An Easy Introduction to the Constellations

A Reference Guide to Exploring the Night Sky with Your Eyes, Binoculars and Telescopes

Cover Image: The constellation Orion, cropped from an original wide field image of the southern Milky Way.

Credit: Akira Fuji.

Original Image: http://www.spacetelescope.org/images/heic0708d/

First Edition, April 2016

For the new astronomers across the world

Contents

Introduction

About this Book

This book combines two previous works, *Easy Things to See With a Small Telescope* and *The Easy Guide to the Night Sky.* They were both written because I primarily wanted to inspire anyone new to the hobby and to provide suggestions on what to observe. I placed a particular focus on the easiest objects with a few challenges thrown in for good measure.

I decided to combine the two because, although they were reasonably successful and well received, I felt it would be more cost effective for readers to only buy one book.

Also, by combining both books, my goal is to first introduce you to the constellations and allow you to discover the things that can be predominantly seen with just your eyes or binoculars. Once you gain some familiarity with these sights, you can go deeper and explore a little more with a small telescope.

The book is divided up as follows:

- The Things You Can See
- Telescope Basics
- Tips 'n Tricks
- Signposts to the Stars
- The Star Charts & Observing Lists
- The Constellation Guide
- Easy Objects for Small Telescopes
- Observation Logs
- Appendix

In the first section, *The Things You Can See,* I'll discuss the different objects you'll be able to observe (including the planets) and review the information to be found on the individual pages for the objects themselves.

In the *Telescope Basics* section, I'll talk a little about the telescopes on the market and which ones might be best suited for you.

The *Tips 'n Tricks* section will give you exactly that – a few pointers and handy hints I've picked up over the years. You'll find that many of these are already in practice by amateur astronomers (like yourself!) all over the world.

Signposts to the Stars is a short section about the two most important constellations for beginners: Ursa Major and Orion. More specifically, it'll show you how you can use these constellations to find others in the sky. If you're very new to astronomy, you might find this useful.

The *Star Charts & Observing Lists* section is where the fun begins. The section begins with a table; simply look for the current time of year (for example, early November) and then look for the current time of night (for example, 10 p.m.) The table will then tell you which star chart to use. (In this example, that would be Chart 7.)

The chart will show you the night sky as it appears at that time and also includes a list of constellations and objects to observe. The lists are not an all-inclusive list as I've focused on the constellations and objects that should be highest over the horizon at that time. Some may be too low to make any observation worthwhile.

The next two sections are the largest. First is *The Constellation Guide*; this goes through the most interesting constellations and includes a little of the mythology, information on its brightest stars and also suggests a few objects to observe. You'll find a general chart of the area as well as simulated depictions of the stars or objects as seen through binoculars.

Again, I've tried to focus on objects that are either visible with just your eyes or binoculars, although a few might require a small telescope and a couple might prove a challenge for absolute beginners. My hope is that, if at first you don't succeed, you'll be inspired to keep trying and return to observe those objects once you've gained a little more experience.

If you own a small telescope (or plan on buying one) you'll find the next section, *Easy Objects for Small Telescopes*, particularly useful. Here you'll find information on over sixty "deep sky" objects. For each object you'll find a general star map, a simulated view through a finderscope and a simulation of what you might see through your eyepiece. (See *The Things You Can See* section on page 13 for more information.)

I always recommend that folks keep a record of their observations (especially if you're new to the hobby) and have included some pages in the *Observation Logs* for you to make some basic notes. You'll find enough pages for you to make notes and sketches for fifty different observations.

Lastly, there's an *Appendix* that contains some useful supplemental information: for example, a list of constellations, the greek alphabet, a basic glossary and recommended resources such as books, software and Facebook groups.

About the Author

Photo by James Bartlett

I've had an interest in astronomy since I was six and although my interest has waxed and waned like the Moon, I've always felt compelled to stop and stare at the stars.

In the late 90's, I discovered the booming frontier of the internet, and like a settler in the Midwest, I quickly staked my claim on it. I started to build a (now-defunct) website called *StarLore*. It was designed to be an online resource for amateur astronomers who wanted to know more about the constellations - and all the stars and deep sky objects to be found within them. It was quite an undertaking.

After the website was featured in the February 2001 edition of *Sky & Telescope* magazine, I began reviewing astronomical websites and software for their rival, *Astronomy*. This was something of a dream come true; I'd been reading the magazine since I was a kid and now my name was regularly appearing in it.

Unfortunately, a financial downturn forced my monthly column to be cut after a few years but I'll always be grateful for the chance to write for the world's best-selling astronomy magazine.

I emigrated from England to the United States in 2004 and spent three years under relatively clear, dark skies in Oklahoma. I then relocated to Kentucky in 2008 and then California in 2013. I now live in the suburbs of Los Angeles; not the most ideal location for astronomy, but there are still a number of naked eye events that are easily visible on any given night.

Also by the Author...

2016 An Astronomical Year is written for everyone with an interest in astronomy and contains information on hundreds of night sky events throughout the year. It was designed for astronomers of all levels and includes details of the lunar phases and eclipses, as well as conjunctions, oppositions, magnitude and apparent diameter changes for the planets and major asteroids.

To date, the 2015 edition has been downloaded nearly 3,000 times, was ranked #1 in Free Kindle Astronomy books, within the Top 10 Paid Kindle Astronomy books and within the Top 50 Free Kindle Non-Fiction books.

It is available in paperback and Kindle editions in the United States, Canada and the United Kingdom. (Please be aware that due to the cost of printing in color, the paperback does not contain images and is purely text only.)

2016 The Night Sky Sights is specifically designed for absolute beginners and casual stargazers without a telescope. The guide highlights over 125 astronomical events in 2016 - all of them visible with just your eyes - and showcases events visible in both the evening and pre-dawn sky as well as those you can see throughout the night.

It is currently available in paperback and Kindle editions in the United States, Canada and the United Kingdom.

The Astronomical Almanac (2016-2020): A Comprehensive Guide to Night Sky Events provides details of thousands of astronomical events from 2016 to the end of 2020. Designed for more experience astronomers, this the guide includes almost daily data and information on the Moon and planets, as well as Pluto, Ceres, Pallas, Juno and Vesta.

To date, the 2015-2019 edition has been downloaded nearly 6,000 times, was ranked #1 in the Free Kindle Astronomy book category, #3 in the Paid Kindle Astronomy book category and within the Top 50 of *all* Free Kindle books in October 2014.

It is available in paperback and Kindle editions worldwide, including the United States, Canada, the United Kingdom and Australia.

The Amateur Astronomer's Notebook: A Journal for Recording and Sketching Astronomical Observations is the perfect way to log your observations of the Moon, stars, planets and deep sky objects. It is available as both a full-size 8.5" by 11" journal and also as a 5" by 8" pocket notebook. The larger edition has room for 150 observing sessions while the pocket edition allows you to record 100 observations.

It is available as a paperback in selected areas. (Full Size Edition: United States, Canada and the United Kingdom. Pocket Edition: United States, Canada and the United Kingdom.)

The Deep Sky Observer's Guide offers you the night sky at your fingertips. As an amateur astronomer, you want to know what's up tonight and you don't always have the time to plan ahead. Maybe the clouds have suddenly parted. Maybe you're at a star party. Maybe you want to challenge yourself with something new but don't know where to start.

The guide can solve these problems in a conveniently sized paperback that easily fits in your back pocket. Take it outside and let the guide suggest any one of over 1,300 deep sky objects, all visible with a small telescope and many accessible via binoculars.

Echoes of Earth – a collection of science fiction, mythological and philosophical short stories that I wrote many, many moons ago. (i.e., in the mid 1990's.)

It is available as a Kindle edition in selected areas. (United States, Canada, the United Kingdom and Australia.)

The Wonder of It All: Your Unique Place Amongst the Sun, Moon, Planets and Stars of the Universe is a book for children and young astronomers everywhere. From our home here on Earth, past the Sun, Moon and planets, this is a journey out to the stars and beyond. A journey of discovery that shows us the beauty and wonder of the cosmos and our special and unique place within it. *The Wonder of It All* will open your child's eyes to the universe and includes notes for parents to help develop an interest in astronomy.

It is currently available in paperback and Kindle editions in the United States, Canada and the United Kingdom.

The Author Online

Amazon US: http://tinyurl.com/rjbamazon-us

Amazon UK: http://tinyurl.com/rjbamazon-uk

Facebook: http://tinyurl.com/rjbfacebook

Twitter: http://tinyurl.com/rjbtwitter

Blog: http://tinyurl.com/theastroyear

Email: astronomywriter@gmail.com

Clear skies,

Richard J. Bartlett

April 30th, 2016

The Things You Can See

About the Deep Sky Objects

There are basically five types of objects you can observe in the night sky:

1. Solar System Objects
2. Stars
3. Star Clusters
4. Nebulae
5. Galaxies

All the objects listed in this book should be easily seen with just your eyes, binoculars or a small telescope. By small, I mean something with an aperture of 70mm (about three inches) to 125mm (about five inches.) Specifically, I've tried to list objects that can be seen with a small 70mm 'scope, but it'll be easier if your 'scope is larger.

Likewise, all the objects are either easily seen with just your eyes (making them very easy to find) or else lie very close to a bright star. That way, you'll be able to easily locate them by first finding the star itself. The object will almost always be within the same binocular or finderscope field of view as the star.

That being said, finding a highlighted object doesn't just depend upon your skill as an observer or how good your equipment is – it will also greatly depend upon the sky conditions at your location. Assuming the skies are clear, your location may also be adversely affected by light pollution. Light from nearby towns and cities can brighten the sky and make it harder to see the fainter objects in the sky.

This is, basically, why you can't see the stars during the daytime. The light of the Sun brightens the sky and makes it impossible to see the stars. The full Moon has a similar effect. This will prove to be important as you hunt for fainter targets, such as variable stars, star clusters, nebulae and galaxies, but bright stars will mostly be unaffected.

I've seen the majority of these objects from the suburbs of Los Angeles (hello, light pollution) but I'm not you, I don't know your personal circumstances and, consequently, I can't guarantee your own personal experiences.

(A word to readers in the United Kingdom: there are two objects you won't be able to see from your latitude and I apologize in advance for that. These are the Butterfly Cluster and Messier 7. Unfortunately, they're too far south to rise over the horizon and I feel bad about that because I'm English myself. If it's any consolation, they barely skim the southern horizon from Los Angeles while observers in the southern hemisphere are lucky enough to see them pass overhead.)

In the first major section, *The Constellation Guide*, I've tried to stick to the objects that can be seen with binoculars as many households will already have a pair available. I own 8x30 and 10x50 binoculars; the first number is the magnification and the second is the width of the "objective lens" – that is, the lenses that point up toward the sky. The 10x50 binoculars provide a slightly higher magnification (10x, in this case) and with 50mm lenses the binoculars are able to gather more light, making it possible to see more as a result.

(10x50 binoculars are a good choice for astronomy, especially if you're just getting started. You can buy larger binoculars but, of course, they'll be heavier and may require a tripod for steady viewing. 10x50 binoculars are still light enough to be used over an extended period of time without your arms getting fatigued.)

My Celestron UpClose G2 10x50's are excellent for observing the night sky. Of course, there are other 10x50's by other manufacturers - Orion, for example, have a binocular stargazing kit for less than a family meal at a restaurant - and a decent pair won't cost an arm and a leg. Even if your binoculars are smaller or older (as a kid, I used my step-grandfather's binoculars from World War I) or you feel your binoculars aren't the best for whatever reason, use them anyway because I always believe that *something* is better than *nothing*.

Not only are binoculars inexpensive, but it's often easier to find objects as they provide a wider field of view compared to a telescope. There's also something to be said for hunting down (and bagging) a target as opposed to letting a GoTo telescope find it for you. (Think of it like learning to ride a bike before learning to ride a motorcycle.)

And, unlike an astronomical telescope that will invert your view, binoculars can be used during the daytime too. (My girlfriend and I take ours to the beach for whale watching!)

Before we get started, let's take a few moments to review the information to be found on the individual pages for the objects.

In the Constellation Guide, I've provided a star chart for each constellation that shows the constellation itself and the stars surrounding it. I've also labelled the stars named in the text to make it easier to identify any objects of interest.

Besides the chart of the general area, most constellations will also have a second page that not only further reviews featured stars and objects, but will have three images to illustrate the objects being discussed.

Some of these images will show an *asterism* to be found within the constellation. An asterism is a pattern to be found within the larger context of the constellation itself. The best (and most famous) example is the asterism of seven bright stars in Ursa Major, the Great Bear. This asterism is famously known as the Plough in the United Kingdom and the Big Dipper in North America.

This pattern is so well known that many people think it's the constellation itself, but in reality, the constellation contains many fainter stars and is much larger than the asterism formed by these seven stars.

Besides asterisms, these images will also show you the simulated view through 10x50 binoculars of the highlighted objects. This is based upon my own Celestron binoculars and your view might vary slightly, but hopefully it'll provide you a good idea of what to expect.

There are a few other things to bear in mind when looking at these simulations.

Firstly, the other deep sky objects are drawn depicting their true size in the sky. In reality, you won't see most of the objects this large in your binoculars unless you're observing under very dark skies. As stated earlier, what you'll see depends upon your equipment, your location, the sky conditions at the time and your own eyesight. The same is true of some of the background stars depicted; how many you'll see will, again, depend upon those same factors.

Also, it's difficult to accurately depict the multiple stars because, in order to show which stars are brightest, it's necessary to make the dots representing the stars larger. So some stars may be depicted as being much closer together – or even overlapping – because at least one of the stars is bright and the pair might be quite close.

All the charts and simulations were created using the *Mobile Observatory* app for Wolfgang Zima. Unfortunately, it's only available for Android devices but I consider it an invaluable tool and it's still the only astronomy app I'll use on a daily basis. (You can find it on Google Play and there's more information at http://zima.co)

Of the text itself, wherever possible I discuss the mythology associated with the constellation and then I give a little information on that constellation's brightest star.

I've discussed these stars as potential objects of interest; for example, many have interesting names that tie in with the constellation itself or have features that set them apart from other stars.

The names of the stars are often Arabic in origin, but you'll notice some may have names like "Gamma Delphini." The first part of the name (gamma, in this example) is a letter of the Greek alphabet. Thousands of years ago, the Greeks assigned letters to the stars based upon their brightness. So the brightest star in the constellation would be assigned the letter Alpha – their equivalent of the letter A. The second brightest would be Beta, then Gamma, Delta and so forth. (There's a table detailing the Greek alphabet in the Appendix on page 263.)

The second part of the name refers to the constellation the star belongs to – in this example, Delphini refers to Delphinus, the Dolphin. (Again, there's a list of constellations in the Appendix on page 263)

The system has been refined and updated since then, but it's still very widely used. After all, not every star in the sky can have a name and this is an easy, convenient alternative.

Some of the stars mentioned will have a number, such as 53 Cancri. Although the brighter stars have these numbers too, they're more commonly associated with the fainter stars as these weren't easily visible to the ancients. However, with binoculars or a small telescope these stars can be more easily seen and catalogued.

The number refers to their position within the constellation with the more westerly stars having a lower designation. For example, the most westerly star in the constellation of Cancer would have the designation 1 Cancri. The number of stars in the constellation will vary, as not all the constellations are the same size and each constellation will have a different number of stars.

Of the other stars, many constellations contain double or multiple stars and variable stars that can be observed with binoculars. I've provided a little more information regarding the nature of these objects on page 23.

Some of the other objects (the star clusters, nebulae and galaxies) may have names, such as the Andromeda Galaxy or the Owl Cluster. More often than not, they're named after whatever the object resembles (such as an owl) or sometimes the constellation it resides in (such as Andromeda.)

Charles Messier, circa 1770. Public domain image.

The book also mentions Messier objects. These are objects that were noted by the French comet hunter Charles Messier in the 18th century. Some deep sky objects can appear distinctly comet-like (especially globular star clusters) and Messier wanted to avoid confusing them with any potential discovery.

There are 110 Messier objects in all and this book contains 58 of them. Some are actually quite clearly not comets – for example, the Pleiades – and many astronomers have wondered why Messier went to the trouble of cataloging them. Other objects, such as the famous Double Cluster or the Owl Cluster, are easily visible and yet Messier didn't list them at all.

One other catalog is also well known and well traversed by astronomers – the New General Catalog, or NGC. This list was compiled in the 19th century and contains nearly 8,000 deep sky objects. However, many of them may be beyond the reach of binoculars or a small 'scope. Incidentally, all of the Messier objects were also assigned an NGC number.

Incidentally, whether we're talking about stars or deep sky objects, all these sights will have a *magnitude*, which is basically a measure of the object's brightness. The lower an object's magnitude, the brighter the object. For example, Sirius, the brightest star in the sky, has a magnitude of -1.4 and most suburban locations will allow you to see stars to about magnitude 4. A good rural location, far away from any light pollution, might provide skies dark enough to see stars down to around magnitude 6 or even fainter. This makes the difference between seeing a few hundred stars and a few thousand!

Obviously, again, the number of stars you see will also depend upon your eyesight and the conditions of the sky above you, but generally speaking a person with good eyesight should be able to see stars to about magnitude 5.5 under clear, dark skies. Binoculars will extend the view out to around magnitude 9 (again, depending on the quality and size of the binoculars) and a telescope will allow you to see much fainter objects still.

The second major section, Easy Objects for Small Telescopes, follows a similar format but has a page dedicated to each of the objects you can see.

At the top of each page you'll find the object name or identifier; again, some stars may have names (such as Sirius) while others may be known by their traditional Greek designation (for example, Sirius is also known as Alpha Canis Majoris.)

On one side of the object page you'll see three graphics. The first is a general map of the area where the object may be found. The image spans 45° of sky and in the center is a circle. This circle denotes the area of sky seen through a 6x finderscope, as depicted in the second image.

The finderscope image is right-side up; in other words, north is at the top of the picture and south is at the bottom. I mention this because some finders will flip the view; please be aware of this as you're looking for your object. Again, both these images were created using the *Mobile Observatory* app by Wolfgang Zima.

The third image on the page attempts to depict the view through your eyepiece. I say *attempts* because I don't know exactly what kind of eyepiece you're using or the field of view you'll get, but this should be a pretty good representation. You'll see the magnification in the top left corner and in the bottom left is an indication of north and east. Your telescope will flip the view; it may just be vertically, or horizontally or both but the view depicted should match many mainstream telescopes.

Again, as with the binocular depictions in the Constellation Guide, there are a few things to bear in mind when looking at these eyepiece views.

With the exception of open clusters, the other deep sky objects are drawn depicting their true size in the sky and it's difficult to accurately depict some of the multiple stars. For the sake of clarity, some of the multiple star images may have large magnifications (e.g., 217x) in the corner. In reality, you can probably get a similar view at about half that magnification.

The eyepiece views were created using *Sky Tools 3*, an excellent program for your PC or Mac that allows you to plan an evening under the stars. As you can see, it also accurately simulates the view through your eyepiece. Like *Mobile Observatory*, I consider it invaluable. (See www.skyhound.com for more information.)

My thanks to Greg Crinklaw for his permission to use these images.

On the other side of the page is some information about the object. You'll see its designation (for example, Gamma Andromedae, Messier 45 etc.), the constellation it belongs to and then we have two co-ordinates, R.A. and Declination.

R.A. is short for Right Ascension and this, coupled with Declination, will give you the exact position of the object in the night sky. Think of it this way – if you were to plot the stars on a globe, then R.A. and Declination would be the sky equivalent of longitude and latitude, respectively.

I debated whether to include them as the objects are easily found once you get to know the constellations. However, some of you may be using a star chart, such as the *Pocket Sky Atlas* by Roger W. Sinnott, and anyone with a GoTo can enter the co-ordinates into their hand controller and the telescope will find it for you. (Most of the objects will be included in the GoTo's database anyway.)

Below the co-ordinates you see what type of object it is – for example, multiple stars and open clusters are the most common. I'll talk about these in a moment.

Next you'll see two sets of stars – one for Location and one as a general Rating. In terms of Location, an object with three stars means it's easily visible to the unaided eye. So, for example, a bright star will have three stars for Location.

Objects with two stars for Location are those that can be found by placing a nearby bright star in the finder and the object will appear in the middle. Objects with one star for the Location are those that may be particularly faint or might require you to have the bright star on the edge of the field of view and the object will appear on the opposite edge.

The general rating is more straightforward. I've given a three star rating to the objects I consider to be unmissable and those I always take time to observe. Objects with two stars still provide something worth seeing while those with one star may be appear faint and best observed under clear, dark skies. (As opposed to the light polluted skies of suburbia.)

There are a lot of objects in astronomy that may seem like a "one star" object at first glance, but remember this: there's a sense of accomplishment that comes from being able to track it down and, as you observe it, consider what you're looking at. Again, I've provided some notes to help with this.

(Again, please bear in mind that I want you to have realistic expectations of what you'll see. Astronomy is awe-inspiring, challenging and fires the imagination.)

Lastly, I've indicated when the object is best seen. You might be able to see the object at other times of the year but this will sometimes require you to observe during the early hours of the morning. So the seasons listed here indicate when the object is at its best visibility in the evening. (This is also why I included the star charts earlier in the book – so you can see what's up in the evening and immediately start observing.)

Below that you'll see some text about the object. This information is predominantly based upon my own personal notes and experiences, both while observing from Oklahoma and also from Los Angeles. Again, my goal was to provide you with a realistic idea of what you might see from either a suburban or a city location. I've read a lot of books written by astronomers lucky enough to observe under clear, dark skies but I understand that this isn't an option for everyone.

Throughout the book, you'll notice I don't have many photos of the objects and there's one very good reason for this: more often than not, they don't accurate depict what you'll see. Too many new astronomers get discouraged because they see beautiful, colorful images in magazines and online and expect to see something similar. Those photographs are created by dedicated individuals who'll spend hours combining many images and fine tuning the final result until it's the best it can be.

I hate to break it to you, but your telescope isn't called Hubble either. I don't want you to be discouraged; the night sky can be stunning and awe inspiring but sometimes you have to use a little imagination and truly take into account what you're looking at. I've tried to do that in the text that accompanies each object.

The Whirlpool Galaxy is visible in a small telescope but it requires dark skies and won't look nearly as good as this image! Taken by the author using Slooh. (www.slooh.com)

Solar System Objects

Technically, objects within the solar system aren't deep sky objects but I wanted to briefly discuss them as they're well worth taking some time to truly appreciate them. As you might expect, there are many other books on the market that discuss observing the planets and one of the best is *The Planet Observer's Handbook* by Fred W. Price.

The five brightest planets – Mercury, Venus, Mars, Jupiter and Saturn – are visible with just your eyes (although Mercury can be tricky) with Venus, Jupiter and Saturn providing particularly good views through a small telescope. With a little more experience, you'll also be able to spot Uranus and Neptune.

Realistically, unless you have a larger binoculars with higher magnification, only Jupiter provides anything of interest for binocular observers. Saturn may appear as a slightly elongated bright gold star when the rings are wide open, but otherwise the remaining planets may only appear as little more than starlike points.

Before you do anything, you'll need to know if any planets are currently visible and this is where the *Mobile Observatory* app or *Sky Tools 3* definitely prove their worth. Alternatively, you can try one of the more basic planetarium programs, such as *Stellarium*, which has a free version available for PC's and Mac's.

Mercury and Venus are both known as *inner planets* because they're closer to the Sun. Consequently, they can never appear opposite the Sun in the sky (unlike the other planets) and can only be seen in the early morning or early evening sky. For example, if you're an early riser you might catch them both in the pre-dawn twilight, but if you're looking for them in the evening, you can catch them in the twilight after sunset.

(You'll need to wait until the planet becomes visible at that particular time of day; the planets won't be visible in both the pre-dawn and evening sky on the same day.)

When you're checking their visibility, you'll need to find out when they reach their greatest elongation from the Sun. This is when the planet appears furthest from the Sun in the sky and is well-placed for observation. Here's where it gets slightly confusing. If the planet is at greatest *eastern* elongation it's visible in the evening sky, but in the west. If it's at greatest *western* elongation, it actually means it's visible in the pre-dawn sky, in the east. (The terms relate to their position in relation to the Sun. For example, greatest western elongation means the planet appears to the west of the Sun in the sky.)

There's a table in the Appendix that lists the elongation dates between 2016 and 2025 (see page 271 for details.) You might need some help finding Mercury as it's often low on the horizon and can be a little faint against the twilight sky. It often appears pink-white and may be twinkling rapidly. Using binoculars will definitely help you. Mercury, like the God it's named after, is fleet-footed and may only be visible for a few weeks at a time.

Venus is a lot easier. In fact, Venus is unmissable. It appears as a brilliant white star, sometimes quite high above the horizon, often for hours after sunset and for months at a time. It's been mistaken for a UFO and provides a beautiful addition to the evening night sky, especially during the holiday season. After the Sun and Moon, it's the third brightest object in the sky.

Venus, October 6th 2012. Image by the author using Slooh (www.slooh.com)

Through a telescope both planets will show phases like the Moon. Venus is closest and will therefore appear largest with Mercury appearing much smaller. You'll only need about 35x to clearly see the white disc of Venus while about 100x will provide a great view.

Mars, February 25th, 2012. Image by the author using Slooh. (www.slooh.com)

Mercury's disc is problematic at low power – you'll probably be able to glimpse it at around 35x but you won't be able to clearly see it until you get to around 80x or 90x. To me, it's always appeared a tan or dull gold color.

All the other planets orbit further away from the Sun and can appear opposite the Sun in the sky. When this happens, the planet is said to be at *opposition* and it's the best time to observe the planet as it will rise at sunset and set at sunrise. It's therefore visible all night and will appear at its brightest and largest.

As with Mercury and Venus, there's a table in the Appendix that lists the opposition dates for Mars, Jupiter and Saturn between 2016 and 2025 (see page 272 for details.)

Unfortunately, Mars only reaches opposition once every couple of years and isn't always conveniently placed for observation. Also, the size of its disc will vary, depending upon how close the planet is to us at the time. To the unaided eye it appears bright and coppery and a small telescope will show a tiny disc at low magnification.

At its best (and with higher magnification) you'll be able to see tiny dark markings on its surface and maybe even the glint of an ice cap at the poles. I've seen markings at 54x but you might need to go higher. A magnification of about 100x seems fairly reliable.

Jupiter and Saturn present, by far, the best views of any planet in the solar system and I promise you will never get bored of viewing them.

Jupiter and its moons, October 7th, 2012. Image by the author using Slooh. (www.slooh.com)

Many people are wowed by Saturn when they see it for the first time. Personally, I love Jupiter. Why? Two reasons: firstly, because you get to see more and secondly, you can start an observing session with Jupiter and then come back to it just a few short hours later and you'll see that things have changed.

For example, even a pair of steadily held binoculars will reveal the planet's four largest moons – Io, Europa, Ganymede and Callisto. These moons are collectively known as the Galilean satellites after the Italian astronomer Galileo Galilei who first discovered them in January 1610. These moons orbit the planet fairly quickly and can be seen to cast shadows against Jupiter's disc and disappear in eclipses behind the planet.

Beyond these four moons you'll also see dark stripes across the disc of the planet. Typically you'll see two (the northern and southern equatorial belts) at around 40x and if you increase the magnification to around 70x or 80x you may also notice dark patches at the poles.

You probably won't see the Great Red Spot. This is the Earth-sized storm that looks like Jupiter's eye staring out into space in many of the photographs of the planet. Despite many years of looking, I was only able to see it with my 130 SLT with the use of a blue eyepiece filter to help bring out the contrast in the planet's features. Even then, if I hadn't known it was currently visible, I probably would have missed it!

Jupiter, January 4th, 2015. Image by the author using Slooh. (www.slooh.com)

Saturn, of course, is famous for its rings and is usually the planet that beginners want to see first. There's something about a person's first view of the planet that often stays with them and many amateur astronomers can clearly recall when they saw the planet for the first time.

You've probably seen a lot of photos of the planet but when you actually see it for yourself, you'll notice a third dimension that you just can't experience any other way. It looks like a Christmas tree ornament, impossibly hanging there in space. It's easy to understand how the first telescopic observers, some four

Saturn, December 27th, 2011. Image by the author using Slooh. (www.slooh.com)

hundred years ago, were completely confused and bewildered by the rings. Nowadays, thanks to missions like the *Voyager* and *Cassini* space probes, we know the rings to be comprised of millions of chunks of ice, some no bigger than a first, others as large as a house, all orbiting Saturn in unison.

Like Jupiter, you'll also be able to see a few moons, most notably Titan, the planet's largest. But unlike the moons of Jupiter, they don't move quickly and the view won't considerably change within a few short hours. However, come back the following night and you'll see some differences.

Beyond that, you may see a gap in the rings (called Cassini's Division, named after the astronomer who discovered it) and maybe some faint markings on the disc, but that's about it. To be honest, for me it's like a childhood friend I've grown apart from. I'll still stop by and visit if it's around, but I don't usually stay for long and I don't find there's much to talk about!

Uranus and Neptune can also be seen but you need to know where to look for them. At low power (say, around 35x) they only appear as tiny, starlike points so even if you're using a GoTo, you'll need to know which are the stars and which is the planet. Having said that, there *is* a way you can tell. Both planets will show a fairly distinct color – a pale turquoise tint for Uranus and a sky blue hue for Neptune.

Uranus might also show a disc; I've possibly seen it at 26x but I've typically needed at least 50x to be sure. Neptune is trickier and I've only seen a very tiny disc at a magnification of 160x or higher.

The same is true for asteroids. You'll need to know exactly what you're looking for and be familiar with the background stars. Unfortunately, unlike the major planets, they're too small to appear as anything but a tiny star-like point in your binoculars or telescope and can often be disappointing to new astronomers.

One last thing: comets. Occasionally, a reasonably bright one might come close and make itself visible to amateur astronomers with binoculars or small telescopes. How bright it will be and how it will appear really depends on the comet itself.

No two comets are alike and none are guaranteed to be a spectacular celestial event. I've observed a number and they typically look like misty spheres with a star-like point (the core of the comet) at its center. Some comets will also show a tail while others might have a greenish tint to them.

Comet C/2013 K5, December 12th, 2012. Image by the author using Slooh. (www.slooh.com)

Be sure to join an astronomical group (be it online or in person) to keep up-to-date with any potential surprise visitors that might be coming around.

Multiple Stars

The double star Kuma, aka Nu Draconis. Image by the author using Slooh. (www.slooh.com)

I hate the phrase "star gazing." For me, it always conjures up an image of someone standing still, staring up at the night sky. Or maybe peering through a telescope at a single bright star. Someone else might walk by. "What are you looking at?" they ask. The observer points up at the sky. "That bright star. That one, right there."

Many non-astronomers think this is what "star gazers" do. It's like when I stopped eating white and red meat during the 1990's. People thought I just ate peas and carrots instead. (Incidentally, I've since returned to the dark side.)

Nothing could be further from the truth, especially given that many stars are not single stars at all. In fact, most of them are *multiple* stars. To the unaided eye, they'll appear to be a single star, but when you observe them with binoculars or a telescope, that star is split in two. Some stars may have three or four components.

There are two kinds of multiple stars: those that are true multiple star systems, where the stars orbit one another, and those that are *optical* doubles and only appear close together due to a chance alignment. In reality, they may be light years apart.

Some multiples have components that are equally bright (such as Kuma) while others show stunning colors (such as the famous Albireo.)

The best thing about multiple stars (beside their abundancy and variety) is that they're largely unaffected by moonlight and light pollution. As long as both components are reasonably bright, they'll still shine through the brightened sky caused by an intrusive Moon.

Variable Stars

On occasion I'll mention a *variable star* that can be observed in the constellation. As the name suggests, these are stars that typically appear to vary in brightness over a period of time. Some may change brightness over a matter of hours, some will take a few days while others (and more commonly) may take tens or even hundreds of days.

For example, one of the most famous variable stars is Algol in the constellation of Perseus (see page 162.) Its normal brightness is magnitude 2.1 but it will fade to magnitude 3.4 for about ten hours and then return to magnitude 2.1 again. In all, it takes 2 days, 20 hours and 49 minutes to complete this cycle. This is known as its *period.*

Why does this happen? Algol is the class example of an *eclipsing* variable. Unseen to our eyes is a smaller, fainter companion that regularly passes in front of the brighter star and dims its light. Hence, the magnitude appears to drop as the star fades. The faint companion takes ten hours to move across the face of the brighter star and then Algol brightens as the eclipse comes to an end.

Not all variable stars are the same. Most have much longer periods and are often red giant stars nearing the end of their lives. Their brightness changes as the star pulsates, like a heart beat, as the star expands and contracts over a regular period of time. Others are irregular and unpredictable while some are classified as recurring novae – stars that suddenly and inexplicably brighten before fading and unpredictably brightening again.

Star Clusters

Messier 41 in the constellation of Canis Major. Image by the author using Slooh. (www.slooh.com)

Star clusters also fall into two categories: open star clusters and globular star clusters. An open star cluster contains tens or hundreds of stars, all of which literally appear clustered together in the same relatively small area of sky. This is not a chance alignment – these stars are genuinely grouped close together in space and are born from the same nebula. Consequently, they're usually quite young and are quite literally, stellar siblings.

Some clusters appear fairly large and only require a low magnification with member stars scattered across the field of view. The famous Hyades, Pleiades and Praesepe open clusters are excellent examples and look great in binoculars. Others are small and compact and may require the higher magnification of a telescope to be properly appreciated. Messier 103 is one example.

The other type of cluster is the globular cluster. These are spherical balls of stars in space and contain thousands of stars within a tightly packed area. They lie thousands of light years away, usually close to the hub of the galaxy, and are very old in comparison to the younger open clusters. It's not unusual for a globular to be over ten billion years old – nearly as old as the universe itself.

Through a small telescope or binoculars a globular can often appear like the head of a comet without the tail. Alternatively, you might think of it as being a faint and fuzzy star. But with the higher magnification of a telescope you may be able to resolve some of the individual stars around the outer edges. You may also notice chains of stars or the cluster may appear to be misshapen. Not all globulars are the same!

The Keystone Cluster in the constellation of Hercules. Image by the author using Slooh. (www.slooh.com)

For my money, the best globular in the northern hemisphere is the Keystone Cluster but Messier 22 is a fine example too.

Whether you're observing open or globular star clusters, you'll get the best views away from the lights but you can still get some great views from the suburbs, even with light pollution.

Nebulae

There are, of course, several categories of nebulae but the most common are clouds of gas and dust in space. These are the birthplaces of stars and can cover an area light years in diameter. The Orion Nebula is the best example of this in the northern hemisphere. It's easily seen with the unaided eye as a tiny, misty patch in the sword of Orion and can be a stunning sight in a telescope.

Unfortunately, there are only a few of that kind in the book as most nebulae are fainter and don't appear close to a bright star. However, there are a few others that might draw your attention.

The Orion Nebula in the constellation of Orion. Image by the author using Slooh. (www.slooh.com)

One is the Ring Nebula, an outstanding example of a planetary nebula. These nebulae are small, faint and may be tricky for an inexperienced observer to locate. However, many will present a planet-like disc when observed through a telescope. The Ring is the best and brightest example and literally looks like a tiny smoke ring in space. It's located in the constellation of Lyra and is best seen in the summer.

The other kind of nebula is a supernova remnant. Again, these are typically faint and only one is easily seen in a small telescope. This is the Crab Nebula and it's the remains of a star that exploded nearly a thousand years ago. The Crab is located in Taurus, the Bull, and is best seen in the autumn and winter.

Galaxies

The Andromeda Galaxy in the constellation of Andromeda. Image by the author using Slooh. (www.slooh.com)

Galaxies come in all different kinds of shapes and sizes but may be disappointing to the beginner. There are a lot of images around showing star-studded spirals in space, but the reality is that you're likely to only see a small, misty patch. Depending upon the object, your equipment and location, you may be able to see some shape and structure with a telescope and patience definitely pays dividends, but don't get your hopes up if you're affected by light pollution.

Also, the vast majority of galaxies are small, faint and can be difficult to locate. With experience, you'll be able to spot a number of them but to begin with there's only one that's easily seen. The Andromeda Galaxy appears as a misty patch with the unaided eye and is conveniently located close to a number of bright stars.

(I've listed a few others in the Constellations Guide, which may be visible with binoculars under clear, dark skies, but they might prove to be a challenge for inexperienced observers.)

Telescope Basics

Refractors vs Reflectors

There are many different types of telescope but they broadly fall into one of two categories: refractors or reflectors.

A refractor is what many people consider to be the classic concept of a telescope. It's basically a tube with a lens at one end and an eyepiece at the other. It sits atop a tripod and is usually the kind found in department and toy stores as a "first telescope" for children.

By the way, there are only a small number of telescope manufacturers I'd consider and, out of those, I've only ever bought from Orion and Celestron. There are others, but both these manufacturers are very well respected in the industry and both produce fine equipment at a reasonable price. If you're looking at a refractor telescope in a department store and it's *not* made by Orion or Celestron, I'd walk away. You might think it's a good buy for the price, but it's probably cheap for a reason.

As you're shopping around, you might come across a telescope that's listed as, for example, a 70mm refractor. This means the lens of the telescope is 70mm in diameter – or just under three inches wide. The size of the lens will dictate how much light is gathered by the telescope and, consequently, how much you'll be able to see as a result.

The 70mm Celestron TravelScope. Image by the author.

Again, stay away from department store telescopes that make ridiculous claims like "magnifies up to 1,000x!" In theory, any telescope can magnify by any amount but the view is worthless because the optics simply won't produce a good quality view. If your optics are imperfect then any imperfections are also magnified by 1,000x.

Also, it's almost a waste of time to try and magnify anything that much. I suspect a lot of inexperienced folk believe that you need a high magnification to see anything – because, after all, the stars and planets are only tiny points of light and those same folks are unaware of everything else you can see at low power.

In reality, only planets need a high magnification and you can still see a lot at 100x, which is well within the range of a decent small 'scope made by a reputable manufacturer.

A good refractor is able to produce a relatively high magnification, which makes it an excellent choice for observing the planets. A refractor also has the benefit of being quite portable as, for technical reasons I won't go into here (I promise to restrict the jargon) the size of the lenses typically range from about 70mm to 100mm in diameter.

Both Orion and Celestron produce some excellent "first 'scope" refractors – especially if you're looking for something portable or you're on a budget. As of this date (November 2015) there are a number of options under $100, such as the Orion 10034 GoScope II or the Celestron 21035 70mm Travel Scope. Both of these come with a backpack that allow you to easily pack up your 'scope and take it wherever you want to go – great for family vacations!

(I personally own the Celestron and have used it under the dark skies of Ojai, California. Talk about reliving childhood memories...)

The other kind of telescope is a reflector. There are, in fact, many different types of reflectors but I'll try to keep this simple and straightforward. Whereas a refractor works by having the light pass through a lens, a reflector works by literally reflecting the light with mirrors.

The light enters through the open end of the telescope tube, called the aperture. It then strikes the large mirror at the bottom of the tube, called the primary mirror, and is reflected back up toward the aperture. Before it can escape again, it hits a much smaller mirror (called the secondary) and is reflected out the side of the tube via the eyepiece.

There's one big advantage to this: a refractor typically requires a long tube because the light isn't reflected and must travel in a straight line. With a reflector, the light is bounced twice and so the required tube length is halved. This has the added benefit of allowing for much larger mirrors and, hence, more light gathering power.

There's one more big difference between a refractor and a reflector. A refractor telescope is almost always found on top of a tripod and there are many reflectors that are similarly mounted. However, there's one very popular mount that's in very common usage today: the Dobsonian.

A Dobsonian mount is relatively simple. The tube sits upon the mount, which, in turn, sits upon a base on the ground. The mount then turns from side to side and allows the telescope to move up and down.

Dobsonian mounts are designed to carry some pretty large beasts and can be permanently grounded. This makes them a favorite with amateur telescope makers who aren't so concerned with portability but would rather have a powerful telescope instead.

(Having said that, you can also have Dobsonians that are relatively small but are still portable, such as Orion's range of SkyQuest XT 'scopes. I owned an XT 4.5 – as seen in this image - and loved it.)

Dobsonians (aka "Dobs") are "light buckets" and are great for deep sky observing but they have one big potential downside – depending on your point of view. Dobs typically aren't motorized.

What does this mean?

Well, when you're looking at an object in the night sky through a telescope, you'll notice that it appears to drift across the field of view as the sky turns above you. Yes, that's right, you will literally see the stars move from east to west.

Having a motorized mount allows your telescope to compensate for this and, consequently, the object you're observing will stay locked within your field of view. This can be critical if you want to try your hand at astrophotography or want to make a detailed sketch.

Reprinted with permission from Orion Telescopes & Binoculars, www.telescope.com

(There are some exceptions to this. For example, Orion produces a range of motorized Dobsonian 'scopes that will guide you to your object.)

GoTo 'Scopes

One more thing... technology has evolved a lot over the past few decades and amateur astronomers have certainly benefitted as a result. One of the advancements has been the development of the "GoTo" telescope.

A GoTo is typically a reflector on a motorized mount that can automatically find objects for you through the use of a computerized hand controller. I think I see a see a light bulb appearing over your head, but before you get too excited, I feel I must tell you this:

Although GoTo's are great - I own one – there's *nothing* that can beat tracking down and "discovering" an object for yourself. A GoTo may be able to place an object in your field of view within seconds, but it's nothing compared to the thrill of using your observational skill (and a decent star chart) to track it down yourself.

The Celestron 130 SLT GoTo. Image by the author.

So, while all of the objects in this book can easily be found with a GoTo, I *strongly* encourage you to go old school and leave the GoTo behind.

I know what you're thinking though... if I'm not recommending a GoTo, why do I own one myself? Well, there's two main reasons. Firstly, and primarily, it's because of light pollution. I live in the suburbs of Los Angeles, where only the brightest of the bright stars can be seen on any clear night. It's extremely difficult to find any deep sky objects under these conditions but the GoTo can find them for me.

Secondly, when I bought my 'scope I wanted something with a motorized mount and they're pretty much all GoTo's now. True, you don't have to use that functionality, but when you have a database of thousands of objects at your fingertips, it's kinda hard to resist it.

(Incidentally, I'd argue that whether you're using a GoTo or not, you still need to know what you're looking for. A GoTo mount will often point the telescope very close to the object but it will rarely place it smack in the middle of your field of view. Sometimes, if you're unlucky, you have to slew the 'scope a little to see your object and if you're not sure what you're looking for, it can be easily missed.)

Which Telescope Should You Buy?

I know I've thrown a lot of information at you and you're probably scratching your head right about now. Don't worry, it gets easier. If you've already bought your telescope, you might even be wondering if you've bought the right one.

In fact, let me address that issue right now. The answer to the question "did I buy the right telescope?" is always "yes. Yes you did." Why? Because anything that allows you to turn your eyes to the stars and to explore the universe is a gift – even if you bought it yourself. When Galileo discovered the four largest moons of Jupiter in 1610 he did it with a simple telescope that makes those department store 'scopes I warned you about look like the Hubble Space Telescope.

The Celestron 76mm FirstScope. Image by the author.

So if you bought your own telescope (even from a department store) and you're wondering if you made the right choice, then don't worry about it. If, on the other hand, you haven't bought one yet (or are looking to buy another – congratulations, you've caught the bug) then here are some recommendations.

<u>Under $100 and/or Portable</u>

You've got some excellent choices for small 'scopes. If you're looking for something as a beginner 'scope, maybe for a child or a young adult, I'd go with either the Orion 10033 FunScope 76mm TableTop Reflector or the Celestron 76mm FirstScope. Both are available for about $70 and can provide great views of the Moon, planets and deep sky objects to get you started. (Almost all of the objects in this book should be within reach of these 'scopes, although that will also depend upon your viewing location and your own eyesight.)

From a parent's point of view, if you buy one of these for little Jimmy and he loses interest after a while (I can't imagine why, but who knows) then you haven't lost much. And as an added bonus, these 'scopes are small and very portable. The only downside is that you'll need something like a table to put them on.

I've personally bought the Celestron FirstScope for my nine year old son and for my girlfriend's five year old daughter; some of the notes in this book are based upon observations made with this 'scope.

Alternatively, you can go with either the Orion 10034 GoScope II or the Celestron 21035 Travel Scope I mentioned earlier. Although they both have a slightly smaller lens (70mm) they're attached to a tripod that can easily be adjusted to suit the observer's height – so no need for a table then.

<u>Under $300</u>

If you're more serious, I highly recommend the Orion SkyQuest XT 4.5 Dobsonian. This is the telescope I owned when I lived in Oklahoma and I loved it. I spent almost every moonless weekend night outside and was able to "discover" something new every night. I've observed almost every object in this book with that 'scope and the only reason why I didn't observe the rest was because I never got around to it. Many of the notes are based upon my observations during that time.

After Oklahoma, I moved to Kentucky where the light pollution from Louisville dampened my spirits. Then nearly six years later, I moved to Los Angeles. I drove from Kentucky to California with all my worldly possessions in my car and, unfortunately, I didn't have room for the 'scope. Consequently, I very regrettably gave it away. In fact, Dawn, if you still have it, you've no excuse for not using it now!

As the name implies, it's a reflector with a mirror that's four and a half inches wide, or about 112mm. It has a handle and weighs about sixteen pounds, so it's portable, but you'll need a small table to put it on. (I used a folding table I bought from a hardware store for about $20 and that worked just fine.)

Under $400

If you're looking for a GoTo – I won't judge – then I recommend the Celestron 130 SLT. This is the 'scope I bought (and still own) after I moved to Los Angeles and, as I said earlier, it's proven to be invaluable under these light polluted skies. The only drawback that is that it's a little large to be taken on a trip. I'd certainly move it for an astronomy weekend, but for a regular family trip I'd take the Travel Scope or the FirstScope instead.

The mirror is 130mm wide, which is just slightly more than five inches. Every one of the objects in this book should be easily visible with this 'scope.

Tips 'n Tricks

Get to Know the Stars

I can't emphasize this enough. Maybe you're already familiar with the night sky and all the constellations, but if you're not, I'm sorry, but there's no way around this. If you don't know your way around the stars then trying to find some of the hidden treasures will be like starting from a random point in the U.S. and trying to get to Disneyland. One easy way to learn the stars is to use a planisphere, like the one below.

Get a Planisphere

I've provided some star charts to help you find your way around the sky but there's something else you can buy which, I believe, is a huge asset to any astronomer. A planisphere. It's basically a disc with the stars plotted upon it that allows you to "dial up" a view of the sky for any night at any time.

This planisphere is actually small enough to fit inside the toolbox I use to store my eyepieces. Image by the author.

It works like this. There are two discs, one attached to the other. On the lower disc is a map of the entire night sky with all the stars that can be seen from your location. Around the edges are the months with marks to indicate the individual dates.

The upper disc is basically a circular mask that covers the sky chart and only allows for a segment of the sky to be visible. The edges of the disc are marked with the twenty-four hours of the day, from midnight to midday and through to midnight again. In order to see a depiction of the night sky on a particular date and at a specific time, you simply move the upper disc so that the desired date and time are aligned. The upper disc will then allow you to see the visible stars and will mask the rest.

I know what you're thinking (and probably saying) – "oh I have an app for that." Yes, you probably do (I have several) – but here's why I don't use the app when I'm observing: firstly, you're relying on your cell phone or tablet battery to keep you going. And if that battery dies and you need your cell phone for the drive home from a dark sky site, you have a potential problem.

Secondly, your eyes need to adjust to the dark (see below) and although many apps will dim the screen, if you hit the wrong button on your device you might find yourself staring at a bright screen instead. And now you've just lost your night vision.

You're probably also saying "how am I supposed to see the planisphere in the dark?" Well, that's what a red flashlight is for (see below.) Ideally, get a planisphere with a white background and black stars (like the star charts I've printed in this book.) This will make it easier for you to see it in the dark.

You can buy a planisphere online from almost any telescope manufacturer, including Orion. Be sure to get the correct one for your latitude otherwise you might still be lost!

Get a Red Flashlight

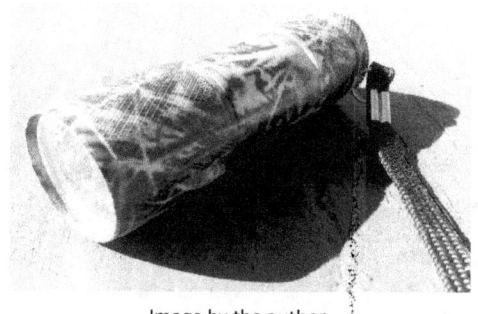

Image by the author.

A red flashlight is an essential tool for any astronomer. Your eyes need time to adjust to the dark (see below) and if you use a regular flashlight, your night vision will be ruined. On top of that, if you're out with a group and you use a regular flashlight, you run the risk of ruining your fellow astronomer's night vision.

That will make them angry. And you don't want to make an amateur astronomer angry.

All that being said, your eyes aren't adversely affected by red light and, many moons ago, astronomers would buy some clear plastic red tape to cover their flashlights. Nowadays you don't have to go to all that bother as a red flashlight is pretty inexpensive.

You can easily buy them online and may even be able to buy them from stores selling camping and hunting equipment. If you can, get one that allows you to adjust the brightness of the beam. This will make it easier for you to view your planisphere, star charts or even (hopefully!) read this book!

Let Your Eyes Adjust to the Dark

Many people are aware that their eyes grow accustomed to the dark. Stand in a lit bedroom at night, with the curtains drawn and then turn out the lights. What do you see? Probably not much if the room is properly dark.

Likewise, if you need to get up in the middle of the night and you turn the light on, what happens? How do your eyes feel? More than likely you'll be dazzled by the sudden burst of light.

It takes your eyes about an hour to fully adjust to the dark and you don't want to lose that sensitivity (hence the red flashlight.) Without that sensitivity you may not be able to see the finer details in some of the deep sky objects (such as nebulae and galaxies) and some of those objects may not be seen at all.

So what do you do while you're waiting? Well, I for one don't wait around. Take some time to enjoy the night sky. If there's a crescent Moon in the sky, I might take a look at that (but with the use of a filter – see below.) *Don't* observe the Moon after your eyes have adapted to the dark as its brightness will probably ruin your night vision again.

Look at the planets, if any are visible. Multiple stars are also a good choice as their light is concentrated in a single, bright point and most are easily seen. (There are a few examples of faint multiple stars but usually you're good to go.)

As my eyes adjust, I also take some time to look at the brighter star clusters, such as the Pleiades or the Praesepe. These are often comprised of many bright stars and can be a very welcome distraction as you wait for your night vision to kick in. I *do* also recommend you revisit them later in the session as you'll frequently see many more of the fainter stars once your eyes have adjusted to the dark.

Beware the Moon

You may notice that I don't talk about the Moon in this book. There's a reason for this. As pretty as it is, the Moon (like an overcast sky) is not my friend. The Moon brightens the sky, just as the Sun does, so instead of the nice, dark sky you'll get on a moonless night, you'll be looking at a deep blue sky instead.

How does this affect your observing? Well, some of the fainter fuzzies (galaxies and nebulae, specifically) will be lost against the glow of the sky. On the plus side, multiple stars are often unaffected as, again, their light is concentrated into a single bright point that can easily shine through.

The waxing crescent Moon, May 30th 2006. Taken with an Orion SkyQuest XT 4.5. Image by the author.

I don't typically observe when the Moon is out, or, if it's a crescent Moon I'll observe if it's relatively close to the horizon (as a crescent Moon often is.) Consequently, I'll often wait until about three or four days after the Moon has turned full. By that time, the Moon won't be rising until some time after sunset, giving me the evening hours to conduct my observing. It stays that way until about two or three days after new Moon, when the Moon sets a few hours after the Sun and I can use the rest of the evening for my observing.

Anything between an evening crescent and just after full Moon is a problem because the Moon is visible for much of the evening.

That's not to say the Moon isn't worth observing, because it is – honestly, it can be stunning – and there are a myriad of books and maps out there that will help you with that. But since this book focuses on the "deep sky" objects of stars, star clusters, nebulae and galaxies, I won't be talking about it here. (I *do* recommend you take the time to enjoy the lunar landscape though!)

Image by the author.

If you're going to look at the Moon, I highly recommend you get yourself a lunar filter for your eyepiece, especially if the Moon is between its half phase and being full. The moonlight can be dazzling when magnified through your telescope and the filter will cut out about 85% of that light, making the Moon far easier to observe. Without it, it's like someone shining a flashlight into your eyes. Again, you can easily buy a good filter online and they're not expensive at all.

Let Your 'Scope Adjust to the Air Temperature

Just as your eyes need time to adjust, so does your telescope. Admittedly, this is something I often neglect to do as I'm so keen to get out there that I frequently skip this step. Also, I'm easily distracted and will often find myself getting sidetracked when I should be putting the 'scope outside.

Why is it important? Well, your telescope is probably kept inside and your home is probably kept at a reasonably warm temperature. When you take it outside, the air temperature will be different. It might be warmer or it might be cooler, but until your telescope reaches the same temperature you might find the images are not as sharp as they could be.

If you can handle the wait, take your telescope outside about an hour before it's time to observe. Look at it this way – it can take about ninety minutes for the sky to get dark after sunset anyway, so if you take it outside within the first thirty minutes after the sun goes down and *then* go out after twilight has vanished and it's truly dark, then you should be fine.

Locate the Object with Binoculars First

If you've never observed the object before, using a pair of binoculars can make the search easier and also help you to familiarize yourself with that particular patch of the sky. Many of the objects in this book can be seen with binoculars and a standard pair of 10x50's should do the trick. As always, there are plenty of options available and you can get a good quality pair online or through major retailers.

If you're interested in binocular astronomy, there a multitude of books out there that are specifically aimed at binocular observers. For example, I can highly recommend Philip Harrington's *Touring the Universe through Binoculars* and Gary Seronik's *Binocular Highlights.* I'll often grab my binoculars when I only have a few objects to observe (or too little time) and have included my notes in the descriptive text for each object.

Likewise, many finderscopes will actually magnify the sky a little – often by about 6x. Although not as powerful as binoculars, you should still be able to see almost all the objects in this book within the finder. I've included a graphic depicting the 6x finderscope view beside the text for each object.

Learn to Calculate the Magnification of Your Eyepieces

This is actually very easy but you need to know two important numbers before you begin: the focal length of your telescope and the focal length of the eyepiece you're using.

What's focal length? Well, it's basically the distance that light must travel from entering your telescope (or eyepiece) in order to reach your eye. It's always measured in millimeters and it can always be found on the telescope or eyepiece itself.

Image by the author.

On the telescope, this information can be found on a label that's often affixed to the tube, either on the side or close to the focuser where the eyepiece is inserted. It'll probably have the make and model of the telescope and the focal length may be represented with the letters FL.

On an eyepiece, it's printed at the top, almost always on the side and sometimes around the lens of the eyepiece itself.

To calculate the magnification of the eyepiece, you simply divide the telescope's focal length by the focal length of the eyepiece. For example, most telescopes come with a 10mm eyepiece (meaning it has a focal length of 10mm.) My Celestron 130 SLT GoTo telescope has a focal length of 650mm so a 10mm eyepiece will give me a magnification of 65x.

Get Yourself a Range of Eyepieces

Obviously then, it's useful to have a small number of eyepieces as this will give you a good range of magnifications to play around with. Most telescopes seem to come with a fairly large eyepiece (for example, a 20mm or 25mm) that will give a low powered view and also a smaller eyepiece (typically a 10mm or a 6mm) that will give a higher powered view.

Personally, I wouldn't go any lower than 6mm because the lens and field of view typically decreases in relation to the focal length. This means you have to put your eye closer to the eyepiece and, in my experience, this actually makes observing a little uncomfortable. (There's a way around that – it's called a barlow and I'll talk about that in a moment.)

Likewise, unless my telescope had a high focal length (say, over 900mm) I wouldn't go over 30mm. (You can, for example, buy 50mm eyepieces but they tend to be a little pricey.)

I'd recommend a 25mm and then adding eyepieces in 5mm increments. I therefore own a 25mm, 20mm, 15mm and 10mm along with a 6mm and a 9mm. (Once you get to 10mm, the sizes tend to decrease in smaller increments.)

You don't have to buy these individually either. Both Orion and Celestron produce eyepiece kits for beginners that will give you a good range to start off with. My only word of caution is to make sure you get the right barrel size for your telescope. The vast majority of telescopes have a focuser that will fit eyepieces with a barrel 1.25 inches wide, but there are some that will take slightly different sizes. If you're not sure, check your telescope's manual or contact the manufacturer before buying.

But why do you need a range of magnification anyway? Surely a high and a low powered eyepiece is enough? Potentially, yes, but you'll soon learn that when it comes to observing deep sky objects, there comes a point when you magnify an object *too* much and it loses its aesthetic appeal. A star cluster is prettier when you can see the whole cluster and it's set against a backdrop of faint stars. Multiple stars (especially those of differing colors) are more attractive when they appear close together, rather than highly magnified and far apart.

Think you only need the two that came with your 'scope? Observe the deep sky objects in this book and then see how you feel!

Some of the eyepieces I own. From left to right, 25mm, 20mm, 15mm, 10mm and a 6mm. Image by the author.

Know the Maximum Useful Magnification of Your Telescope

Unfortunately, as I mentioned in the Introduction, your telescope is probably not going to produce a good quality image at 1,000x. In fact, I'd be impressed if you can even squeeze 1,000x out of your 'scope. There are a number of reasons for this, but suffice it to say, every telescope has its maximum useful magnification and it's important to know this before you start observing (or start buying eyepieces that go beyond this limit.)

Quite simply, the theoretical maximum magnification is 50x the aperture of your telescope in inches or about twice the aperture in millimeters. For example, my Orion SkyQuest XT 4.5 Dobsonian had an aperture of 4½ inches so the theoretical maximum was 225x. (4½ inches in millimeters is slightly more than 114, so that would be 228x.)

However, I've found that this isn't always the case. Realistically, I've found the views can be unreliable beyond half this amount. Sometimes they're better, but more often they're not. (For example, the only time I've seen the Great Red Spot on Jupiter with my Celestron 130 SLT is when I had it right at the maximum of 260x. However, I don't tend to notice much improvement in many faint and fuzzy deep sky objects.)

Therefore I'd go for 25x the aperture of your telescope in inches or – more conveniently – the aperture of your telescope in millimeters. So I usually stick to 130x or less on my Celestron 130 SLT and have never been disappointed as a result.

Invest in a Barlow Lens

I've recommended that you buy yourself a range of eyepieces and although you can buy these for a reasonable price, some can be quite expensive. Also, the smaller the focal length of your eyepiece, the smaller your field of view, which can make it hard to observe.

The solution? A 2x barlow lens. This is basically a short tube that slots into your telescope's focuser, where the eyepiece normally goes. You then pop your eyepiece into the barlow and *voila!* Your magnification is instantly doubled. And the best part is your field of view remains the same.

For example, a 20mm eyepiece in my Celestron 130 SLT will give me a magnification of about 32½x with a nice, wide, comfortable field of view. But if I'm using a 2x barlow, I double up to 65x and still be comfortable with the same field of view.

A 2x barlow. Image by the author.

Barlows are very easily obtained, are reasonably priced and the advantages are obvious. Better magnification. A more enjoyable observing experience. Potentially few eyepieces needed so you won't have to carry as much equipment around. It's easier on the eyes *and* the wallet.

Buy a Toolbox for Your Equipment

Yes, a toolbox. Or a fishing tackle box. One of those boxes you buy in hardware stores to keep small tools, screws and odds and ends does the job quite nicely.

Mine has a lower and an upper compartment; in the lower I keep the things I might only need to use once or a few times in a session: my GoTo power cable, a pocket night sky guide and a mini planisphere.

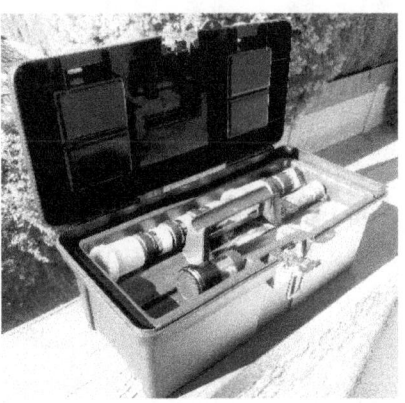

In the upper compartment, I keep the items I'll use frequently, such as my eyepieces, my barlow, my red flashlight and my filters.

And all I have to do is carry it outside. Trust me, it's invaluable.

Image by the author.

Start with a Lower Power Eyepiece and Work Your Way Up

When I'm observing, I always start at a low magnification and then increase it. For starters, it's easier to find your target and many star clusters are best observed at low magnification anyway. If the object is easily seen, then I'll gradually increase the power to see how the view changes.

If, however, the target isn't seen, then I'll check to make sure I'm looking in the right place and then up the power to something closer to 100x. This is also true of multiple stars that can't be split at low magnification. For example, I may only see a single, bright star at 35x so I'll up the power to around 100x and see if my luck improves. If the star's split, I'll start decreasing the magnification to find the point where it becomes a single star again. Quite frequently, I've gotten down to 35x again and still been able to split the star – but only because I now know where to look for the companion.

Focus on the Stars

If you're staring at something faint and fuzzy (for example, a globular star cluster) you'll often find it difficult to properly focus the view. Instead of trying to focus on the object, focus on one of the nearby stars instead. And if you don't see any in the same field of view, try moving the 'scope slightly so that some are visible. It's far easier to focus on a single point of light than something grey, misty and ill-defined!

The Colors of Multiple Stars

I've found that star colors can be very subjective and may even seem to vary from night to night. There are a number of reasons for this. Unless the colors are quite strong (such as the famous double star Albireo) what you see might vary depending upon the sky conditions, the equipment you're using, the height of the star in the sky and, frankly simply, your own eyesight.

So if I've described a blue-white star and you're seeing just white – or maybe even pale gold or yellow – then there's nothing wrong. It just means that, like almost everything in life, you're having a slightly different experience. (If you read the observations of others, you'll often find a wide range of colors being described.)

If you're having difficulty seeing any color at all, try de-focusing the view. As the star loses clarity, it will appear larger, spreading the light over a wider area in your field of view and sometimes making the color easier to see.

Use Averted Vision

What's averted vision? Basically, it's observing the object out of the corner of your eye. Without getting too technical, your peripheral vision is more sensitive to light (and movement) than your direct vision. It's thought this enabled our ancestors to detect potential nearby predators before they became a more immediate threat.

If you can't see an object, try using averted vision. If that doesn't work, try tapping the telescope tube gently to make the view move as averted vision is also good for detecting movement.

Once you've found it, looking at the object with averted vision can reveal detail not visible with direct observation. However, there's a trade-off.

From personal experience, it seems that while averted vision may help you to see a faint object, it's usually lacking in color, whereas direct vision may show shades of green or blue. The center of your eye is better suited for providing a good, all-round color view while it seems the edge of your vision is best suited for detection and motion.

Keep an Astronomical Journal

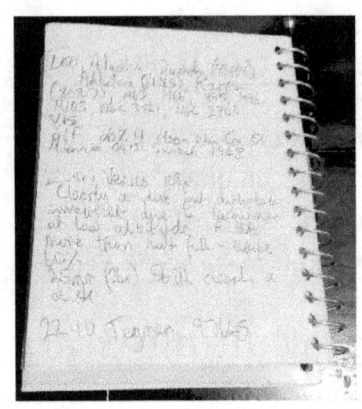

One of my journals, complete with chicken-scratch handwriting. Image by the author.

As a kid I used to keep one and wrote an entry for every object I observed on every night. Well, almost. I admit to skipping it sometimes. There's a couple of good reasons for this. Firstly, it's a great way to see keep track of which objects you've seen and to feel a real sense of accomplishment as your tally increases.

Secondly, I use it to plan my session. Before I go outside, I'll make a list of all the objects I want to observe, along with any additional information I'll need – for example, the object's catalogue number or co-ordinates. As I'm going outside, I'll also make a note of the weather, sunset time and Moon information such as its phase, illumination and rise or set time.

Lastly, it's simply a good reference tool. It's a rare object that only gets observed once and you'll return to many of your favorite objects over and over again. However, I don't re-read my notes before I observe that object again. Why? Because it can actually prejudice your viewing experience and influence your expectations.

I've also found it sometimes spoils the experience a little. There's nothing better than observing an object for the first time and discovering something new and when you're starting out there's literally hundreds of objects to discover. If you don't review your notes beforehand, it's easy to forget what you've seen and then you go to the eyepiece and rediscover that object all over again. (There's an exception to this rule. Sometimes I'll observe an object but make a separate note to re-observe it later, looking for a specific detail.)

Similarly, unless you're having difficulty observing an object, try to avoid the notes of others beforehand. I know this sounds contrary to the concept of this book, but I'm hoping the text will give you an idea of what to expect while encouraging you to have your own personal experience. And that's really the point of astronomy. It's a personal experience for everyone. Reading someone else's opinion of an object beforehand can color that experience and, I feel, you lose something as a result.

Compare your notes after you're done for the night, not before. That way you can find out if that star was truly multiple or if you were just imagining it. Or you'll read about a detail you missed and you'll want to go back and observe that object again.

Go, observe, keep a journal and compare notes later. You'll enjoy it far more that way.

Signposts to the Stars

Ursa Major

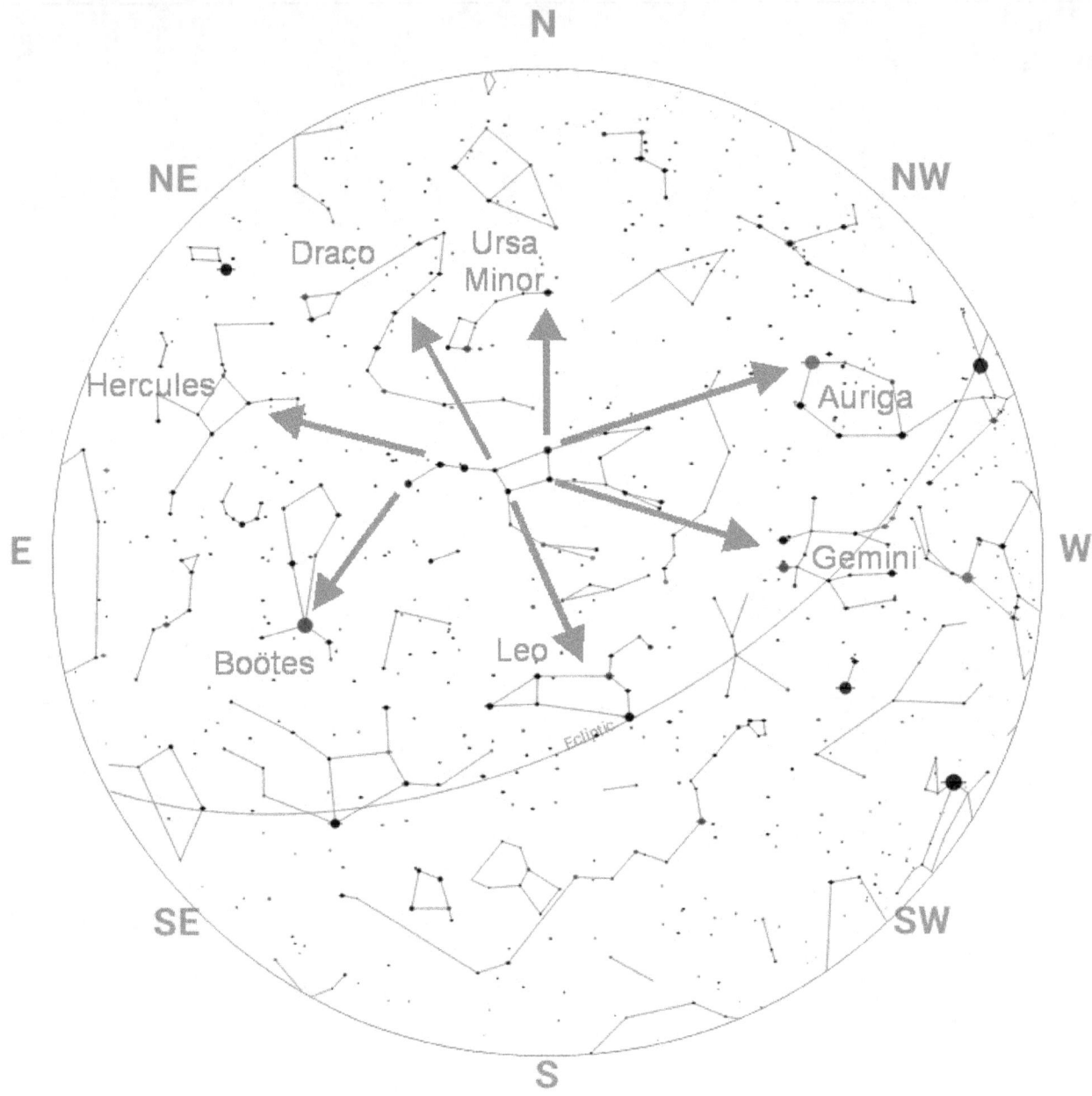

You can use the brightest stars in Ursa Major, commonly known as the Big Dipper (or the Plough in the United Kingdom) to find a number of other constellations. The table below lists these constellations.

Object	Page
Auriga	112
Boötes	114
Draco	142

Object	Page
Gemini	144
Hercules	146

Object	Page
Leo	148
Ursa Minor	178

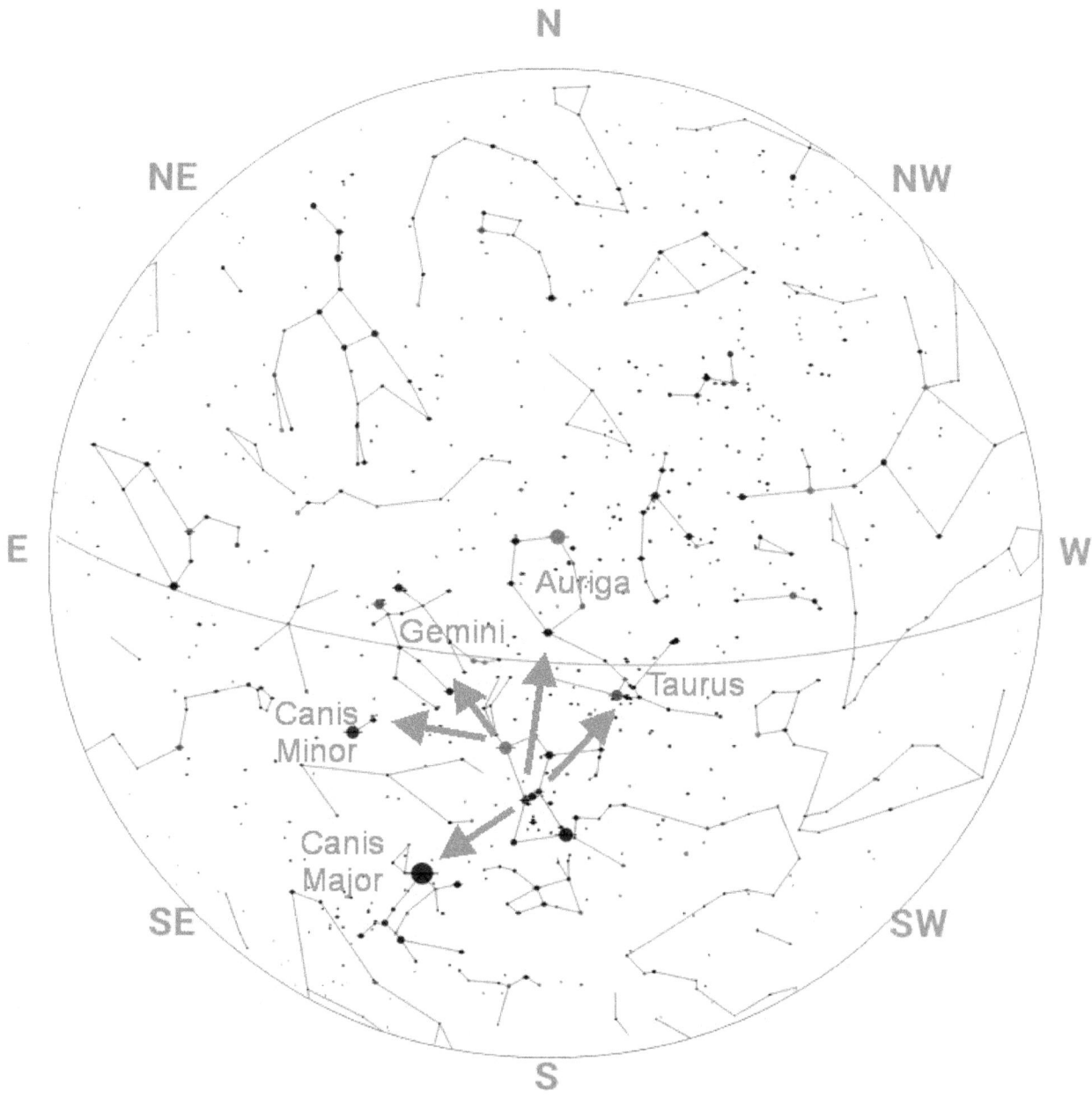

You can use Orion to find a number of other constellations. The table below lists these constellations.

Object	Page	Object	Page	Object	Page
Auriga	112	Canis Minor	122	Taurus	174
Canis Major	120	Gemini	144		

Star Charts & Observing Lists

Star Chart Tables

If observing during daylight savings time, first deduct one hour and then refer to the corresponding chart number. For example, for 10pm daylight savings time in early August, use chart 18.

	6pm	7pm	8pm	9pm	10pm	11pm
Early January	1	2	3	4	5	6
Late January	2	3	4	5	6	7
Early February	3	4	5	6	7	8
Late February	4	5	6	7	8	9
Early March	5	6	7	8	9	10
Late March	6	7	8	9	10	11
Early April	7	8	9	10	11	12
Late April	8	9	10	11	12	13
Early May	9	10	11	12	13	14
Late May	10	11	12	13	14	15
Early June	11	12	13	14	15	16
Late June	12	13	14	15	16	17
Early July	13	14	15	16	17	18
Late July	14	15	16	17	18	19
Early August	15	16	17	18	19	20
Late August	16	17	18	19	20	21
Early September	17	18	19	20	21	22
Late September	18	19	20	21	22	23
Early October	19	20	21	22	23	24
Late October	20	21	22	23	24	1
Early November	21	22	23	24	1	2
Late November	22	23	24	1	2	3
Early December	23	24	1	2	3	4
Late December	24	1	2	3	4	5

If observing during daylight savings time, first deduct one hour and then refer to the corresponding chart number. For example, for 2am daylight savings time in early July, use chart 20.

	12am	1am	2am	3am	4am	5am	6am
Early January	7	8	9	10	11	12	13
Late January	8	9	10	11	12	13	14
Early February	9	10	11	12	13	14	15
Late February	10	11	12	13	14	15	16
Early March	11	12	13	14	15	16	17
Late March	12	13	14	15	16	17	18
Early April	13	14	15	16	17	18	19
Late April	14	15	16	17	18	19	20
Early May	15	16	17	18	19	20	21
Late May	16	17	18	19	20	21	22
Early June	17	18	19	20	21	22	23
Late June	18	19	20	21	22	23	24
Early July	19	20	21	22	23	24	1
Late July	20	21	22	23	24	1	2
Early August	21	22	23	24	1	2	3
Late August	22	23	24	1	2	3	4
Early September	23	24	1	2	3	4	5
Late September	24	1	2	3	4	5	6
Early October	1	2	3	4	5	6	7
Late October	2	3	4	5	6	7	8
Early November	3	4	5	6	7	8	9
Late November	4	5	6	7	8	9	10
Early December	5	6	7	8	9	10	11
Late December	6	7	8	9	10	11	12

Chart 1

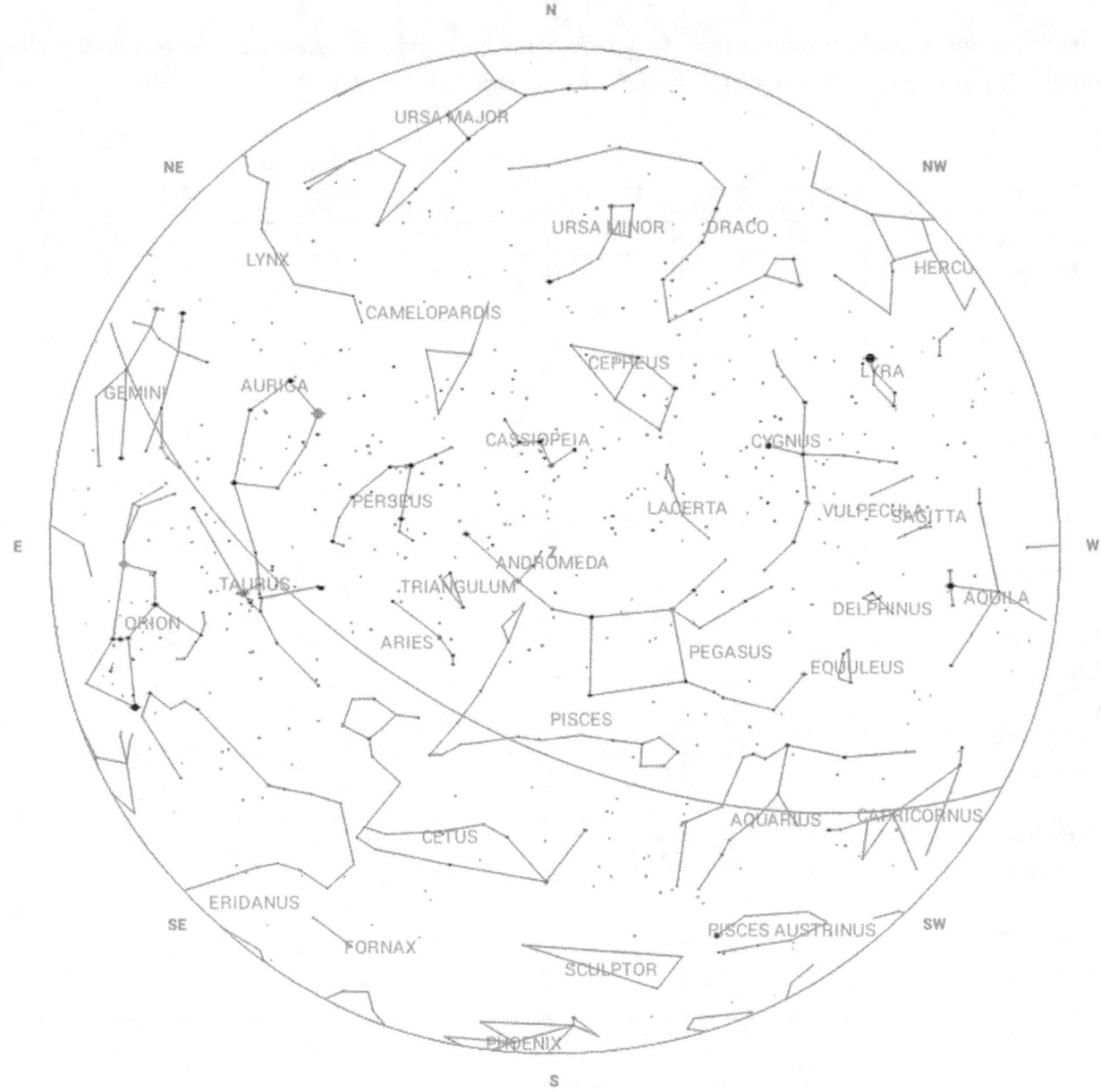

The following table shows the constellations at their best visibility at this time.

Object	Page	Object	Page	Object	Page
Andromeda	104	Cygnus	138	Perseus	162
Aquarius	106	Delphinus	140	Pisces	164
Aries	110	Draco	142	Sagitta	166
Auriga	112	Equuleus	160	Taurus	174
Cassiopeia	126	Lyra	154	Ursa Minor	178
Cepheus	128	Pegasus	160	Vulpecula	166
Cetus	130				

The following table shows telescopic objects at their best visibility (listed by constellation. NB: Not all visible objects may be listed; there may be some close to the horizon which might be visible but not at their best.)

	Constellation	R.A.	Dec.	Type	Location	Rating	Page
Pi Andromedae	Andromeda	00h 37m	+33° 43'	Multiple Star	★ ★	★ ★	184
Andromeda Galaxy	Andromeda	00h 43m	+41° 16'	Spiral Galaxy	★ ★	★ ★	185
NGC 752	Andromeda	01h 58m	+37° 52'	Open Cluster	★ ★	★ ★	186
Almach	Andromeda	02h 04m	+42° 20'	Multiple Star	★ ★ ★	★ ★ ★	187
Mesarthim	Aries	01h 53m	+19° 18'	Multiple Star	★ ★ ★	★ ★ ★	192
Lambda Arietis	Aries	01h 58m	+23° 36'	Multiple Star	★ ★ ★	★ ★ ★	193
Messier 37	Auriga	05h 52m	+32° 33'	Open Cluster	★	★ ★ ★	204
Achird	Cassiopeia	00h 49m	+57° 49'	Multiple Star	★ ★ ★	★ ★ ★	188
Owl Cluster	Cassiopeia	01h 20m	+58° 17'	Open Cluster	★	★ ★ ★	189
Messier 103	Cassiopeia	01h 33m	+60° 39'	Open Cluster	★	★	190
NGC 663	Cassiopeia	01h 46m	+61° 13'	Open Cluster	★ ★	★ ★ ★	191
Albireo	Cygnus	19h 31m	+27° 58'	Multiple Star	★ ★ ★	★ ★ ★	241
Messier 29	Cygnus	20h 24m	+38° 30'	Open Cluster	★	★ ★	242
Gamma Delphini	Delphinus	20h 47m	+16° 08'	Multiple Star	★ ★ ★	★ ★	246
Kuma	Draco	17h 32m	+55° 11'	Multiple Star	★ ★ ★	★ ★ ★	233
Messier 35	Gemini	06h 09m	+24° 21'	Open Cluster	★ ★	★ ★ ★	205
Double Double	Lyra	18h 44m	+39° 40'	Multiple Star	★ ★ ★	★ ★ ★	235
Sheliak	Lyra	18h 50m	+33° 22'	Multiple Star	★ ★ ★	★ ★	236
Ring Nebula	Lyra	18h 54m	+33° 02'	Planetary Nebula	★	★ ★ ★	237
Meissa	Orion	05h 35m	+09° 56'	Multiple Star	★ ★ ★	★ ★ ★	200
Messier 15	Pegasus	21h 30m	+12° 10'	Globular Cluster	★	★ ★	247
Double Cluster	Perseus	02h 22m	+57° 09'	Open Clusters	★ ★ ★	★ ★ ★	194
Messier 34	Perseus	02h 42m	+42° 45'	Open Cluster	★ ★	★ ★	195
Zeta Sagittae	Sagitta	19h 49m	+19° 09'	Multiple Star	★ ★	★ ★	243
The Pleiades	Taurus	03h 48m	+24° 10'	Open Cluster	★ ★ ★	★ ★ ★	197
Crab Nebula	Taurus	05h 35m	+22° 01'	Supernova Remnant	★	★	198
Polaris	Ursa Minor	02h 32m	+89° 16'	Multiple Star	★ ★ ★	★	196
Coathanger Cluster	Vulpecula	19h 26m	+20° 13'	Open Cluster	★ ★	★ ★	239
Dumbbell Nebula	Vulpecula	20h 00m	+22° 43'	Planetary Nebula	★	★ ★	240

Chart 2

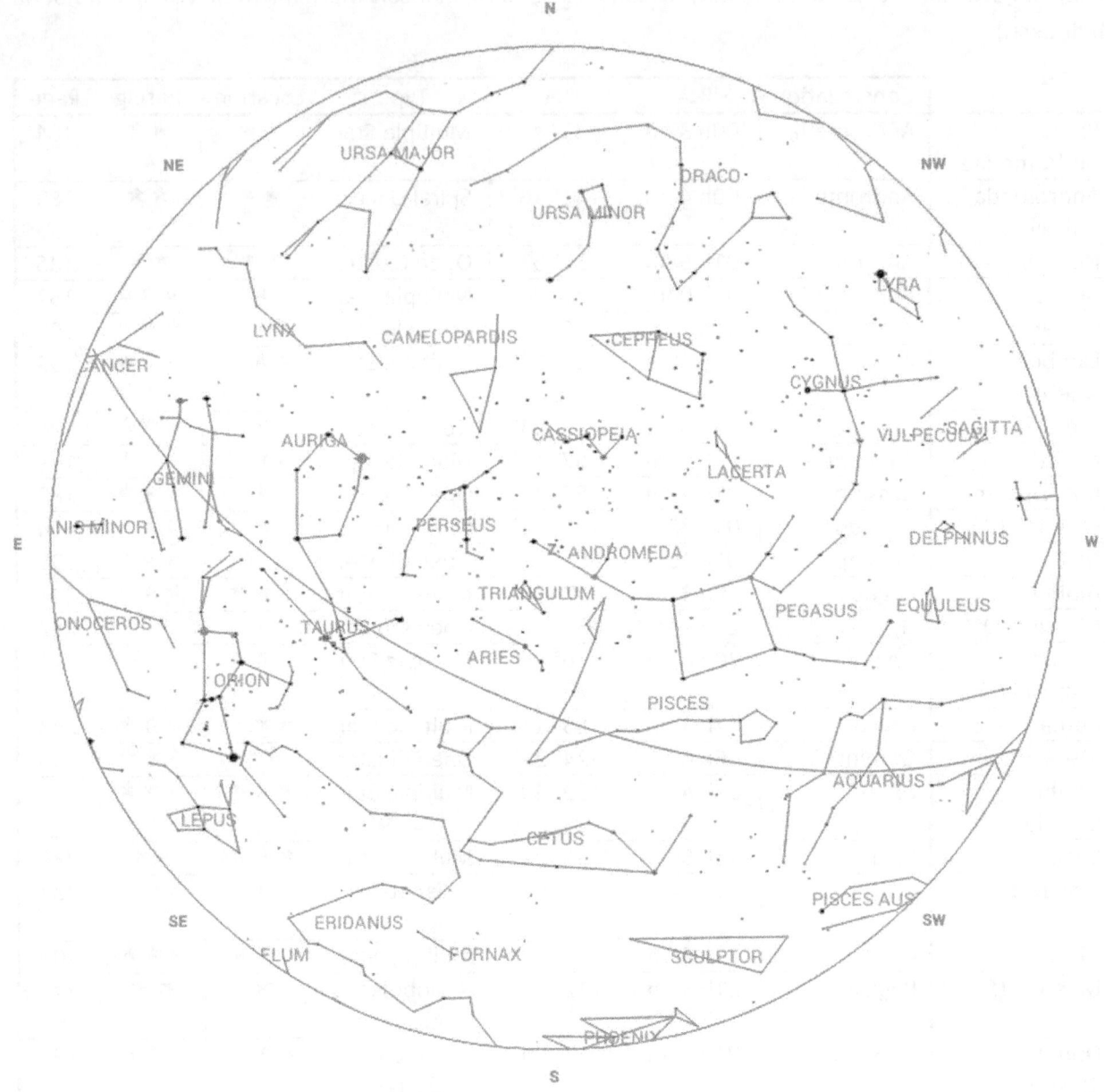

The following table shows the constellations at their best visibility at this time.

Object	Page	Object	Page	Object	Page
Andromeda	104	Cygnus	138	Pegasus	160
Aries	110	Delphinus	140	Perseus	162
Auriga	112	Draco	142	Pisces	164
Cassiopeia	126	Equuleus	160	Taurus	174
Cepheus	128	Gemini	144	Ursa Minor	178
Cetus	130	Orion	158		

The following table shows telescopic objects at their best visibility (listed by constellation. NB: Not all visible objects may be listed; there may be some close to the horizon which might be visible but not at their best.)

	Constellation	R.A.	Dec.	Type	Location	Rating	Page
Pi Andromedae	Andromeda	00h 37m	+33° 43'	Multiple Star	★★	★★	184
Andromeda Galaxy	Andromeda	00h 43m	+41° 16'	Spiral Galaxy	★★	★★	185
NGC 752	Andromeda	01h 58m	+37° 52'	Open Cluster	★★	★★	186
Almach	Andromeda	02h 04m	+42° 20'	Multiple Star	★★★	★★★	187
Mesarthim	Aries	01h 53m	+19° 18'	Multiple Star	★★★	★★★	192
Lambda Arietis	Aries	01h 58m	+23° 36'	Multiple Star	★★★	★★★	193
Messier 37	Auriga	05h 52m	+32° 33'	Open Cluster	★	★★★	204
Achird	Cassiopeia	00h 49m	+57° 49'	Multiple Star	★★★	★★★	188
Owl Cluster	Cassiopeia	01h 20m	+58° 17'	Open Cluster	★	★★★	189
Messier 103	Cassiopeia	01h 33m	+60° 39'	Open Cluster	★	★	190
NGC 663	Cassiopeia	01h 46m	+61° 13'	Open Cluster	★★	★★★	191
Albireo	Cygnus	19h 31m	+27° 58'	Multiple Star	★★★	★★★	241
Messier 29	Cygnus	20h 24m	+38° 30'	Open Cluster	★	★★	242
Gamma Delphini	Delphinus	20h 47m	+16° 08'	Multiple Star	★★★	★★	246
Kuma	Draco	17h 32m	+55° 11'	Multiple Star	★★★	★★★	233
Messier 35	Gemini	06h 09m	+24° 21'	Open Cluster	★★	★★★	205
Mintaka	Orion	05h 32m	-00° 18'	Multiple Star	★★★	★★	199
Meissa	Orion	05h 35m	+09° 56'	Multiple Star	★★★	★★★	200
Orion Nebula	Orion	05h 35m	-05° 23'	Nebula	★★★	★★★	201
Sigma Orionis	Orion	05h 39m	-02° 36'	Multiple Star	★★★	★★★	202
Messier 15	Pegasus	21h 30m	+12° 10'	Globular Cluster	★	★★	247
Double Cluster	Perseus	02h 22m	+57° 09'	Open Clusters	★★★	★★★	194
Messier 34	Perseus	02h 42m	+42° 45'	Open Cluster	★★	★★	195
The Pleiades	Taurus	03h 48m	+24° 10'	Open Cluster	★★★	★★★	197
Crab Nebula	Taurus	05h 35m	+22° 01'	Supernova Remnant	★	★	198
Polaris	Ursa Minor	02h 32m	+89° 16'	Multiple Star	★★★	★	196

Chart 3

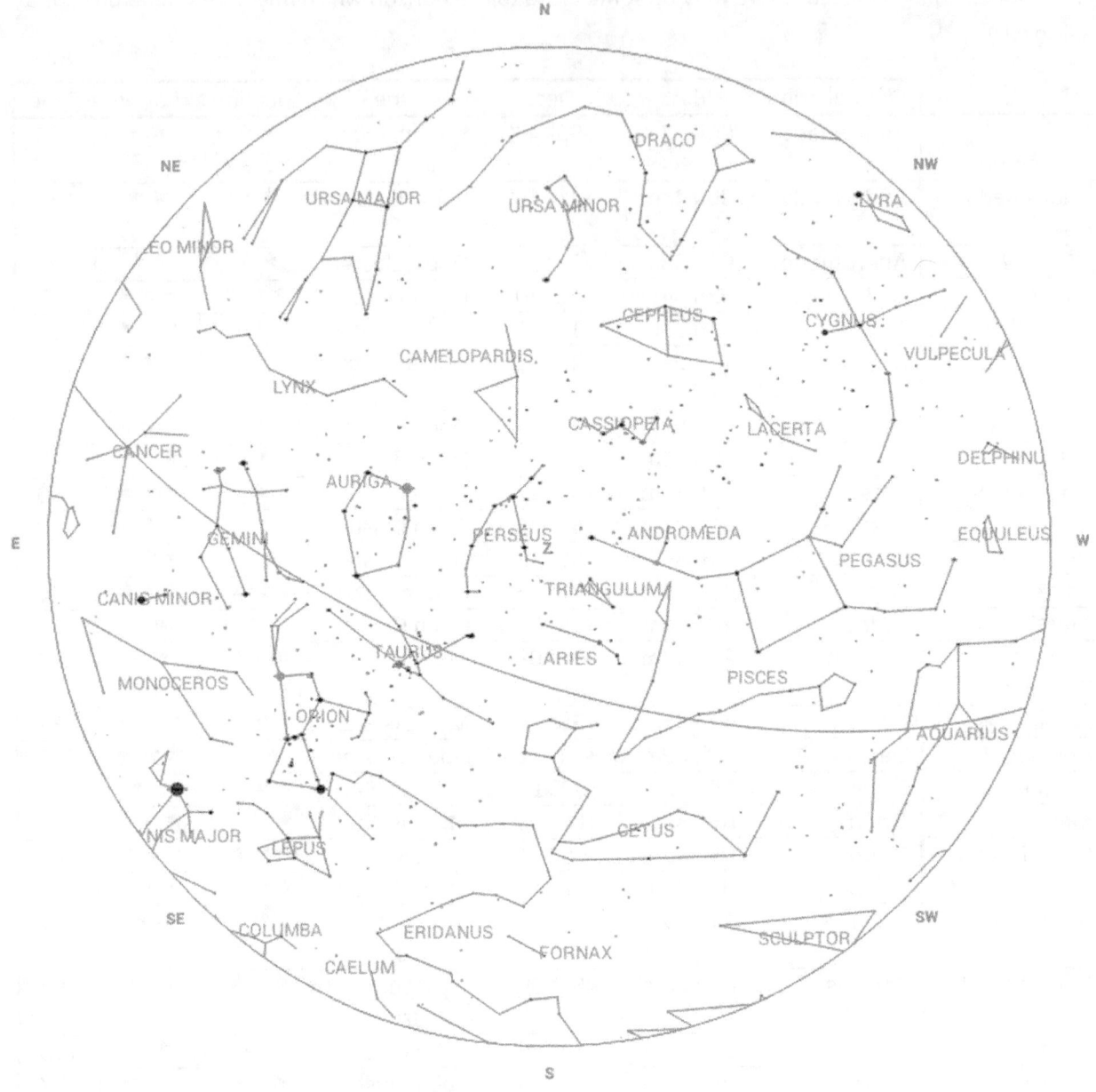

The following table shows the constellations at their best visibility at this time.

Object	Page
Andromeda	104
Aries	110
Auriga	112
Cassiopeia	126
Cepheus	128

Object	Page
Cetus	130
Gemini	144
Orion	158
Pegasus	160

Object	Page
Perseus	162
Pisces	164
Taurus	174
Ursa Minor	178

The following table shows telescopic objects at their best visibility (listed by constellation. NB: Not all visible objects may be listed; there may be some close to the horizon which might be visible but not at their best.)

	Constellation	R.A.	Dec.	Type	Location	Rating	Page
Pi Andromedae	Andromeda	00h 37m	+33° 43'	Multiple Star	★★	★★	184
Andromeda Galaxy	Andromeda	00h 43m	+41° 16'	Spiral Galaxy	★★	★★	185
NGC 752	Andromeda	01h 58m	+37° 52'	Open Cluster	★★	★★	186
Almach	Andromeda	02h 04m	+42° 20'	Multiple Star	★★★	★★★	187
Mesarthim	Aries	01h 53m	+19° 18'	Multiple Star	★★★	★★★	192
Lambda Arietis	Aries	01h 58m	+23° 36'	Multiple Star	★★★	★★★	193
Messier 37	Auriga	05h 52m	+32° 33'	Open Cluster	★	★★★	204
Achird	Cassiopeia	00h 49m	+57° 49'	Multiple Star	★★★	★★★	188
Owl Cluster	Cassiopeia	01h 20m	+58° 17'	Open Cluster	★	★★★	189
Messier 103	Cassiopeia	01h 33m	+60° 39'	Open Cluster	★	★	190
NGC 663	Cassiopeia	01h 46m	+61° 13'	Open Cluster	★★	★★★	191
Messier 35	Gemini	06h 09m	+24° 21'	Open Cluster	★★	★★★	205
Castor	Gemini	07h 35m	+31° 53'	Multiple Star	★★★	★★★	206
Mintaka	Orion	05h 32m	-00° 18'	Multiple Star	★★★	★★	199
Meissa	Orion	05h 35m	+09° 56'	Multiple Star	★★★	★★★	200
Orion Nebula	Orion	05h 35m	-05° 23'	Nebula	★★★	★★★	201
Sigma Orionis	Orion	05h 39m	-02° 36'	Multiple Star	★★★	★★★	202
Double Cluster	Perseus	02h 22m	+57° 09'	Open Clusters	★★★	★★★	194
Messier 34	Perseus	02h 42m	+42° 45'	Open Cluster	★★	★★	195
The Pleiades	Taurus	03h 48m	+24° 10'	Open Cluster	★★★	★★★	197
Crab Nebula	Taurus	05h 35m	+22° 01'	Supernova Remnant	★	★	198
Polaris	Ursa Minor	02h 32m	+89° 16'	Multiple Star	★★★	★	196

Chart 4

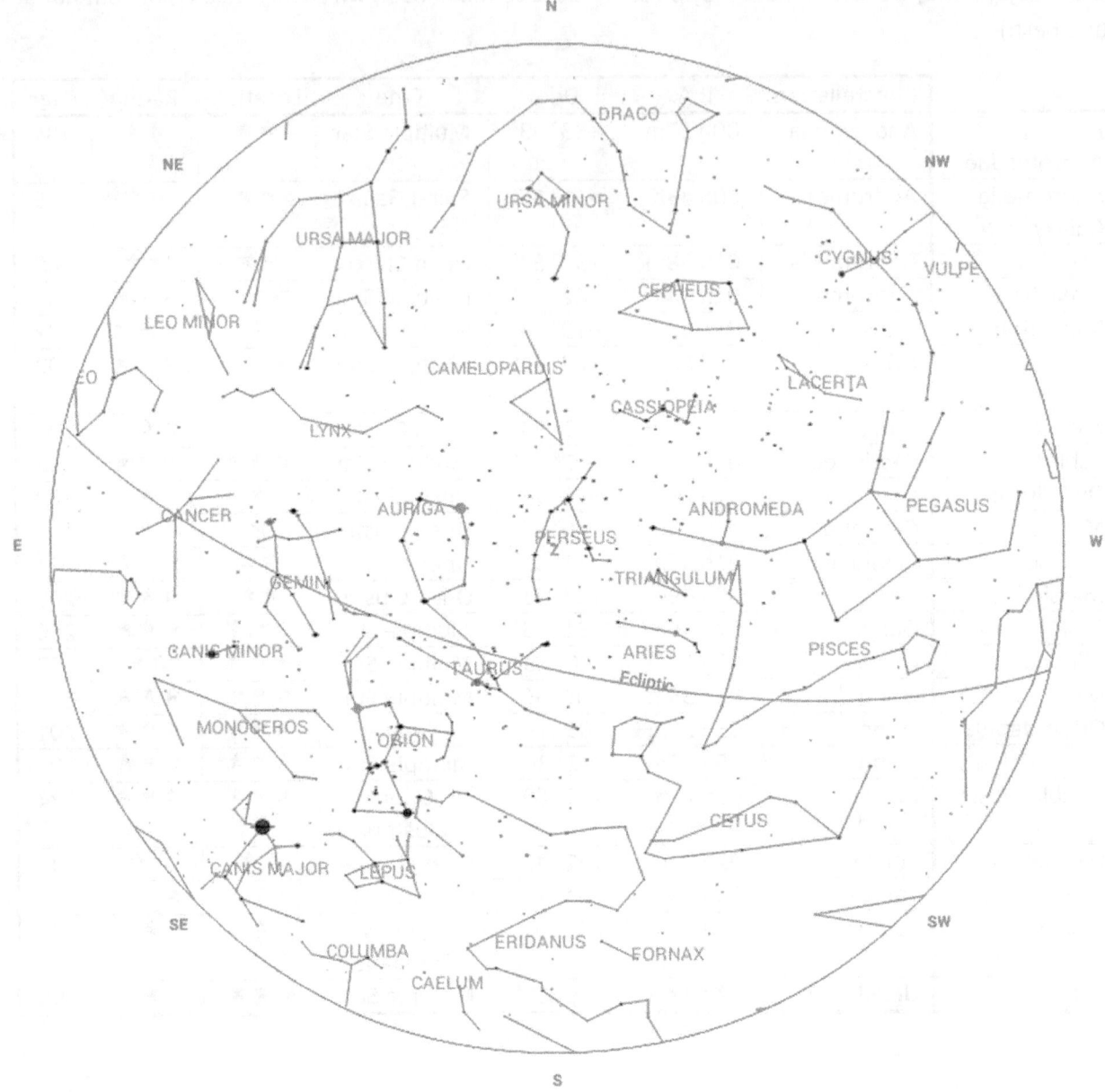

The following table shows the constellations at their best visibility at this time.

Object	Page
Andromeda	104
Aries	110
Auriga	112
Cancer	116
Canis Minor	122

Object	Page
Cassiopeia	126
Cepheus	128
Cetus	130
Gemini	144
Lepus	150

Object	Page
Orion	158
Perseus	162
Pisces	164
Taurus	174
Ursa Minor	178

The following table shows telescopic objects at their best visibility (listed by constellation. NB: Not all visible objects may be listed; there may be some close to the horizon which might be visible but not at their best.)

	Constellation	R.A.	Dec.	Type	Location	Rating	Page
Pi Andromedae	Andromeda	00h 37m	+33° 43'	Multiple Star	★★	★★	184
Andromeda Galaxy	Andromeda	00h 43m	+41° 16'	Spiral Galaxy	★★	★★	185
NGC 752	Andromeda	01h 58m	+37° 52'	Open Cluster	★★	★★	186
Almach	Andromeda	02h 04m	+42° 20'	Multiple Star	★★★	★★★	187
Mesarthim	Aries	01h 53m	+19° 18'	Multiple Star	★★★	★★★	192
Lambda Arietis	Aries	01h 58m	+23° 36'	Multiple Star	★★★	★★★	193
Messier 37	Auriga	05h 52m	+32° 33'	Open Cluster	★	★★★	204
Praesepe	Cancer	08h 40m	+19° 40'	Open Cluster	★★	★★★	212
Iota Cancri	Cancer	08h 47m	+28° 46'	Multiple Star	★★★	★★★	213
Procyon	Canis Minor	07h 39m	+05° 13'	Multiple Star	★★★	★	210
Achird	Cassiopeia	00h 49m	+57° 49'	Multiple Star	★★★	★★★	188
Owl Cluster	Cassiopeia	01h 20m	+58° 17'	Open Cluster	★	★★★	189
Messier 103	Cassiopeia	01h 33m	+60° 39'	Open Cluster	★	★	190
NGC 663	Cassiopeia	01h 46m	+61° 13'	Open Cluster	★★	★★★	191
Messier 35	Gemini	06h 09m	+24° 21'	Open Cluster	★★	★★★	205
Castor	Gemini	07h 35m	+31° 53'	Multiple Star	★★★	★★★	206
Gamma Leporis	Lepus	05h 44m	-22° 27'	Multiple Star	★★★	★★	203
Beta Monocerotis	Monoceros	06h 29m	-07° 02'	Multiple Star	★★	★★	207
Mintaka	Orion	05h 32m	-00° 18'	Multiple Star	★★★	★★	199
Meissa	Orion	05h 35m	+09° 56'	Multiple Star	★★★	★★★	200
Orion Nebula	Orion	05h 35m	-05° 23'	Nebula	★★★	★★★	201
Sigma Orionis	Orion	05h 39m	-02° 36'	Multiple Star	★★★	★★★	202
Double Cluster	Perseus	02h 22m	+57° 09'	Open Clusters	★★★	★★★	194
Messier 34	Perseus	02h 42m	+42° 45'	Open Cluster	★★	★★	195
The Pleiades	Taurus	03h 48m	+24° 10'	Open Cluster	★★★	★★★	197
Crab Nebula	Taurus	05h 35m	+22° 01'	Supernova Remnant	★	★	198
Polaris	Ursa Minor	02h 32m	+89° 16'	Multiple Star	★★★	★	196

Chart 5

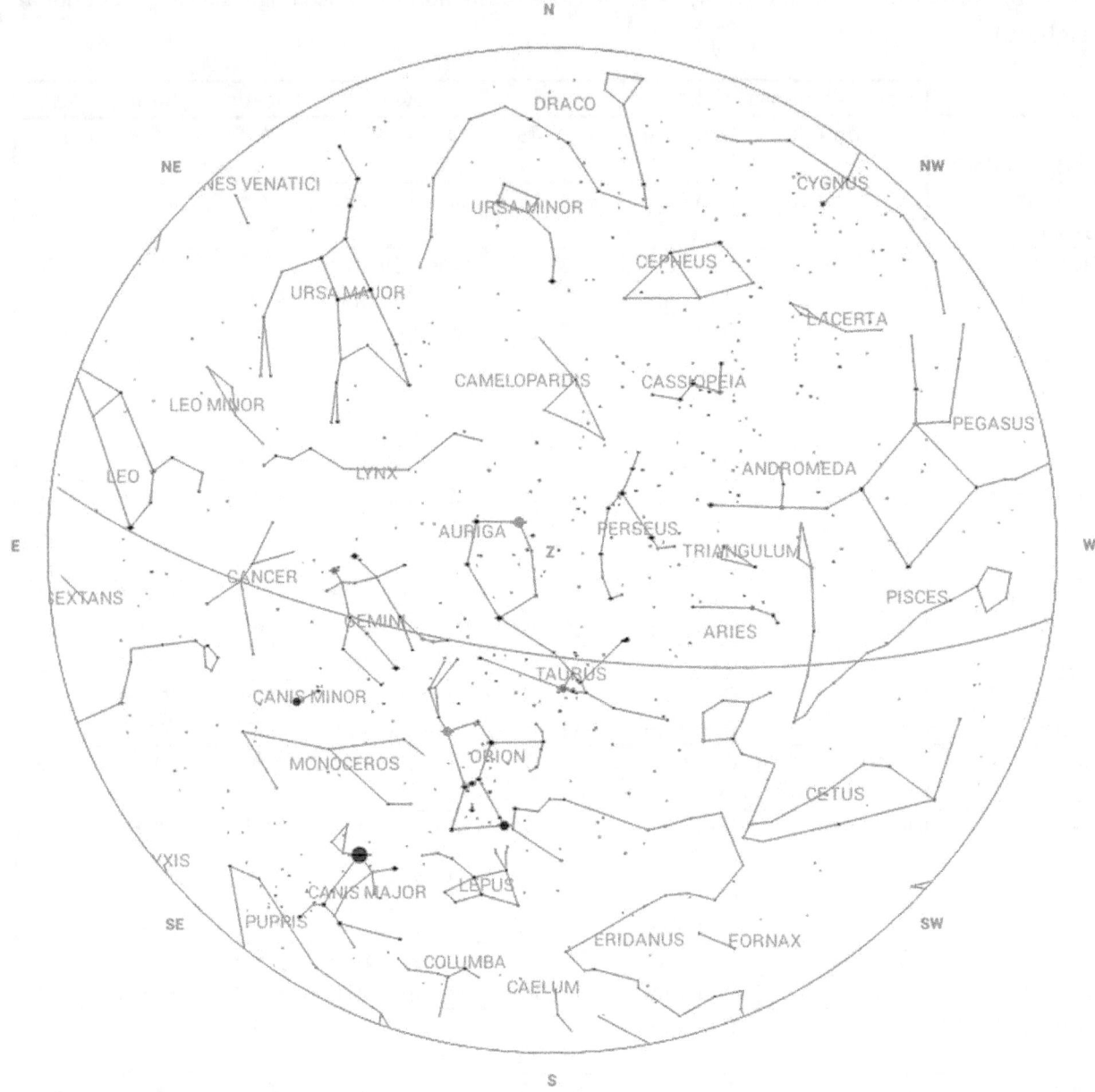

The following table shows the constellations at their best visibility at this time.

Object	Page
Andromeda	104
Aries	110
Auriga	112
Cancer	116
Canis Major	120

Object	Page
Canis Minor	122
Cassiopeia	126
Cepheus	128
Gemini	144
Lepus	150

Object	Page
Orion	158
Perseus	162
Taurus	174
Ursa Minor	178

The following table shows telescopic objects at their best visibility (listed by constellation. NB: Not all visible objects may be listed; there may be some close to the horizon which might be visible but not at their best.)

	Constellation	R.A.	Dec.	Type	Location	Rating	Page
Pi Andromedae	Andromeda	00h 37m	+33° 43'	Multiple Star	★★	★★	184
Andromeda Galaxy	Andromeda	00h 43m	+41° 16'	Spiral Galaxy	★★	★★	185
NGC 752	Andromeda	01h 58m	+37° 52'	Open Cluster	★★	★★	186
Almach	Andromeda	02h 04m	+42° 20'	Multiple Star	★★★	★★★	187
Mesarthim	Aries	01h 53m	+19° 18'	Multiple Star	★★★	★★★	192
Lambda Arietis	Aries	01h 58m	+23° 36'	Multiple Star	★★★	★★★	193
Messier 37	Auriga	05h 52m	+32° 33'	Open Cluster	★	★★★	204
Praesepe	Cancer	08h 40m	+19° 40'	Open Cluster	★★	★★★	212
Iota Cancri	Cancer	08h 47m	+28° 46'	Multiple Star	★★★	★★★	213
Messier 67	Cancer	08h 51m	+11° 49'	Open Cluster	★	★★	214
Sirius	Canis Major	06h 45m	-16° 43'	Multiple Star	★★★	★	208
Procyon	Canis Minor	07h 39m	+05° 13'	Multiple Star	★★★	★	210
Achird	Cassiopeia	00h 49m	+57° 49'	Multiple Star	★★★	★★★	188
Owl Cluster	Cassiopeia	01h 20m	+58° 17'	Open Cluster	★	★★★	189
Messier 103	Cassiopeia	01h 33m	+60° 39'	Open Cluster	★	★	190
NGC 663	Cassiopeia	01h 46m	+61° 13'	Open Cluster	★★	★★★	191
Messier 35	Gemini	06h 09m	+24° 21'	Open Cluster	★★	★★★	205
Castor	Gemini	07h 35m	+31° 53'	Multiple Star	★★★	★★★	206
Gamma Leporis	Lepus	05h 44m	-22° 27'	Multiple Star	★★★	★★	203
Beta Monocerotis	Monoceros	06h 29m	-07° 02'	Multiple Star	★★	★★	207
Mintaka	Orion	05h 32m	-00° 18'	Multiple Star	★★★	★★	199
Meissa	Orion	05h 35m	+09° 56'	Multiple Star	★★★	★★★	200
Orion Nebula	Orion	05h 35m	-05° 23'	Nebula	★★★	★★★	201
Sigma Orionis	Orion	05h 39m	-02° 36'	Multiple Star	★★★	★★★	202
Double Cluster	Perseus	02h 22m	+57° 09'	Open Clusters	★★★	★★★	194
Messier 34	Perseus	02h 42m	+42° 45'	Open Cluster	★★	★★	195
The Pleiades	Taurus	03h 48m	+24° 10'	Open Cluster	★★★	★★★	197
Crab Nebula	Taurus	05h 35m	+22° 01'	Supernova Remnant	★	★	198
Polaris	Ursa Minor	02h 32m	+89° 16'	Multiple Star	★★★	★	196

Chart 6

The following table shows the constellations at their best visibility at this time.

Object	Page
Andromeda	104
Aries	110
Auriga	112
Cancer	116
Canis Major	120

Object	Page
Canis Minor	122
Cassiopeia	126
Cepheus	128
Gemini	144
Leo	148

Object	Page
Lepus	150
Orion	158
Perseus	162
Taurus	174
Ursa Minor	178

The following table shows telescopic objects at their best visibility (listed by constellation. NB: Not all visible objects may be listed; there may be some close to the horizon which might be visible but not at their best.)

	Constellation	R.A.	Dec.	Type	Location	Rating	Page
Pi Andromedae	Andromeda	00h 37m	+33° 43'	Multiple Star	★★	★★	184
Andromeda Galaxy	Andromeda	00h 43m	+41° 16'	Spiral Galaxy	★★	★★	185
NGC 752	Andromeda	01h 58m	+37° 52'	Open Cluster	★★	★★	186
Almach	Andromeda	02h 04m	+42° 20'	Multiple Star	★★★	★★★	187
Mesarthim	Aries	01h 53m	+19° 18'	Multiple Star	★★★	★★★	192
Lambda Arietis	Aries	01h 58m	+23° 36'	Multiple Star	★★★	★★★	193
Messier 37	Auriga	05h 52m	+32° 33'	Open Cluster	★	★★★	204
Praesepe	Cancer	08h 40m	+19° 40'	Open Cluster	★★	★★★	212
Iota Cancri	Cancer	08h 47m	+28° 46'	Multiple Star	★★★	★★★	213
Messier 67	Cancer	08h 51m	+11° 49'	Open Cluster	★	★★	214
Sirius	Canis Major	06h 45m	-16° 43'	Multiple Star	★★★	★	208
Messier 41	Canis Major	06h 46m	-20° 45'	Open Cluster	★★	★★★	209
Procyon	Canis Minor	07h 39m	+05° 13'	Multiple Star	★★★	★	210
Achird	Cassiopeia	00h 49m	+57° 49'	Multiple Star	★★★	★★★	188
Owl Cluster	Cassiopeia	01h 20m	+58° 17'	Open Cluster	★	★★★	189
Messier 103	Cassiopeia	01h 33m	+60° 39'	Open Cluster	★	★	190
NGC 663	Cassiopeia	01h 46m	+61° 13'	Open Cluster	★★	★★★	191
Messier 35	Gemini	06h 09m	+24° 21'	Open Cluster	★★	★★★	205
Castor	Gemini	07h 35m	+31° 53'	Multiple Star	★★★	★★★	206
Regulus	Leo	10h 08m	+11° 58'	Multiple Star	★★★	★	215
Adhafera	Leo	10h 17m	+23° 25'	Multiple Star	★★★	★★	216
Algieba	Leo	10h 20m	+19° 51'	Multiple Star	★★★	★★★	217
Gamma Leporis	Lepus	05h 44m	-22° 27'	Multiple Star	★★★	★★	203
Beta Monocerotis	Monoceros	06h 29m	-07° 02'	Multiple Star	★★	★★	207
Mintaka	Orion	05h 32m	-00° 18'	Multiple Star	★★★	★★	199
Meissa	Orion	05h 35m	+09° 56'	Multiple Star	★★★	★★★	200
Orion Nebula	Orion	05h 35m	-05° 23'	Nebula	★★★	★★★	201
Sigma Orionis	Orion	05h 39m	-02° 36'	Multiple Star	★★★	★★★	202
Double Cluster	Perseus	02h 22m	+57° 09'	Open Clusters	★★★	★★★	194
Messier 34	Perseus	02h 42m	+42° 45'	Open Cluster	★★	★★	195
Messier 93	Puppis	07h 44m	-23° 51'	Open Cluster	★	★★	211
The Pleiades	Taurus	03h 48m	+24° 10'	Open Cluster	★★★	★★★	197
Crab Nebula	Taurus	05h 35m	+22° 01'	Supernova Remnant	★	★	198
Polaris	Ursa Minor	02h 32m	+89° 16'	Multiple Star	★★★	★	196

Chart 7

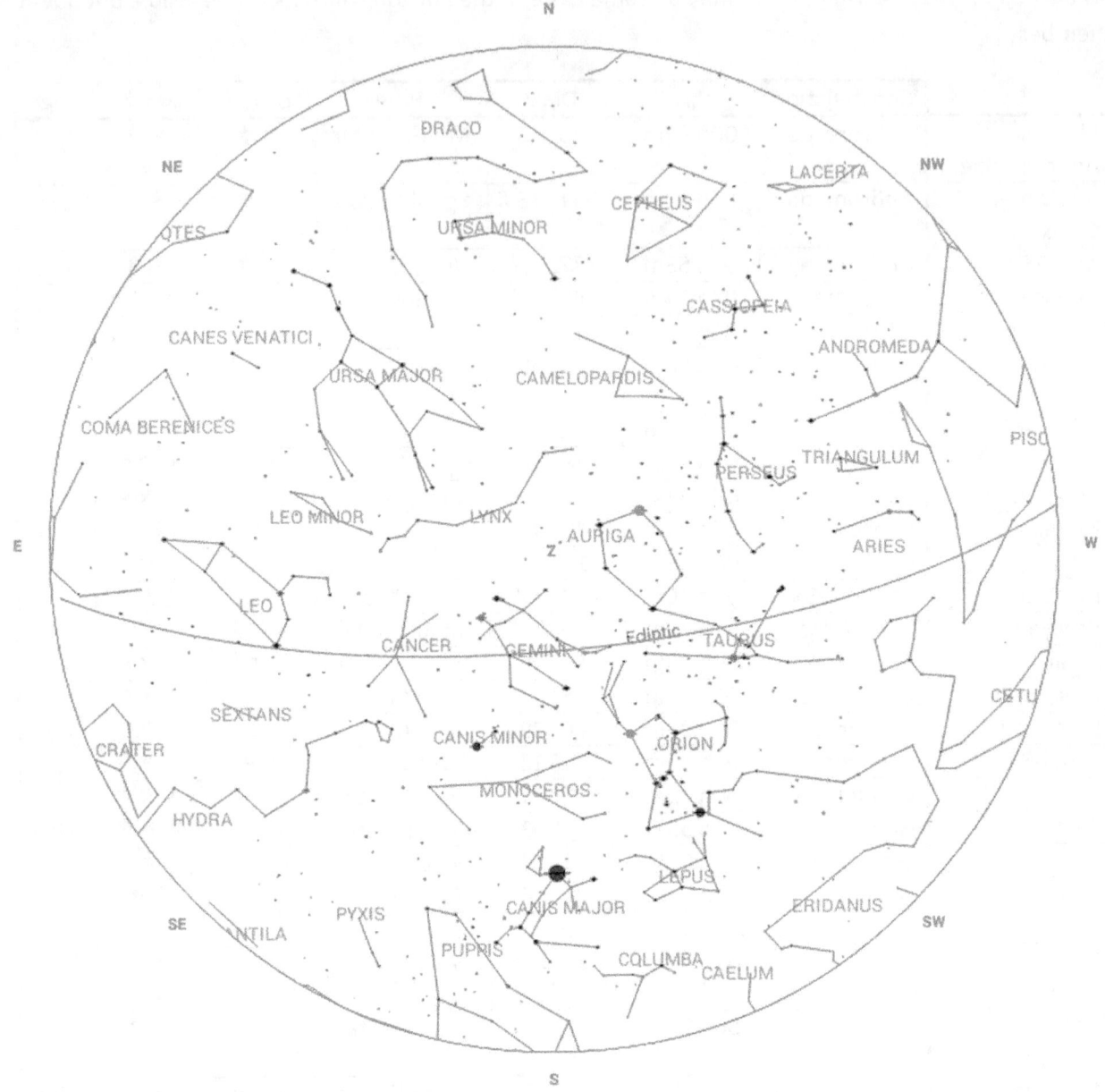

The following table shows the constellations at their best visibility at this time.

Object	Page
Aries	110
Auriga	112
Cancer	116
Canes Venatici	64
Canis Major	120
Canis Minor	122

Object	Page
Cassiopeia	126
Cepheus	128
Gemini	144
Leo	148
Lepus	150

Object	Page
Orion	158
Perseus	162
Taurus	174
Ursa Major	176
Ursa Minor	178

The following table shows telescopic objects at their best visibility (listed by constellation. NB: Not all visible objects may be listed; there may be some close to the horizon which might be visible but not at their best.)

	Constellation	R.A.	Dec.	Type	Location	Rating	Page
Andromeda Galaxy	Andromeda	00h 43m	+41° 16′	Spiral Galaxy	★★	★★	185
NGC 752	Andromeda	01h 58m	+37° 52′	Open Cluster	★★	★★	186
Almach	Andromeda	02h 04m	+42° 20′	Multiple Star	★★★	★★★	187
Mesarthim	Aries	01h 53m	+19° 18′	Multiple Star	★★★	★★★	192
Lambda Arietis	Aries	01h 58m	+23° 36′	Multiple Star	★★★	★★★	193
Messier 37	Auriga	05h 52m	+32° 33′	Open Cluster	★	★★★	204
Praesepe	Cancer	08h 40m	+19° 40′	Open Cluster	★★	★★★	212
Iota Cancri	Cancer	08h 47m	+28° 46′	Multiple Star	★★★	★★★	213
Messier 67	Cancer	08h 51m	+11° 49′	Open Cluster	★	★★	214
Sirius	Canis Major	06h 45m	-16° 43′	Multiple Star	★★★	★	208
Messier 41	Canis Major	06h 46m	-20° 45′	Open Cluster	★★	★★★	209
Procyon	Canis Minor	07h 39m	+05° 13′	Multiple Star	★★★	★	210
Achird	Cassiopeia	00h 49m	+57° 49′	Multiple Star	★★★	★★★	188
Owl Cluster	Cassiopeia	01h 20m	+58° 17′	Open Cluster	★	★★★	189
Messier 103	Cassiopeia	01h 33m	+60° 39′	Open Cluster	★	★	190
NGC 663	Cassiopeia	01h 46m	+61° 13′	Open Cluster	★★	★★★	191
Messier 35	Gemini	06h 09m	+24° 21′	Open Cluster	★★	★★★	205
Castor	Gemini	07h 35m	+31° 53′	Multiple Star	★★★	★★★	206
Regulus	Leo	10h 08m	+11° 58′	Multiple Star	★★★	★	215
Adhafera	Leo	10h 17m	+23° 25′	Multiple Star	★★★	★★	216
Algieba	Leo	10h 20m	+19° 51′	Multiple Star	★★★	★★★	217
Denebola	Leo	11h 49m	+14° 34′	Multiple Star	★★★	★★	218
Gamma Leporis	Lepus	05h 44m	-22° 27′	Multiple Star	★★★	★★	203
Beta Monocerotis	Monoceros	06h 29m	-07° 02′	Multiple Star	★★	★★	207
Mintaka	Orion	05h 32m	-00° 18′	Multiple Star	★★★	★★	199
Meissa	Orion	05h 35m	+09° 56′	Multiple Star	★★★	★★★	200
Orion Nebula	Orion	05h 35m	-05° 23′	Nebula	★★★	★★★	201
Sigma Orionis	Orion	05h 39m	-02° 36′	Multiple Star	★★★	★★★	202
Double Cluster	Perseus	02h 22m	+57° 09′	Open Clusters	★★★	★★★	194
Messier 34	Perseus	02h 42m	+42° 45′	Open Cluster	★★	★★	195
Messier 93	Puppis	07h 44m	-23° 51′	Open Cluster	★	★★	211
The Pleiades	Taurus	03h 48m	+24° 10′	Open Cluster	★★★	★★★	197
Crab Nebula	Taurus	05h 35m	+22° 01′	Supernova Remnant	★	★	198
Mizar & Alcor	Ursa Major	13h 24m	+54° 56′	Multiple Star	★★★	★★★	221
Polaris	Ursa Minor	02h 32m	+89° 16′	Multiple Star	★★★	★	196

Chart 8

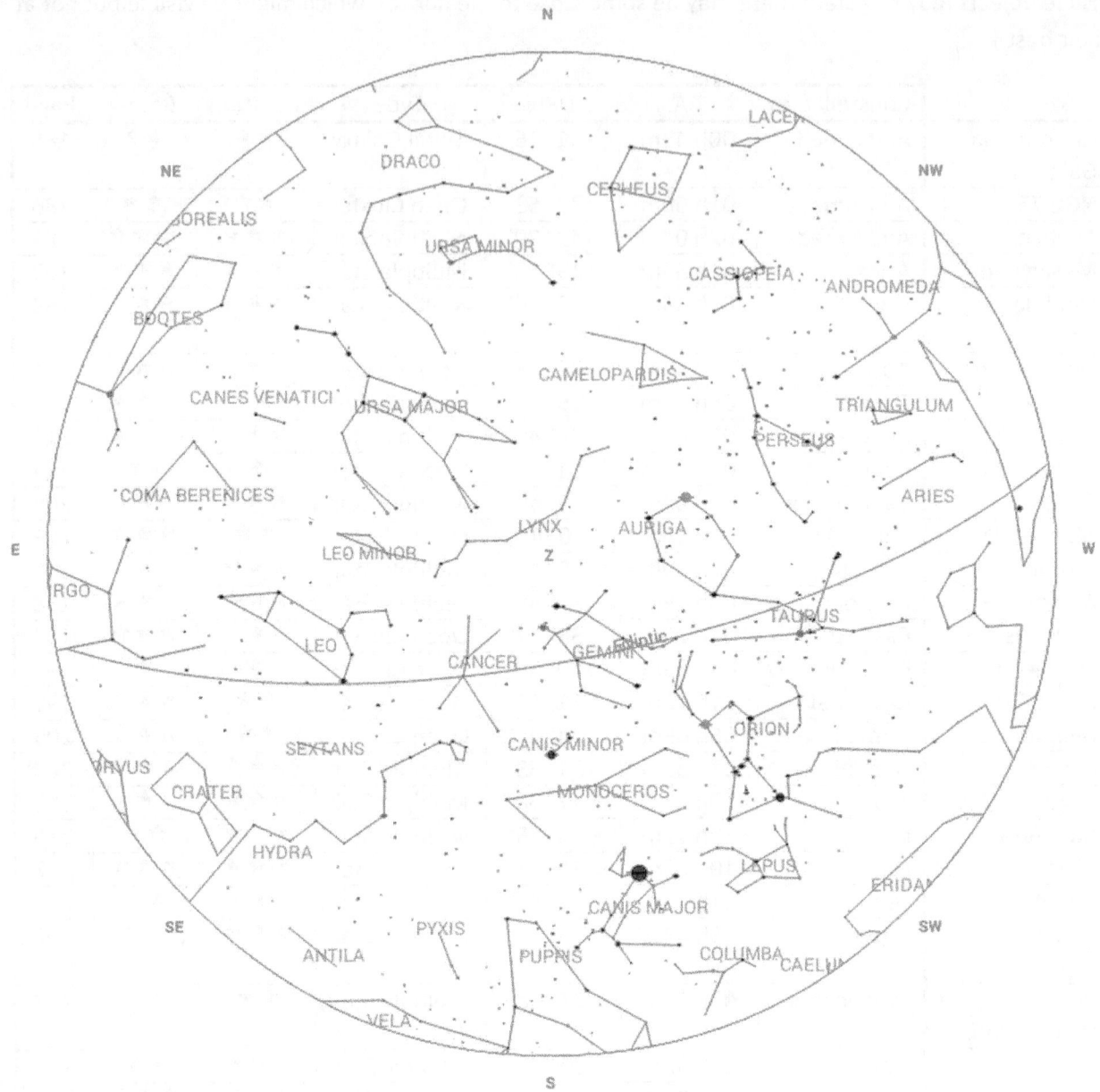

The following table shows the constellations at their best visibility at this time.

Object	Page
Auriga	112
Cancer	116
Canes Venatici	118
Canis Major	120
Canis Minor	122

Object	Page
Cassiopeia	126
Gemini	144
Leo	148
Lepus	150
Orion	158

Object	Page
Perseus	162
Taurus	174
Ursa Major	176
Ursa Minor	178

The following table shows telescopic objects at their best visibility (listed by constellation. NB: Not all visible objects may be listed; there may be some close to the horizon which might be visible but not at their best.)

	Constellation	R.A.	Dec.	Type	Location	Rating	Page
Messier 37	Auriga	05h 52m	+32° 33'	Open Cluster	★	★★★	204
Praesepe	Cancer	08h 40m	+19° 40'	Open Cluster	★★	★★★	212
Iota Cancri	Cancer	08h 47m	+28° 46'	Multiple Star	★★★	★★★	213
Messier 67	Cancer	08h 51m	+11° 49'	Open Cluster	★	★★	214
Cor Caroli	Canes Venatici	12h 56m	+38° 19'	Multiple Star	★★★	★★★	220
Sirius	Canis Major	06h 45m	-16° 43'	Multiple Star	★★★	★	208
Messier 41	Canis Major	06h 46m	-20° 45'	Open Cluster	★★	★★★	209
Procyon	Canis Minor	07h 39m	+05° 13'	Multiple Star	★★★	★	210
Achird	Cassiopeia	00h 49m	+57° 49'	Multiple Star	★★★	★★★	188
Owl Cluster	Cassiopeia	01h 20m	+58° 17'	Open Cluster	★	★★★	189
Messier 103	Cassiopeia	01h 33m	+60° 39'	Open Cluster	★	★	190
NGC 663	Cassiopeia	01h 46m	+61° 13'	Open Cluster	★★	★★★	191
Messier 35	Gemini	06h 09m	+24° 21'	Open Cluster	★★	★★★	205
Castor	Gemini	07h 35m	+31° 53'	Multiple Star	★★★	★★★	206
Regulus	Leo	10h 08m	+11° 58'	Multiple Star	★★★	★	215
Adhafera	Leo	10h 17m	+23° 25'	Multiple Star	★★★	★★	216
Algieba	Leo	10h 20m	+19° 51'	Multiple Star	★★★	★★★	217
Denebola	Leo	11h 49m	+14° 34'	Multiple Star	★★★	★★	218
Gamma Leporis	Lepus	05h 44m	-22° 27'	Multiple Star	★★★	★★	203
Beta Monocerotis	Monoceros	06h 29m	-07° 02'	Multiple Star	★★	★★	207
Mintaka	Orion	05h 32m	-00° 18'	Multiple Star	★★★	★★	199
Meissa	Orion	05h 35m	+09° 56'	Multiple Star	★★★	★★★	200
Orion Nebula	Orion	05h 35m	-05° 23'	Nebula	★★★	★★★	201
Sigma Orionis	Orion	05h 39m	-02° 36'	Multiple Star	★★★	★★★	202
Double Cluster	Perseus	02h 22m	+57° 09'	Open Clusters	★★★	★★★	194
Messier 34	Perseus	02h 42m	+42° 45'	Open Cluster	★★	★★	195
Messier 93	Puppis	07h 44m	-23° 51'	Open Cluster	★	★★	211
The Pleiades	Taurus	03h 48m	+24° 10'	Open Cluster	★★★	★★★	197
Crab Nebula	Taurus	05h 35m	+22° 01'	Supernova Remnant	★	★	198
Mizar & Alcor	Ursa Major	13h 24m	+54° 56'	Multiple Star	★★★	★★★	221
Polaris	Ursa Minor	02h 32m	+89° 16'	Multiple Star	★★★	★	196

Chart 9

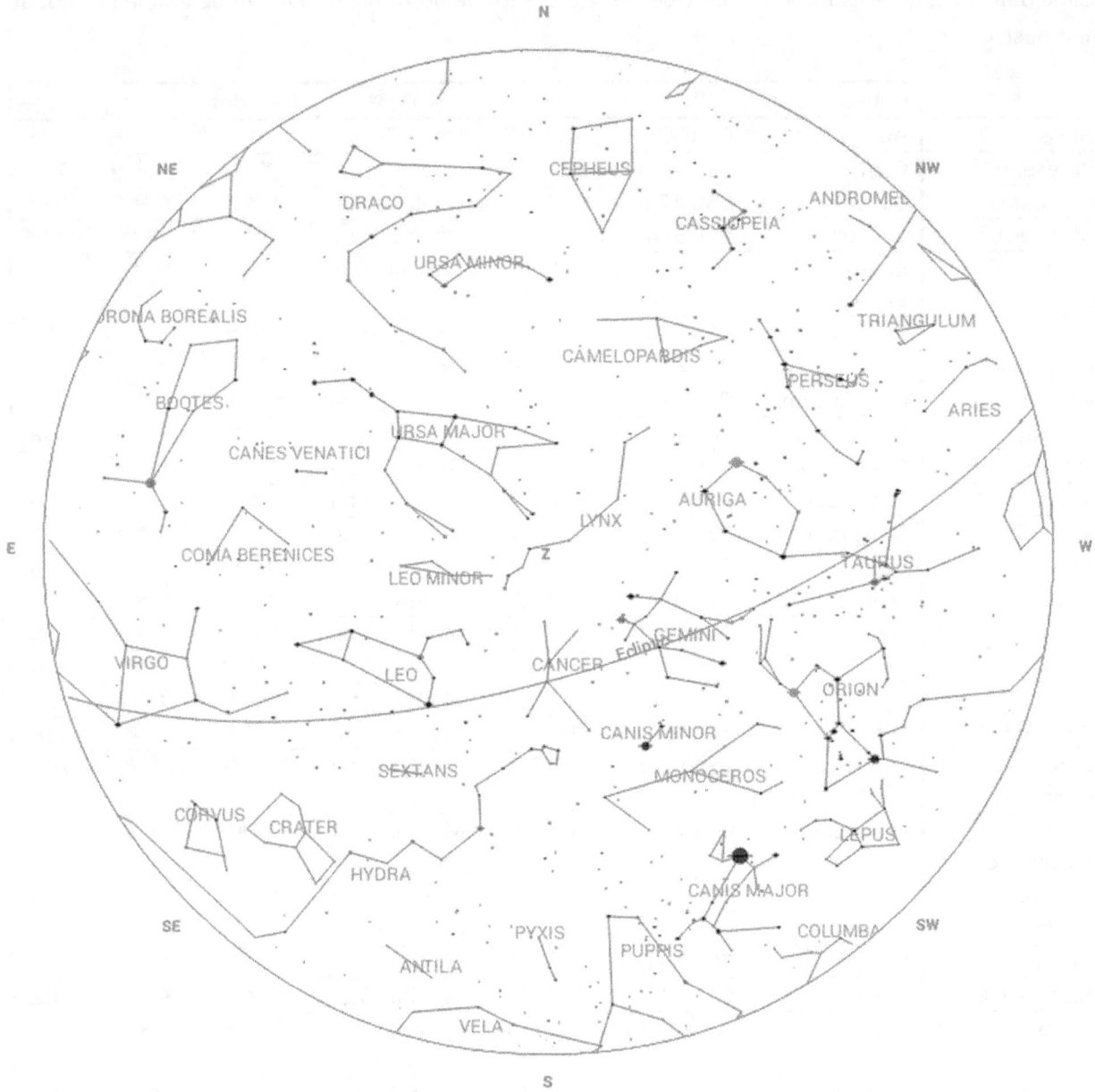

The following table shows the constellations at their best visibility at this time.

Object	Page
Auriga	112
Boötes	114
Cancer	116
Canes Venatici	118
Canis Major	120

Object	Page
Canis Minor	122
Cassiopeia	126
Coma Berenices	132
Gemini	144
Leo	148

Object	Page
Orion	158
Perseus	162
Taurus	174
Ursa Major	176
Ursa Minor	178

The following table shows telescopic objects at their best visibility (listed by constellation. NB: Not all visible objects may be listed; there may be some close to the horizon which might be visible but not at their best.)

	Constellation	R.A.	Dec.	Type	Location	Rating	Page
Messier 37	Auriga	05h 52m	+32° 33'	Open Cluster	★	★★★	204
Arcturus	Boötes	14h 16m	+19° 11'	Multiple Star	★★★	★	222
Praesepe	Cancer	08h 40m	+19° 40'	Open Cluster	★★	★★★	212
Iota Cancri	Cancer	08h 47m	+28° 46'	Multiple Star	★★★	★★★	213
Messier 67	Cancer	08h 51m	+11° 49'	Open Cluster	★	★★	214
Cor Caroli	Canes Venatici	12h 56m	+38° 19'	Multiple Star	★★★	★★★	220
Sirius	Canis Major	06h 45m	-16° 43'	Multiple Star	★★★	★	208
Messier 41	Canis Major	06h 46m	-20° 45'	Open Cluster	★★	★★★	209
Procyon	Canis Minor	07h 39m	+05° 13'	Multiple Star	★★★	★	210
Achird	Cassiopeia	00h 49m	+57° 49'	Multiple Star	★★★	★★★	188
Owl Cluster	Cassiopeia	01h 20m	+58° 17'	Open Cluster	★	★★★	189
Messier 103	Cassiopeia	01h 33m	+60° 39'	Open Cluster	★	★	190
NGC 663	Cassiopeia	01h 46m	+61° 13'	Open Cluster	★★	★★★	191
Messier 35	Gemini	06h 09m	+24° 21'	Open Cluster	★★	★★★	205
Castor	Gemini	07h 35m	+31° 53'	Multiple Star	★★★	★★★	206
Regulus	Leo	10h 08m	+11° 58'	Multiple Star	★★★	★	215
Adhafera	Leo	10h 17m	+23° 25'	Multiple Star	★★★	★★	216
Algieba	Leo	10h 20m	+19° 51'	Multiple Star	★★★	★★★	217
Denebola	Leo	11h 49m	+14° 34'	Multiple Star	★★★	★★	218
Beta Monocerotis	Monoceros	06h 29m	-07° 02'	Multiple Star	★★	★★	207
Mintaka	Orion	05h 32m	-00° 18'	Multiple Star	★★★	★★	199
Meissa	Orion	05h 35m	+09° 56'	Multiple Star	★★★	★★★	200
Orion Nebula	Orion	05h 35m	-05° 23'	Nebula	★★★	★★★	201
Sigma Orionis	Orion	05h 39m	-02° 36'	Multiple Star	★★★	★★★	202
Double Cluster	Perseus	02h 22m	+57° 09'	Open Clusters	★★★	★★★	194
Messier 34	Perseus	02h 42m	+42° 45'	Open Cluster	★★	★★	195
Messier 93	Puppis	07h 44m	-23° 51'	Open Cluster	★	★★	211
The Pleiades	Taurus	03h 48m	+24° 10'	Open Cluster	★★★	★★★	197
Crab Nebula	Taurus	05h 35m	+22° 01'	Supernova Remnant	★	★	198
Mizar & Alcor	Ursa Major	13h 24m	+54° 56'	Multiple Star	★★★	★★★	221
Polaris	Ursa Minor	02h 32m	+89° 16'	Multiple Star	★★★	★	196

Chart 10

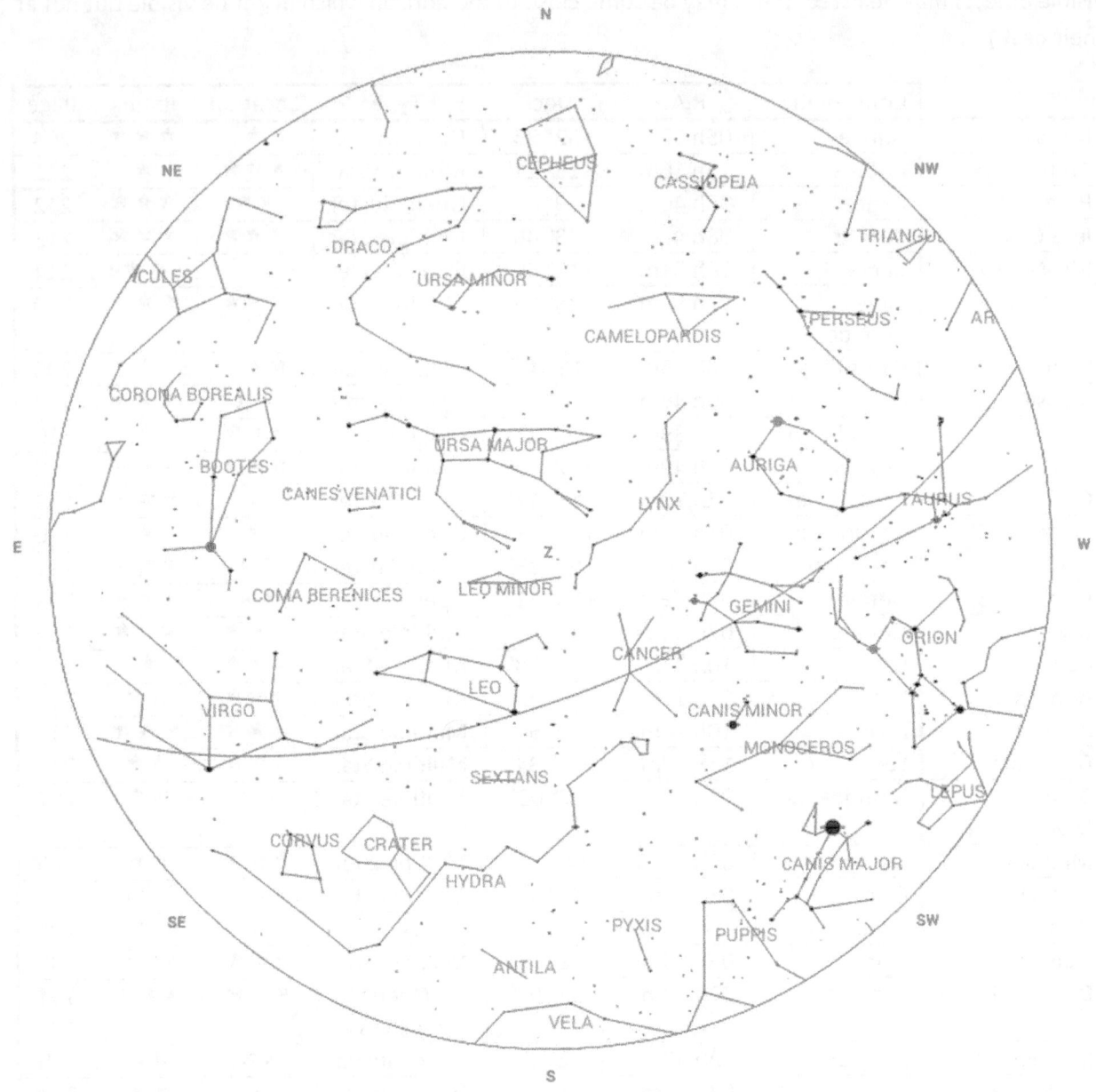

The following table shows the constellations at their best visibility at this time.

Object	Page	Object	Page	Object	Page
Auriga	112	Coma Berenices	132	Taurus	174
Boötes	114	Gemini	144	Ursa Major	176
Cancer	116	Leo	148	Ursa Minor	178
Canes Venatici	118	Orion	158		
Canis Minor	122	Perseus	162		

The following table shows telescopic objects at their best visibility (listed by constellation. NB: Not all visible objects may be listed; there may be some close to the horizon which might be visible but not at their best.)

	Constellation	R.A.	Dec.	Type	Location	Rating	Page
Messier 37	Auriga	05h 52m	+32° 33'	Open Cluster	★	★★★	204
Arcturus	Boötes	14h 16m	+19° 11'	Multiple Star	★★★	★	222
Delta Boötis	Boötes	15h 16m	+33° 19'	Multiple Star	★★★	★★★	223
Praesepe	Cancer	08h 40m	+19° 40'	Open Cluster	★★	★★★	212
Iota Cancri	Cancer	08h 47m	+28° 46'	Multiple Star	★★★	★★★	213
Messier 67	Cancer	08h 51m	+11° 49'	Open Cluster	★	★★	214
Cor Caroli	Canes Venatici	12h 56m	+38° 19'	Multiple Star	★★★	★★★	220
Procyon	Canis Minor	07h 39m	+05° 13'	Multiple Star	★★★	★	210
Messier 35	Gemini	06h 09m	+24° 21'	Open Cluster	★★	★★★	205
Castor	Gemini	07h 35m	+31° 53'	Multiple Star	★★★	★★★	206
Regulus	Leo	10h 08m	+11° 58'	Multiple Star	★★★	★	215
Adhafera	Leo	10h 17m	+23° 25'	Multiple Star	★★★	★★	216
Algieba	Leo	10h 20m	+19° 51'	Multiple Star	★★★	★★★	217
Denebola	Leo	11h 49m	+14° 34'	Multiple Star	★★★	★★	218
Beta Monocerotis	Monoceros	06h 29m	-07° 02'	Multiple Star	★★	★★	207
Mintaka	Orion	05h 32m	-00° 18'	Multiple Star	★★★	★★	199
Meissa	Orion	05h 35m	+09° 56'	Multiple Star	★★★	★★★	200
Sigma Orionis	Orion	05h 39m	-02° 36'	Multiple Star	★★★	★★★	202
Double Cluster	Perseus	02h 22m	+57° 09'	Open Clusters	★★★	★★★	194
Messier 34	Perseus	02h 42m	+42° 45'	Open Cluster	★★	★★	195
Messier 93	Puppis	07h 44m	-23° 51'	Open Cluster	★	★★	211
The Pleiades	Taurus	03h 48m	+24° 10'	Open Cluster	★★★	★★★	197
Crab Nebula	Taurus	05h 35m	+22° 01'	Supernova Remnant	★	★	198
Mizar & Alcor	Ursa Major	13h 24m	+54° 56'	Multiple Star	★★★	★★★	221
Polaris	Ursa Minor	02h 32m	+89° 16'	Multiple Star	★★★	★	196

Chart 11

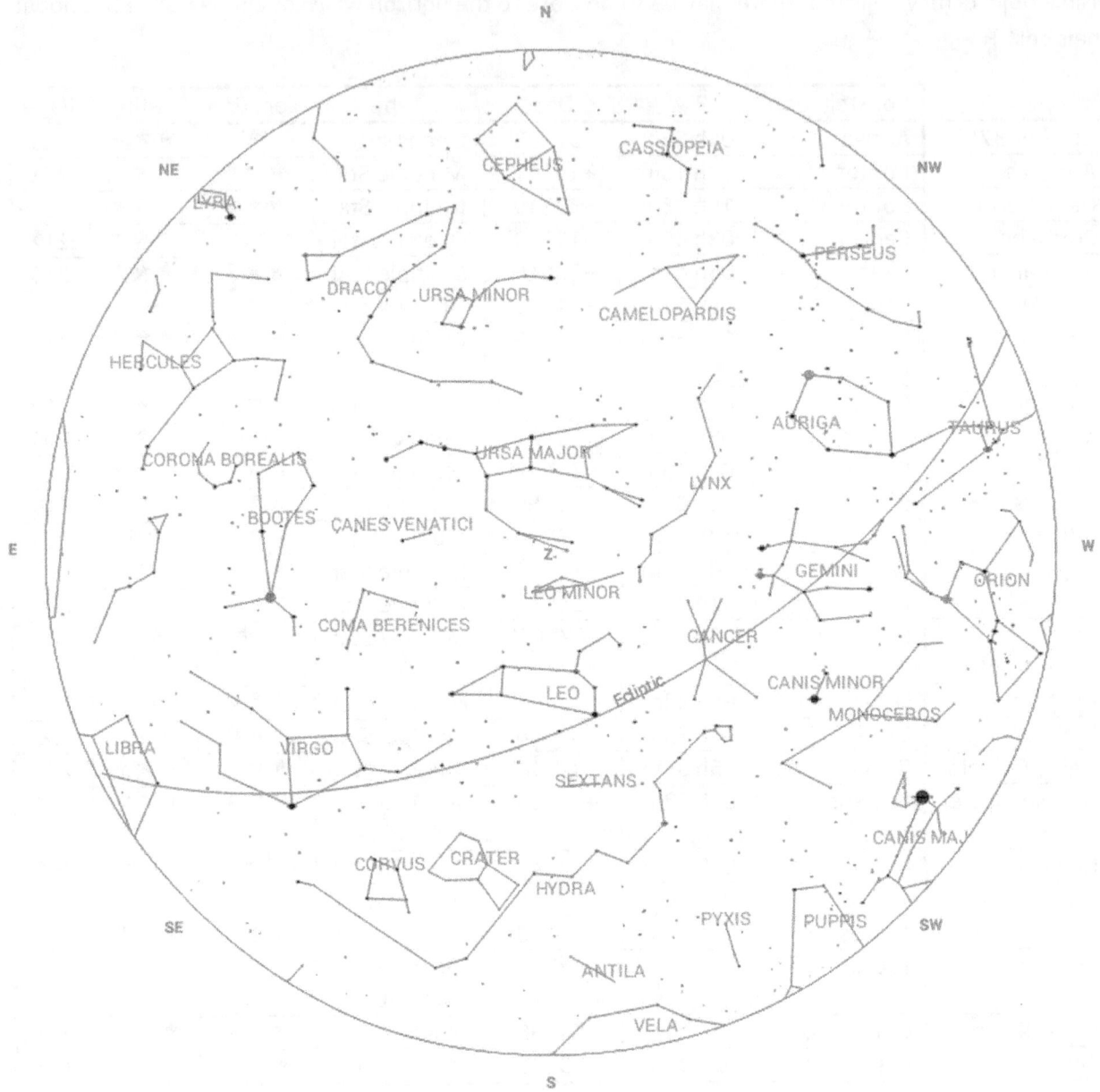

The following table shows the constellations at their best visibility at this time.

Object	Page
Auriga	112
Boötes	114
Cancer	116
Canes Venatici	118
Canis Minor	122

Object	Page
Coma Berenices	132
Corona Borealis	134
Corvus	136
Crater	136
Draco	142

Object	Page
Gemini	144
Leo	148
Ursa Major	176
Ursa Minor	178
Virgo	180

The following table shows telescopic objects at their best visibility (listed by constellation. NB: Not all visible objects may be listed; there may be some close to the horizon which might be visible but not at their best.)

	Constellation	R.A.	Dec.	Type	Location	Rating	Page
Messier 37	Auriga	05h 52m	+32° 33'	Open Cluster	★	★★★	204
Arcturus	Boötes	14h 16m	+19° 11'	Multiple Star	★★★	★	222
Delta Boötis	Boötes	15h 16m	+33° 19'	Multiple Star	★★★	★★★	223
Praesepe	Cancer	08h 40m	+19° 40'	Open Cluster	★★	★★★	212
Iota Cancri	Cancer	08h 47m	+28° 46'	Multiple Star	★★★	★★★	213
Messier 67	Cancer	08h 51m	+11° 49'	Open Cluster	★	★★	214
Cor Caroli	Canes Venatici	12h 56m	+38° 19'	Multiple Star	★★★	★★★	220
Procyon	Canis Minor	07h 39m	+05° 13'	Multiple Star	★★★	★	210
Algorab	Corvus	12h 30m	-16° 31'	Multiple Star	★★★	★★	219
Kuma	Draco	17h 32m	+55° 11'	Multiple Star	★★★	★★★	233
Messier 35	Gemini	06h 09m	+24° 21'	Open Cluster	★★	★★★	205
Castor	Gemini	07h 35m	+31° 53'	Multiple Star	★★★	★★★	206
Keystone Cluster	Hercules	16h 42m	+36° 28'	Globular Cluster	★★	★★★	231
Regulus	Leo	10h 08m	+11° 58'	Multiple Star	★★★	★	215
Adhafera	Leo	10h 17m	+23° 25'	Multiple Star	★★★	★★	216
Algieba	Leo	10h 20m	+19° 51'	Multiple Star	★★★	★★★	217
Denebola	Leo	11h 49m	+14° 34'	Multiple Star	★★★	★★	218
Beta Monocerotis	Monoceros	06h 29m	-07° 02'	Multiple Star	★★	★★	207
Mizar & Alcor	Ursa Major	13h 24m	+54° 56'	Multiple Star	★★★	★★★	221
Polaris	Ursa Minor	02h 32m	+89° 16'	Multiple Star	★★★	★	196

Chart 12

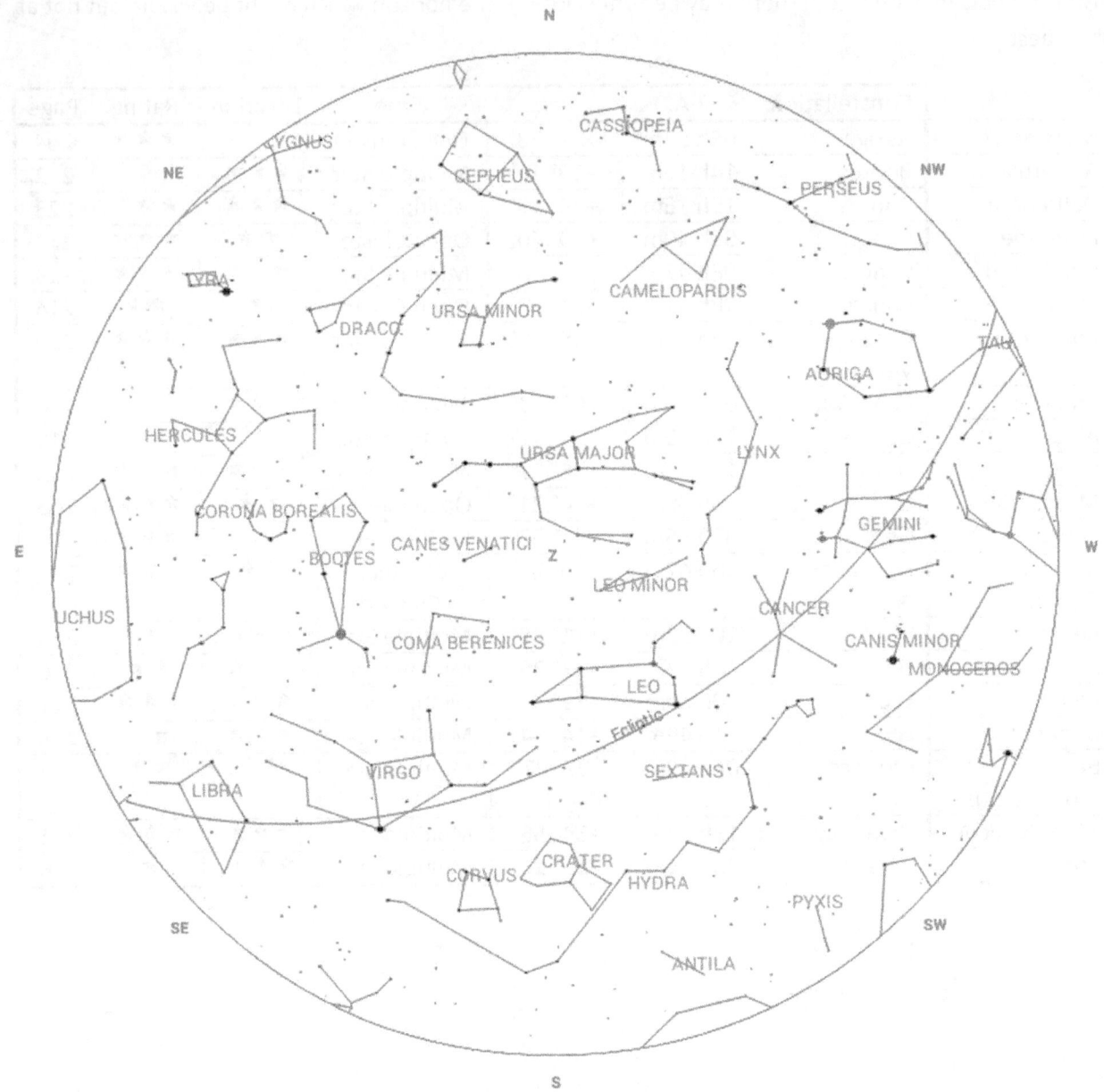

The following table shows the constellations at their best visibility at this time.

Object	Page
Auriga	112
Boötes	114
Cancer	116
Canes Venatici	118
Canis Minor	122
Coma Berenices	132

Object	Page
Corona Borealis	134
Corvus	136
Crater	136
Draco	142
Gemini	144

Object	Page
Hercules	146
Leo	148
Ursa Major	176
Ursa Minor	178
Virgo	180

The following table shows telescopic objects at their best visibility (listed by constellation. NB: Not all visible objects may be listed; there may be some close to the horizon which might be visible but not at their best.)

	Constellation	R.A.	Dec.	Type	Location	Rating	Page
Arcturus	Boötes	14h 16m	+19° 11'	Multiple Star	★ ★ ★	★	222
Delta Boötis	Boötes	15h 16m	+33° 19'	Multiple Star	★ ★ ★	★ ★ ★	223
Praesepe	Cancer	08h 40m	+19° 40'	Open Cluster	★ ★	★ ★ ★	212
Iota Cancri	Cancer	08h 47m	+28° 46'	Multiple Star	★ ★ ★	★ ★ ★	213
Messier 67	Cancer	08h 51m	+11° 49'	Open Cluster	★	★ ★	214
Cor Caroli	Canes Venatici	12h 56m	+38° 19'	Multiple Star	★ ★ ★	★ ★ ★	220
Procyon	Canis Minor	07h 39m	+05° 13'	Multiple Star	★ ★ ★	★	210
Algorab	Corvus	12h 30m	-16° 31'	Multiple Star	★ ★ ★	★ ★	219
Kuma	Draco	17h 32m	+55° 11'	Multiple Star	★ ★ ★	★ ★ ★	233
Messier 35	Gemini	06h 09m	+24° 21'	Open Cluster	★ ★	★ ★ ★	205
Castor	Gemini	07h 35m	+31° 53'	Multiple Star	★ ★ ★	★ ★ ★	206
Keystone Cluster	Hercules	16h 42m	+36° 28'	Globular Cluster	★ ★	★ ★ ★	231
Rho Herculis	Hercules	17h 24m	+37° 08'	Multiple Star	★ ★	★ ★	232
Regulus	Leo	10h 08m	+11° 58'	Multiple Star	★ ★ ★	★	215
Adhafera	Leo	10h 17m	+23° 25'	Multiple Star	★ ★ ★	★ ★	216
Algieba	Leo	10h 20m	+19° 51'	Multiple Star	★ ★ ★	★ ★ ★	217
Denebola	Leo	11h 49m	+14° 34'	Multiple Star	★ ★ ★	★ ★	218
Mizar & Alcor	Ursa Major	13h 24m	+54° 56'	Multiple Star	★ ★ ★	★ ★ ★	221
Polaris	Ursa Minor	02h 32m	+89° 16'	Multiple Star	★ ★ ★	★	196

Chart 13

The following table shows the constellations at their best visibility at this time.

Object	Page
Boötes	114
Cancer	116
Canes Venatici	118
Coma Berenices	132
Corona Borealis	134

Object	Page
Corvus	136
Crater	136
Draco	142
Hercules	146
Leo	148

Object	Page
Lyra	154
Ursa Major	176
Ursa Minor	178
Virgo	180

The following table shows telescopic objects at their best visibility (listed by constellation. NB: Not all visible objects may be listed; there may be some close to the horizon which might be visible but not at their best.)

	Constellation	R.A.	Dec.	Type	Location	Rating	Page
Arcturus	Boötes	14h 16m	+19° 11'	Multiple Star	★★★	★	222
Delta Boötis	Boötes	15h 16m	+33° 19'	Multiple Star	★★★	★★★	223
Praesepe	Cancer	08h 40m	+19° 40'	Open Cluster	★★	★★★	212
Iota Cancri	Cancer	08h 47m	+28° 46'	Multiple Star	★★★	★★★	213
Messier 67	Cancer	08h 51m	+11° 49'	Open Cluster	★	★★	214
Cor Caroli	Canes Venatici	12h 56m	+38° 19'	Multiple Star	★★★	★★★	220
Algorab	Corvus	12h 30m	-16° 31'	Multiple Star	★★★	★★	219
Kuma	Draco	17h 32m	+55° 11'	Multiple Star	★★★	★★★	233
Messier 35	Gemini	06h 09m	+24° 21'	Open Cluster	★★	★★★	205
Castor	Gemini	07h 35m	+31° 53'	Multiple Star	★★★	★★★	206
Keystone Cluster	Hercules	16h 42m	+36° 28'	Globular Cluster	★★	★★★	231
Rho Herculis	Hercules	17h 24m	+37° 08'	Multiple Star	★★	★★	232
Regulus	Leo	10h 08m	+11° 58'	Multiple Star	★★★	★	215
Adhafera	Leo	10h 17m	+23° 25'	Multiple Star	★★★	★★	216
Algieba	Leo	10h 20m	+19° 51'	Multiple Star	★★★	★★★	217
Denebola	Leo	11h 49m	+14° 34'	Multiple Star	★★★	★★	218
Zuben Elgenubi	Libra	14h 51m	-16° 00'	Multiple Star	★★★	★★	224
Double Double	Lyra	18h 44m	+39° 40'	Multiple Star	★★★	★★★	235
Sheliak	Lyra	18h 50m	+33° 22'	Multiple Star	★★★	★★	236
Mizar & Alcor	Ursa Major	13h 24m	+54° 56'	Multiple Star	★★★	★★★	221
Polaris	Ursa Minor	02h 32m	+89° 16'	Multiple Star	★★★	★	196

Chart 14

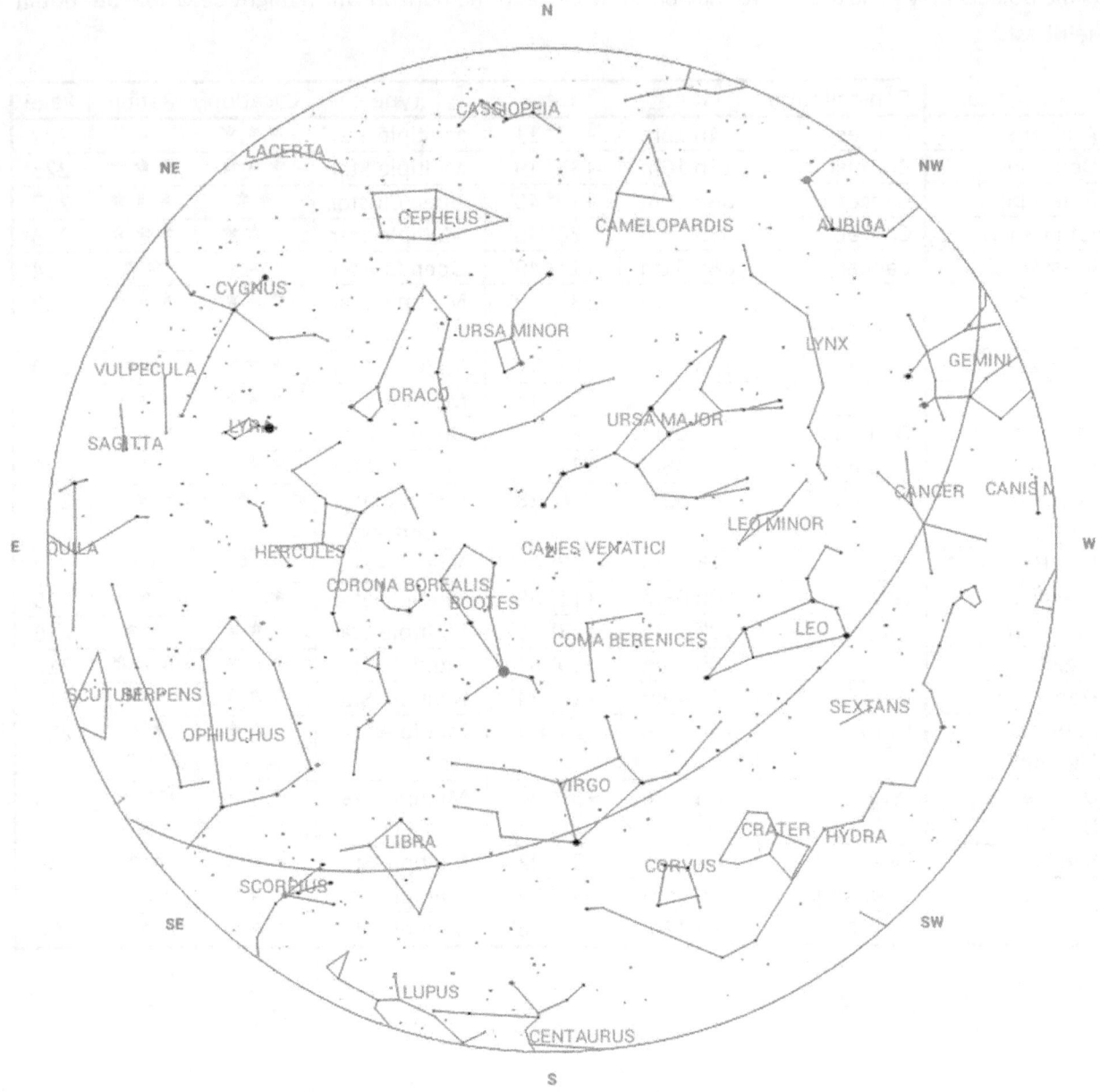

The following table shows the constellations at their best visibility at this time.

Object	Page
Boötes	114
Canes Venatici	118
Cepheus	128
Coma Berenices	132
Corona Borealis	134

Object	Page
Corvus	136
Crater	136
Draco	142
Hercules	146
Leo	148

Object	Page
Libra	152
Lyra	154
Ursa Major	176
Ursa Minor	178
Virgo	180

The following table shows telescopic objects at their best visibility (listed by constellation. NB: Not all visible objects may be listed; there may be some close to the horizon which might be visible but not at their best.)

	Constellation	R.A.	Dec.	Type	Location	Rating	Page
Arcturus	Boötes	14h 16m	+19° 11'	Multiple Star	★★★	★	222
Delta Boötis	Boötes	15h 16m	+33° 19'	Multiple Star	★★★	★★★	223
Iota Cancri	Cancer	08h 47m	+28° 46'	Multiple Star	★★★	★★★	213
Cor Caroli	Canes Venatici	12h 56m	+38° 19'	Multiple Star	★★★	★★★	220
Algorab	Corvus	12h 30m	-16° 31'	Multiple Star	★★★	★★	219
Albireo	Cygnus	19h 31m	+27° 58'	Multiple Star	★★★	★★★	241
Kuma	Draco	17h 32m	+55° 11'	Multiple Star	★★★	★★★	233
Keystone Cluster	Hercules	16h 42m	+36° 28'	Globular Cluster	★★	★★★	231
Rho Herculis	Hercules	17h 24m	+37° 08'	Multiple Star	★★	★★	232
Regulus	Leo	10h 08m	+11° 58'	Multiple Star	★★★	★	215
Adhafera	Leo	10h 17m	+23° 25'	Multiple Star	★★★	★★	216
Algieba	Leo	10h 20m	+19° 51'	Multiple Star	★★★	★★★	217
Denebola	Leo	11h 49m	+14° 34'	Multiple Star	★★★	★★	218
Zuben Elgenubi	Libra	14h 51m	-16° 00'	Multiple Star	★★★	★★	224
Double Double	Lyra	18h 44m	+39° 40'	Multiple Star	★★★	★★★	235
Sheliak	Lyra	18h 50m	+33° 22'	Multiple Star	★★★	★★	236
Ring Nebula	Lyra	18h 54m	+33° 02'	Multiple Star	★	★★★	237
Mizar & Alcor	Ursa Major	13h 24m	+54° 56'	Multiple Star	★★★	★★★	221
Polaris	Ursa Minor	02h 32m	+89° 16'	Multiple Star	★★★	★	196

Chart 15

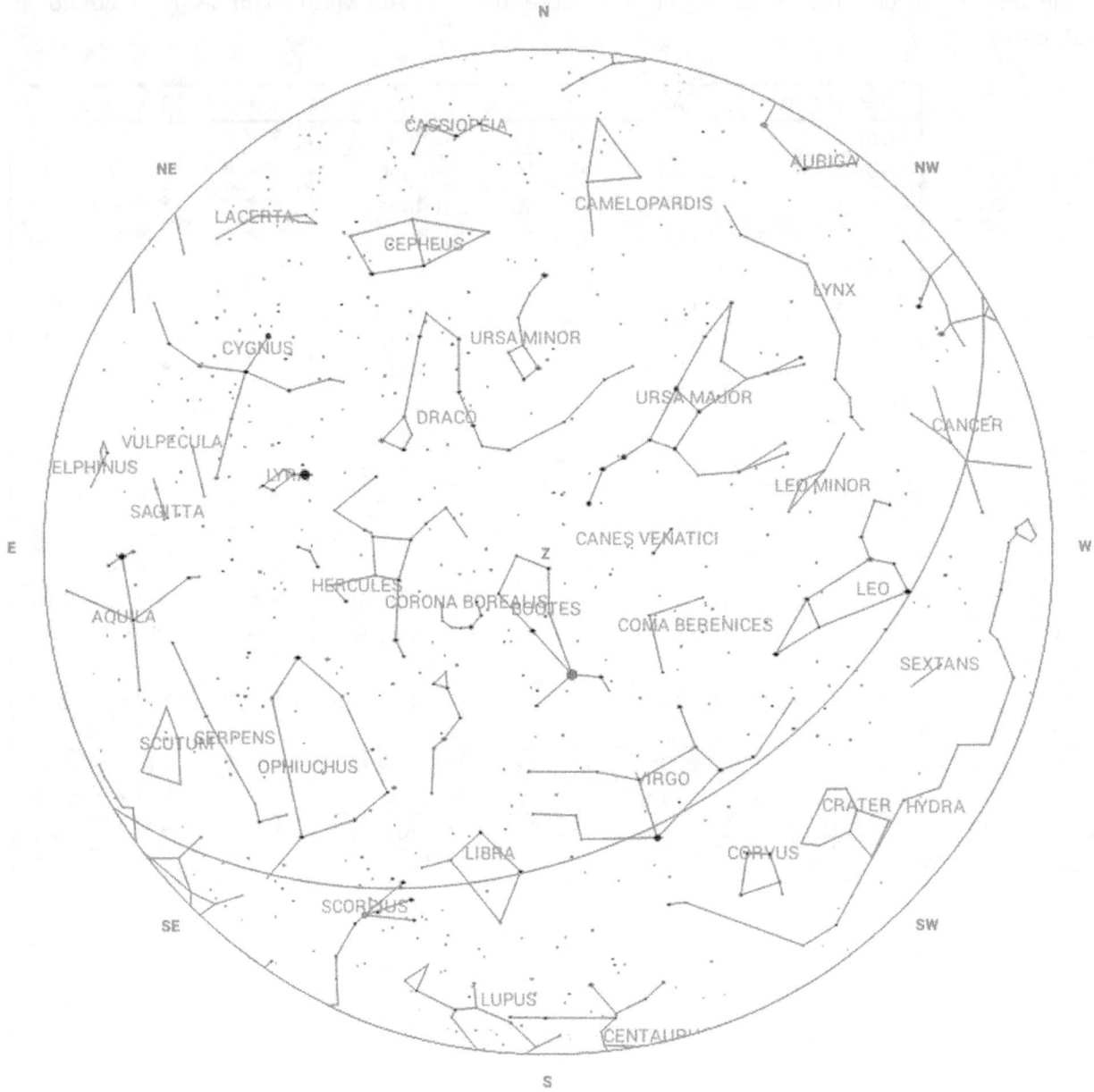

The following table shows the constellations at their best visibility at this time.

Object	Page
Boötes	114
Canes Venatici	118
Cepheus	128
Coma Berenices	132
Corona Borealis	134
Corvus	136

Object	Page
Cygnus	138
Draco	142
Hercules	146
Leo	148
Libra	152
Lyra	154

Object	Page
Ophiuchus	156
Serpens	172
Ursa Major	176
Ursa Minor	178
Virgo	180
Vulpecula	166

The following table shows telescopic objects at their best visibility (listed by constellation. NB: Not all visible objects may be listed; there may be some close to the horizon which might be visible but not at their best.)

	Constellation	R.A.	Dec.	Type	Location	Rating	Page
Arcturus	Boötes	14h 16m	+19° 11'	Multiple Star	★★★	★	222
Delta Boötis	Boötes	15h 16m	+33° 19'	Multiple Star	★★★	★★★	223
Cor Caroli	Canes Venatici	12h 56m	+38° 19'	Multiple Star	★★★	★★★	220
Algorab	Corvus	12h 30m	-16° 31'	Multiple Star	★★★	★★	219
Albireo	Cygnus	19h 31m	+27° 58'	Multiple Star	★★★	★★★	241
Messier 29	Cygnus	20h 24m	+38° 30'	Open Cluster	★	★★	242
Kuma	Draco	17h 32m	+55° 11'	Multiple Star	★★★	★★★	233
Keystone Cluster	Hercules	16h 42m	+36° 28'	Globular Cluster	★★	★★★	231
Rho Herculis	Hercules	17h 24m	+37° 08'	Multiple Star	★★	★★	232
Regulus	Leo	10h 08m	+11° 58'	Multiple Star	★★★	★	215
Adhafera	Leo	10h 17m	+23° 25'	Multiple Star	★★★	★★	216
Algieba	Leo	10h 20m	+19° 51'	Multiple Star	★★★	★★★	217
Denebola	Leo	11h 49m	+14° 34'	Multiple Star	★★★	★★	218
Zuben Elgenubi	Libra	14h 51m	-16° 00'	Multiple Star	★★★	★★	224
Double Double	Lyra	18h 44m	+39° 40'	Multiple Star	★★★	★★★	235
Sheliak	Lyra	18h 50m	+33° 22'	Multiple Star	★★★	★★	236
Ring Nebula	Lyra	18h 54m	+33° 02'	Multiple Star	★	★★★	237
Zeta Sagittae	Sagitta	19h 49m	+19° 09'	Multiple Star	★★	★★	243
Graffias	Scorpius	16h 05m	-19° 48'	Multiple Star	★★★	★★★	225
Jabbah	Scorpius	16h 12m	-19° 28'	Multiple Star	★★	★★★	226
Messier 80	Scorpius	16h 17m	-22° 59'	Globular Cluster	★	★	227
Messier 4	Scorpius	16h 24m	-26° 32'	Globular Cluster	★	★★	228
Mizar & Alcor	Ursa Major	13h 24m	+54° 56'	Multiple Star	★★★	★★★	221
Polaris	Ursa Minor	02h 32m	+89° 16'	Multiple Star	★★★	★	196
Coathanger Cluster	Vulpecula	19h 26m	+20° 13'	Open Cluster	★★	★★	239

Chart 16

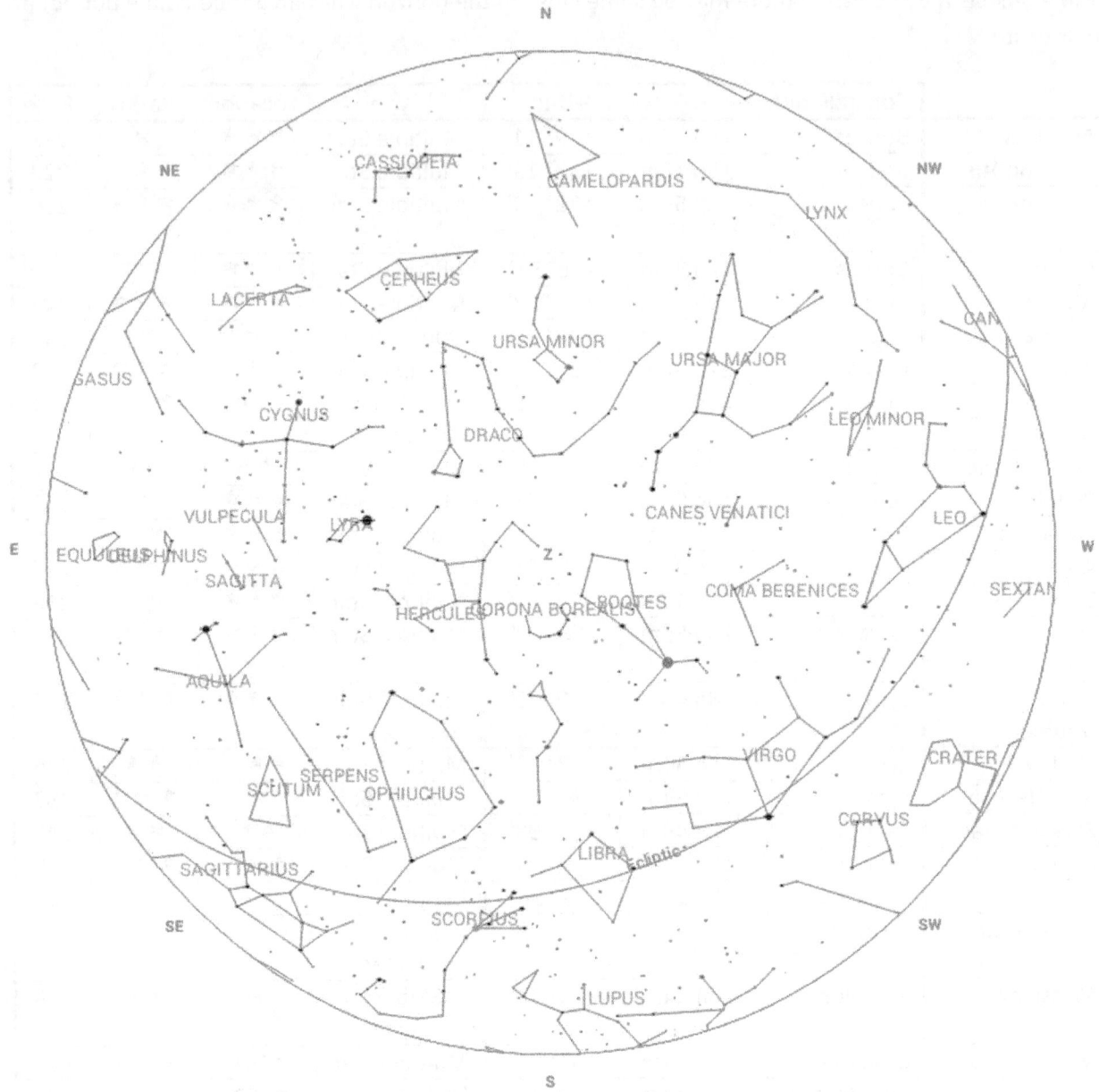

The following table shows the constellations at their best visibility at this time.

Object	Page
Boötes	114
Canes Venatici	118
Cepheus	128
Coma Berenices	132
Corona Borealis	134
Cygnus	138
Delphinus	140

Object	Page
Draco	142
Hercules	146
Leo	148
Libra	152
Lyra	154
Ophiuchus	156
Sagitta	166

Object	Page
Scorpius	170
Serpens	172
Ursa Major	176
Ursa Minor	178
Virgo	180
Vulpecula	166

The following table shows telescopic objects at their best visibility (listed by constellation. NB: Not all visible objects may be listed; there may be some close to the horizon which might be visible but not at their best.)

	Constellation	R.A.	Dec.	Type	Location	Rating	Page
Arcturus	Boötes	14h 16m	+19° 11′	Multiple Star	★★★	★	222
Delta Boötis	Boötes	15h 16m	+33° 19′	Multiple Star	★★★	★★★	223
Cor Caroli	Canes Venatici	12h 56m	+38° 19′	Multiple Star	★★★	★★★	220
Albireo	Cygnus	19h 31m	+27° 58′	Multiple Star	★★★	★★★	241
Messier 29	Cygnus	20h 24m	+38° 30′	Open Cluster	★	★★	242
Gamma Delphini	Delphinus	20h 47m	+16° 08′	Multiple Star	★★★	★★	246
Kuma	Draco	17h 32m	+55° 11′	Multiple Star	★★★	★★★	233
Keystone Cluster	Hercules	16h 42m	+36° 28′	Globular Cluster	★★	★★★	231
Rho Herculis	Hercules	17h 24m	+37° 08′	Multiple Star	★★	★★	232
Regulus	Leo	10h 08m	+11° 58′	Multiple Star	★★★	★	215
Adhafera	Leo	10h 17m	+23° 25′	Multiple Star	★★★	★★	216
Algieba	Leo	10h 20m	+19° 51′	Multiple Star	★★★	★★★	217
Denebola	Leo	11h 49m	+14° 34′	Multiple Star	★★★	★★	218
Zuben Elgenubi	Libra	14h 51m	-16° 00′	Multiple Star	★★★	★★	224
Double Double	Lyra	18h 44m	+39° 40′	Multiple Star	★★★	★★★	235
Sheliak	Lyra	18h 50m	+33° 22′	Multiple Star	★★★	★★	236
Ring Nebula	Lyra	18h 54m	+33° 02′	Multiple Star	★	★★★	237
Zeta Sagittae	Sagitta	19h 49m	+19° 09′	Multiple Star	★★	★★	243
Messier 71	Sagitta	19h 54m	+18° 47′	Globular Cluster	★	★	244
Graffias	Scorpius	16h 05m	-19° 48′	Multiple Star	★★★	★★★	225
Jabbah	Scorpius	16h 12m	-19° 28′	Multiple Star	★★	★★★	226
Messier 80	Scorpius	16h 17m	-22° 59′	Globular Cluster	★	★	227
Messier 4	Scorpius	16h 24m	-26° 32′	Globular Cluster	★	★★	228
Wild Duck Cluster	Scutum	18h 51m	-06° 16′	Open Cluster	★★	★★	238
Mizar & Alcor	Ursa Major	13h 24m	+54° 56′	Multiple Star	★★★	★★★	221
Polaris	Ursa Minor	02h 32m	+89° 16′	Multiple Star	★★★	★	196
Coathanger Cluster	Vulpecula	19h 26m	+20° 13′	Open Cluster	★★	★★	239
Dumbbell Nebula	Vulpecula	20h 00m	+22° 43′	Planetary Nebula	★	★★	240

Chart 17

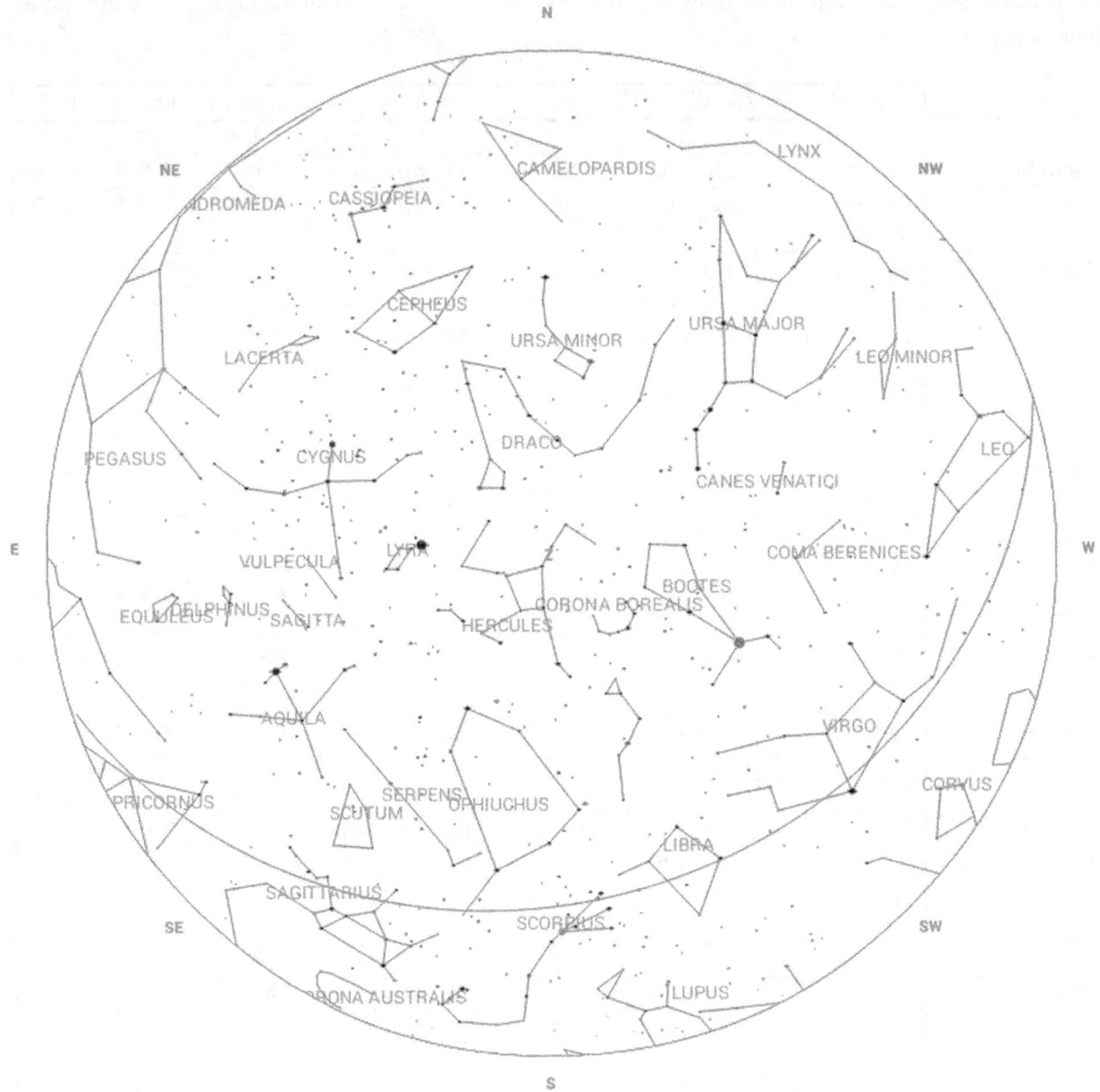

The following table shows the constellations at their best visibility at this time.

Object	Page
Aquila	108
Boötes	114
Canes Venatici	118
Cepheus	128
Coma Berenices	132
Corona Borealis	134
Cygnus	138

Object	Page
Delphinus	140
Draco	142
Hercules	146
Libra	152
Lyra	154
Ophiuchus	156
Sagitta	166

Object	Page
Sagittarius	168
Scorpius	170
Serpens	172
Ursa Major	176
Ursa Minor	178
Virgo	180
Vulpecula	166

The following table shows telescopic objects at their best visibility (listed by constellation. NB: Not all visible objects may be listed; there may be some close to the horizon which might be visible but not at their best.)

	Constellation	R.A.	Dec.	Type	Location	Rating	Page
Arcturus	Boötes	14h 16m	+19° 11'	Multiple Star	★★★	★	222
Delta Boötis	Boötes	15h 16m	+33° 19'	Multiple Star	★★★	★★★	223
Cor Caroli	Canes Venatici	12h 56m	+38° 19'	Multiple Star	★★★	★★★	220
Albireo	Cygnus	19h 31m	+27° 58'	Multiple Star	★★★	★★★	241
Messier 29	Cygnus	20h 24m	+38° 30'	Open Cluster	★	★★	242
Gamma Delphini	Delphinus	20h 47m	+16° 08'	Multiple Star	★★★	★★	246
Kuma	Draco	17h 32m	+55° 11'	Multiple Star	★★★	★★★	233
Keystone Cluster	Hercules	16h 42m	+36° 28'	Globular Cluster	★★	★★★	231
Rho Herculis	Hercules	17h 24m	+37° 08'	Multiple Star	★★	★★	232
Denebola	Leo	11h 49m	+14° 34'	Multiple Star	★★★	★★	218
Zuben Elgenubi	Libra	14h 51m	-16° 00'	Multiple Star	★★★	★★	224
Double Double	Lyra	18h 44m	+39° 40'	Multiple Star	★★★	★★★	235
Sheliak	Lyra	18h 50m	+33° 22'	Multiple Star	★★★	★★	236
Ring Nebula	Lyra	18h 54m	+33° 02'	Multiple Star	★	★★★	237
Zeta Sagittae	Sagitta	19h 49m	+19° 09'	Multiple Star	★★	★★	243
Messier 71	Sagitta	19h 54m	+18° 47'	Globular Cluster	★	★	244
Messier 22	Sagittarius	18h 36m	-23° 54'	Globular Cluster	★★	★★	234
Graffias	Scorpius	16h 05m	-19° 48'	Multiple Star	★★★	★★★	225
Jabbah	Scorpius	16h 12m	-19° 28'	Multiple Star	★★	★★★	226
Messier 80	Scorpius	16h 17m	-22° 59'	Globular Cluster	★	★	227
Messier 4	Scorpius	16h 24m	-26° 32'	Globular Cluster	★	★★	228
Butterfly Cluster	Scorpius	17h 40m	-32° 15'	Open Cluster	★★	★★★	229
Messier 7	Scorpius	17h 54m	-34° 48'	Open Cluster	★★	★★	230
Wild Duck Cluster	Scutum	18h 51m	-06° 16'	Open Cluster	★★	★★	238
Mizar & Alcor	Ursa Major	13h 24m	+54° 56'	Multiple Star	★★★	★★★	221
Polaris	Ursa Minor	02h 32m	+89° 16'	Multiple Star	★★★	★	196
Coathanger Cluster	Vulpecula	19h 26m	+20° 13'	Open Cluster	★★	★★	239
Dumbbell Nebula	Vulpecula	20h 00m	+22° 43'	Planetary Nebula	★	★★	240

Chart 18

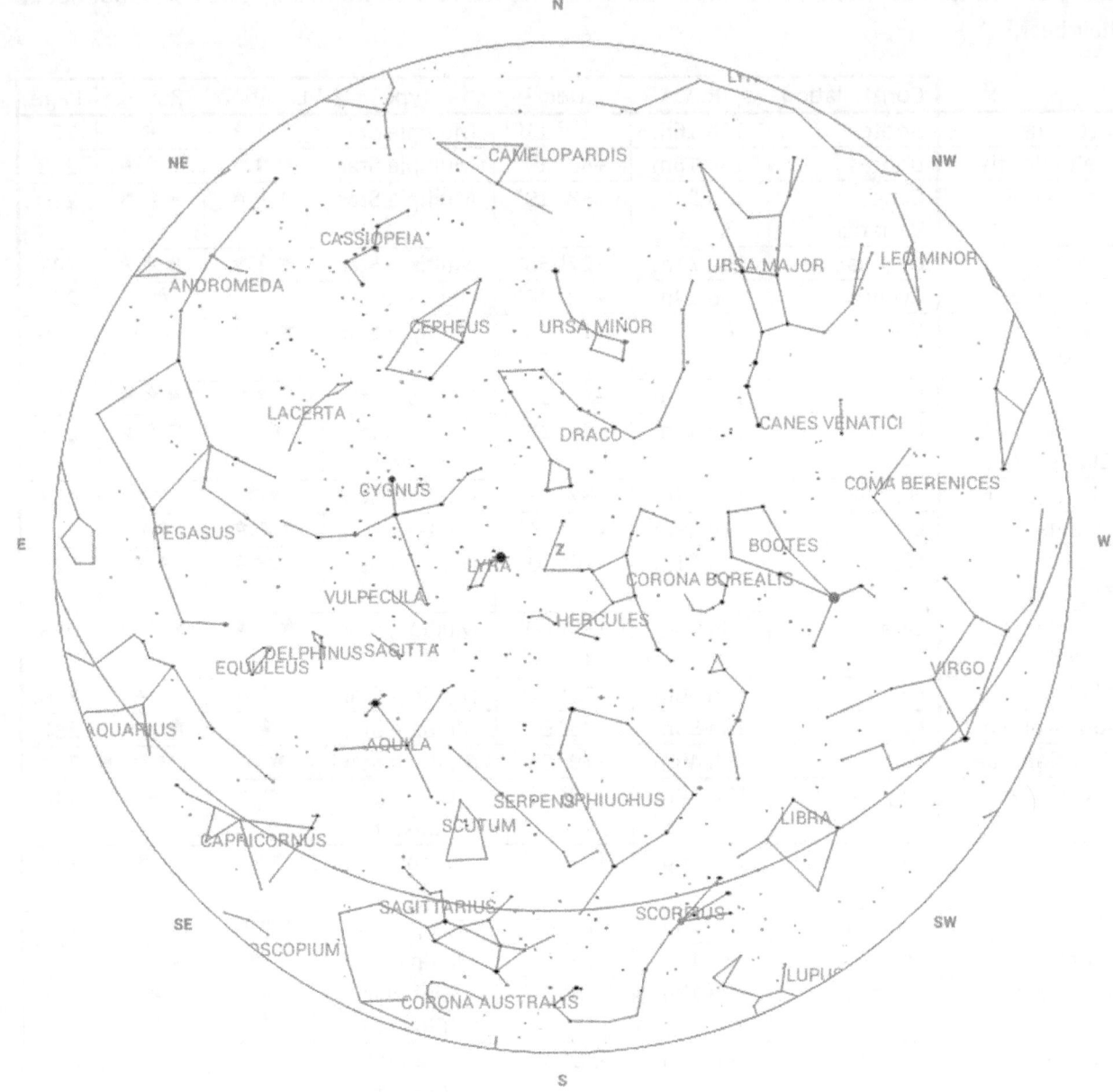

The following table shows the constellations at their best visibility at this time.

Object	Page	Object	Page	Object	Page
Aquila	108	Delphinus	140	Sagitta	166
Boötes	114	Draco	142	Sagittarius	168
Canes Venatici	118	Equuleus	160	Scorpius	170
Cassiopeia	126	Hercules	146	Serpens	172
Cepheus	128	Libra	152	Ursa Major	176
Coma Berenices	132	Lyra	154	Ursa Minor	178
Corona Borealis	134	Ophiuchus	156	Vulpecula	166
Cygnus	138				

The following table shows telescopic objects at their best visibility (listed by constellation. NB: Not all visible objects may be listed; there may be some close to the horizon which might be visible but not at their best.)

	Constellation	R.A.	Dec.	Type	Location	Rating	Page
Arcturus	Boötes	14h 16m	+19° 11'	Multiple Star	★★★	★	222
Delta Boötis	Boötes	15h 16m	+33° 19'	Multiple Star	★★★	★★★	223
Cor Caroli	Canes Venatici	12h 56m	+38° 19'	Multiple Star	★★★	★★★	220
Achird	Cassiopeia	00h 49m	+57° 49'	Multiple Star	★★★	★★★	188
Owl Cluster	Cassiopeia	01h 20m	+58° 17'	Open Cluster	★	★★★	189
NGC 663	Cassiopeia	01h 46m	+61° 13'	Open Cluster	★★	★★★	191
Albireo	Cygnus	19h 31m	+27° 58'	Multiple Star	★★★	★★★	241
Messier 29	Cygnus	20h 24m	+38° 30'	Open Cluster	★	★★	242
Gamma Delphini	Delphinus	20h 47m	+16° 08'	Multiple Star	★★★	★★	246
Kuma	Draco	17h 32m	+55° 11'	Multiple Star	★★★	★★★	233
Keystone Cluster	Hercules	16h 42m	+36° 28'	Globular Cluster	★★	★★★	231
Rho Herculis	Hercules	17h 24m	+37° 08'	Multiple Star	★★	★★	232
Double Double	Lyra	18h 44m	+39° 40'	Multiple Star	★★★	★★★	235
Sheliak	Lyra	18h 50m	+33° 22'	Multiple Star	★★★	★★	236
Ring Nebula	Lyra	18h 54m	+33° 02'	Multiple Star	★	★★★	237
Zeta Sagittae	Sagitta	19h 49m	+19° 09'	Multiple Star	★★	★★	243
Messier 71	Sagitta	19h 54m	+18° 47'	Globular Cluster	★	★	244
Messier 22	Sagittarius	18h 36m	-23° 54'	Globular Cluster	★★	★★	234
Graffias	Scorpius	16h 05m	-19° 48'	Multiple Star	★★★	★★★	225
Jabbah	Scorpius	16h 12m	-19° 28'	Multiple Star	★★	★★★	226
Messier 80	Scorpius	16h 17m	-22° 59'	Globular Cluster	★	★	227
Messier 4	Scorpius	16h 24m	-26° 32'	Globular Cluster	★	★★	228
Butterfly Cluster	Scorpius	17h 40m	-32° 15'	Open Cluster	★★	★★★	229
Messier 7	Scorpius	17h 54m	-34° 48'	Open Cluster	★★	★★	230
Wild Duck Cluster	Scutum	18h 51m	-06° 16'	Open Cluster	★★	★★	238
Mizar & Alcor	Ursa Major	13h 24m	+54° 56'	Multiple Star	★★★	★★★	221
Polaris	Ursa Minor	02h 32m	+89° 16'	Multiple Star	★★★	★	196
Coathanger Cluster	Vulpecula	19h 26m	+20° 13'	Open Cluster	★★	★★	239
Dumbbell Nebula	Vulpecula	20h 00m	+22° 43'	Planetary Nebula	★	★★	240

Chart 19

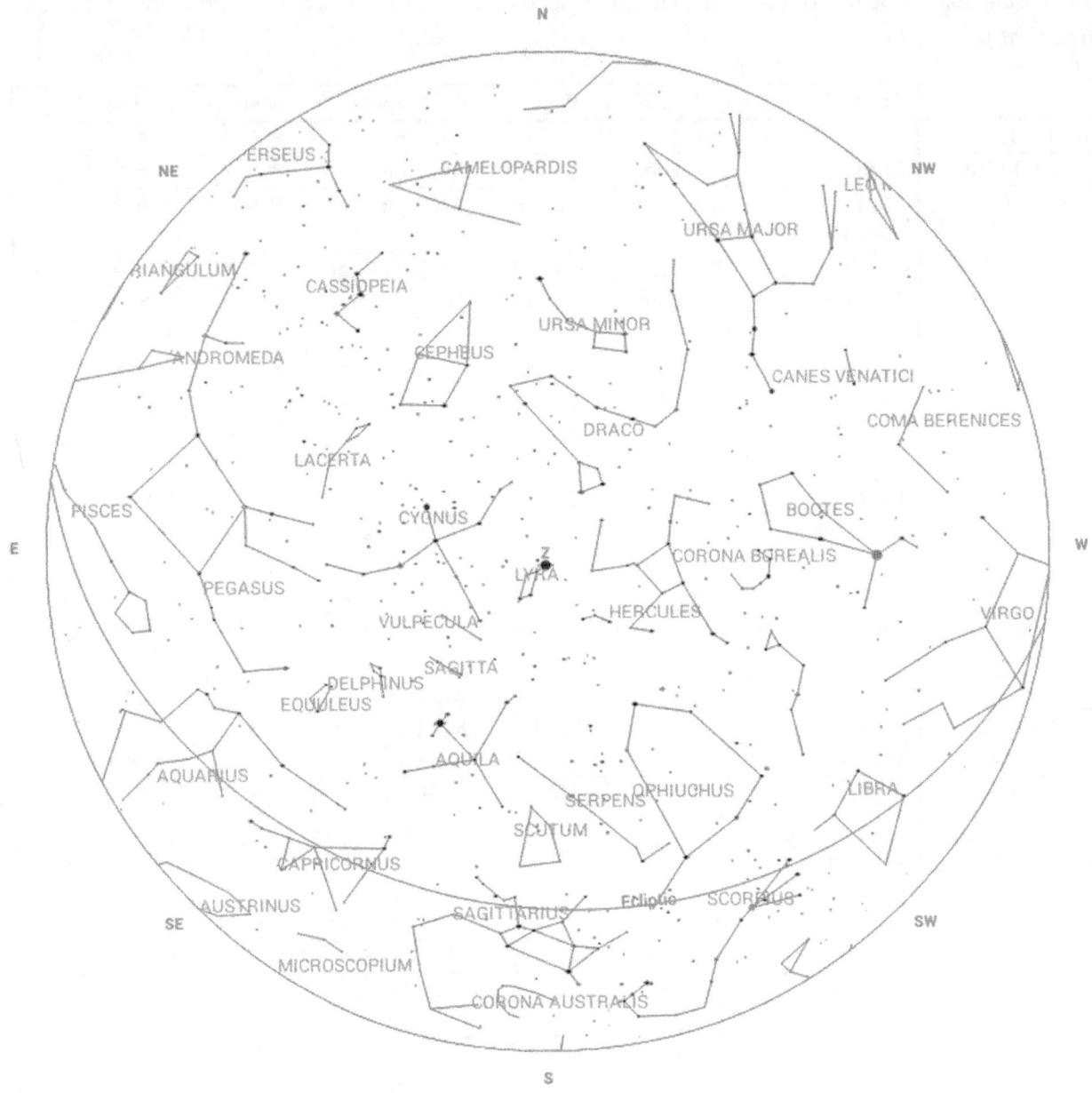

The following table shows the constellations at their best visibility at this time.

Object	Page
Andromeda	104
Aquila	108
Boötes	114
Canes Venatici	118
Capricornus	124
Cassiopeia	126
Cepheus	128
Corona Borealis	134

Object	Page
Cygnus	138
Delphinus	140
Draco	142
Equuleus	160
Hercules	146
Lyra	154
Ophiuchus	156
Pegasus	160

Object	Page
Sagitta	166
Sagittarius	168
Scorpius	170
Serpens	172
Ursa Major	176
Ursa Minor	178
Vulpecula	166

The following table shows telescopic objects at their best visibility (listed by constellation. NB: Not all visible objects may be listed; there may be some close to the horizon which might be visible but not at their best.)

	Constellation	R.A.	Dec.	Type	Location	Rating	Page
Andromeda Galaxy	Andromeda	00h 43m	+41° 16'	Spiral Galaxy	★★	★★	185
Al Giedi	Capricornus	20h 18m	-12° 33'	Multiple Star	★★★	★★	245
Achird	Cassiopeia	00h 49m	+57° 49'	Multiple Star	★★★	★★★	188
Owl Cluster	Cassiopeia	01h 20m	+58° 17'	Open Cluster	★	★★★	189
Messier 103	Cassiopeia	01h 33m	+60° 39'	Open Cluster	★	★	190
NGC 663	Cassiopeia	01h 46m	+61° 13'	Open Cluster	★★	★★★	191
Albireo	Cygnus	19h 31m	+27° 58'	Multiple Star	★★★	★★★	241
Messier 29	Cygnus	20h 24m	+38° 30'	Open Cluster	★	★★	242
Gamma Delphini	Delphinus	20h 47m	+16° 08'	Multiple Star	★★★	★★	246
Kuma	Draco	17h 32m	+55° 11'	Multiple Star	★★★	★★★	233
Keystone Cluster	Hercules	16h 42m	+36° 28'	Globular Cluster	★★	★★★	231
Rho Herculis	Hercules	17h 24m	+37° 08'	Multiple Star	★★	★★	232
Double Double	Lyra	18h 44m	+39° 40'	Multiple Star	★★★	★★★	235
Sheliak	Lyra	18h 50m	+33° 22'	Multiple Star	★★★	★★	236
Ring Nebula	Lyra	18h 54m	+33° 02'	Multiple Star	★	★★★	237
Messier 15	Pegasus	21h 30m	+12° 10'	Globular Cluster	★	★★	247
Zeta Sagittae	Sagitta	19h 49m	+19° 09'	Multiple Star	★★	★★	243
Messier 71	Sagitta	19h 54m	+18° 47'	Globular Cluster	★	★	244
Messier 22	Sagittarius	18h 36m	-23° 54'	Globular Cluster	★★	★★	234
Graffias	Scorpius	16h 05m	-19° 48'	Multiple Star	★★★	★★★	225
Jabbah	Scorpius	16h 12m	-19° 28'	Multiple Star	★★	★★★	226
Messier 4	Scorpius	16h 24m	-26° 32'	Globular Cluster	★	★★	228
Butterfly Cluster	Scorpius	17h 40m	-32° 15'	Open Cluster	★★	★★★	229
Messier 7	Scorpius	17h 54m	-34° 48'	Open Cluster	★★	★★	230
Wild Duck Cluster	Scutum	18h 51m	-06° 16'	Open Cluster	★★	★★	238
Mizar & Alcor	Ursa Major	13h 24m	+54° 56'	Multiple Star	★★★	★★★	221
Polaris	Ursa Minor	02h 32m	+89° 16'	Multiple Star	★★★	★	196
Coathanger Cluster	Vulpecula	19h 26m	+20° 13'	Open Cluster	★★	★★	239
Dumbbell Nebula	Vulpecula	20h 00m	+22° 43'	Planetary Nebula	★	★★	240

Chart 20

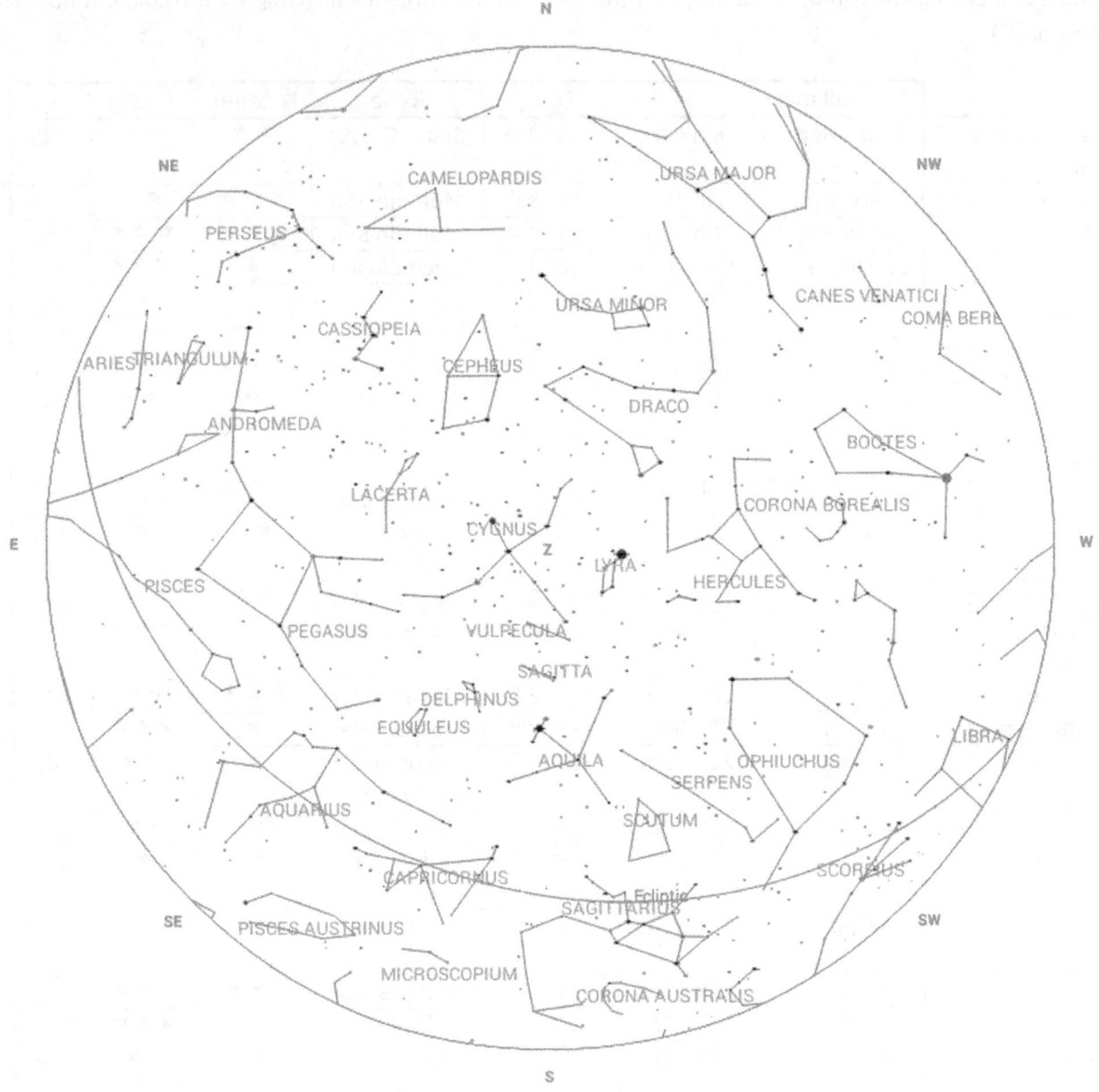

The following table shows the constellations at their best visibility at this time.

Object	Page
Andromeda	104
Aquarius	106
Aquila	108
Boötes	114
Capricornus	124
Cassiopeia	126
Cepheus	128
Corona Borealis	134

Object	Page
Cygnus	138
Delphinus	140
Draco	142
Equuleus	160
Hercules	146
Lyra	154
Ophiuchus	156

Object	Page
Pegasus	160
Sagitta	166
Sagittarius	168
Serpens	172
Ursa Major	176
Ursa Minor	178
Vulpecula	166

The following table shows telescopic objects at their best visibility (listed by constellation. NB: Not all visible objects may be listed; there may be some close to the horizon which might be visible but not at their best.)

	Constellation	R.A.	Dec.	Type	Location	Rating	Page
Pi Andomedae	Andromeda	00h 37m	+33° 43'	Multiple Star	★★	★★	184
Andromeda Galaxy	Andromeda	00h 43m	+41° 16'	Spiral Galaxy	★★	★★	185
NGC 752	Andromeda	01h 58m	+37° 52'	Open Cluster	★★	★★	186
Almach	Andromeda	02h 04m	+42° 20'	Multiple Star	★★★	★★★	187
Arcturus	Boötes	14h 16m	+19° 11'	Multiple Star	★★★	★	222
Delta Boötis	Boötes	15h 16m	+33° 19'	Multiple Star	★★★	★★★	223
Al Giedi	Capricornus	20h 18m	-12° 33'	Multiple Star	★★★	★★	245
Achird	Cassiopeia	00h 49m	+57° 49'	Multiple Star	★★★	★★★	188
Owl Cluster	Cassiopeia	01h 20m	+58° 17'	Open Cluster	★	★★★	189
Messier 103	Cassiopeia	01h 33m	+60° 39'	Open Cluster	★	★	190
NGC 663	Cassiopeia	01h 46m	+61° 13'	Open Cluster	★★	★★★	191
Albireo	Cygnus	19h 31m	+27° 58'	Multiple Star	★★★	★★★	241
Messier 29	Cygnus	20h 24m	+38° 30'	Open Cluster	★	★★	242
Gamma Delphini	Delphinus	20h 47m	+16° 08'	Multiple Star	★★★	★★	246
Kuma	Draco	17h 32m	+55° 11'	Multiple Star	★★★	★★★	233
Keystone Cluster	Hercules	16h 42m	+36° 28'	Globular Cluster	★★	★★★	231
Rho Herculis	Hercules	17h 24m	+37° 08'	Multiple Star	★★	★★	232
Double Double	Lyra	18h 44m	+39° 40'	Multiple Star	★★★	★★★	235
Sheliak	Lyra	18h 50m	+33° 22'	Multiple Star	★★★	★★	236
Ring Nebula	Lyra	18h 54m	+33° 02'	Multiple Star	★	★★★	237
Messier 15	Pegasus	21h 30m	+12° 10'	Globular Cluster	★	★★	247
Zeta Sagittae	Sagitta	19h 49m	+19° 09'	Multiple Star	★★	★★	243
Messier 71	Sagitta	19h 54m	+18° 47'	Globular Cluster	★	★	244
Messier 22	Sagittarius	18h 36m	-23° 54'	Globular Cluster	★★	★★	234
Wild Duck Cluster	Scutum	18h 51m	-06° 16'	Open Cluster	★★	★★	238
Mizar & Alcor	Ursa Major	13h 24m	+54° 56'	Multiple Star	★★★	★★★	221
Polaris	Ursa Minor	02h 32m	+89° 16'	Multiple Star	★★★	★	196
Coathanger Cluster	Vulpecula	19h 26m	+20° 13'	Open Cluster	★★	★★	239
Dumbbell Nebula	Vulpecula	20h 00m	+22° 43'	Planetary Nebula	★	★★	240

Chart 21

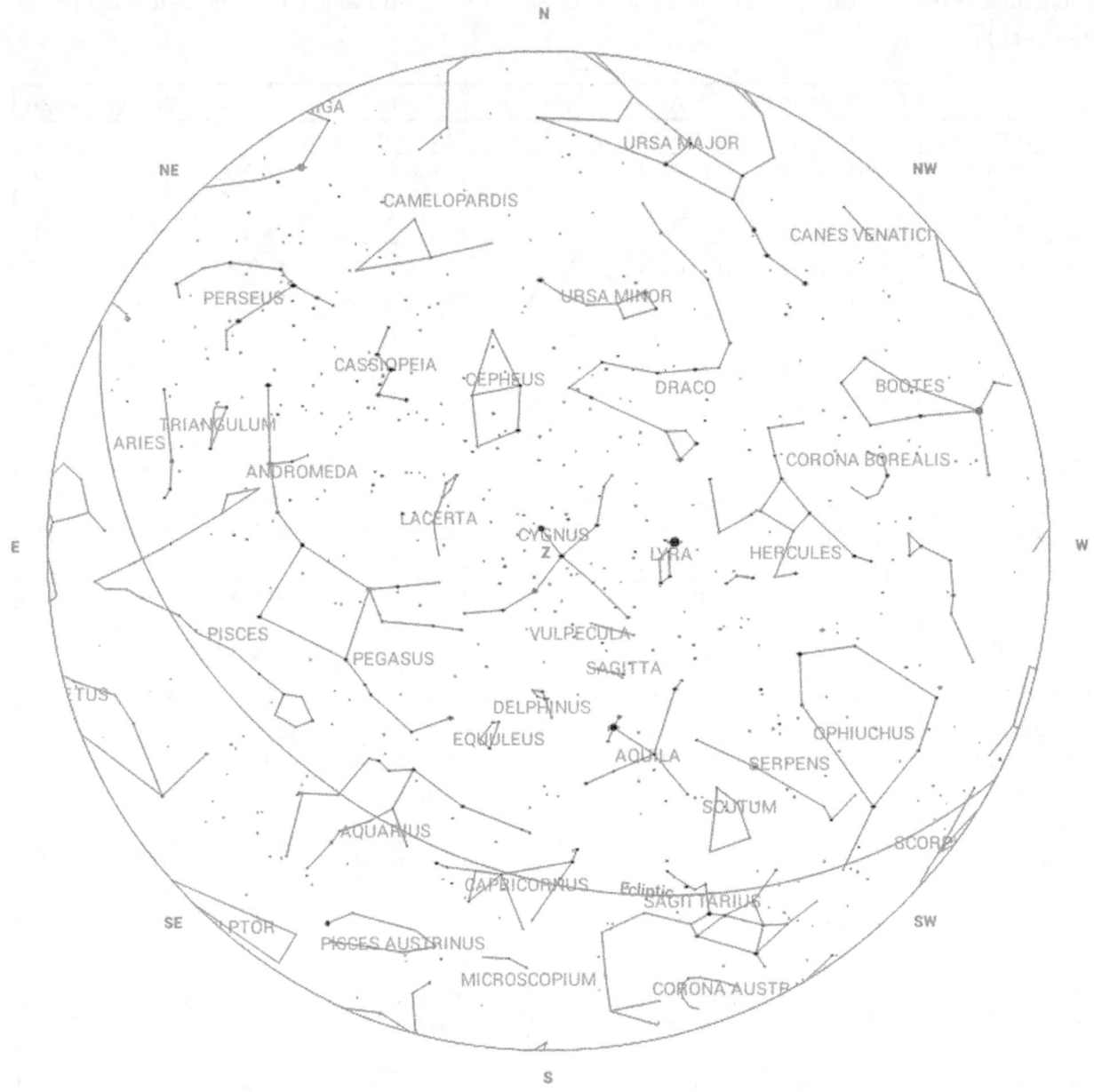

The following table shows the constellations at their best visibility at this time.

Object	Page	Object	Page	Object	Page
Andromeda	104	Cygnus	138	Perseus	162
Aquarius	106	Delphinus	140	Pisces	164
Aquila	108	Draco	142	Sagitta	166
Aries	110	Equuleus	160	Sagittarius	168
Capricornus	124	Hercules	146	Serpens	172
Cassiopeia	126	Lyra	154	Ursa Major	176
Cepheus	128	Ophiuchus	156	Ursa Minor	178
Corona Borealis	134	Pegasus	160	Vulpecula	166

The following table shows telescopic objects at their best visibility (listed by constellation. NB: Not all visible objects may be listed; there may be some close to the horizon which might be visible but not at their best.)

	Constellation	R.A.	Dec.	Type	Location	Rating	Page
Pi Andomedae	Andromeda	00h 37m	+33° 43'	Multiple Star	★★	★★	184
Andromeda Galaxy	Andromeda	00h 43m	+41° 16'	Spiral Galaxy	★★	★★	185
NGC 752	Andromeda	01h 58m	+37° 52'	Open Cluster	★★	★★	186
Almach	Andromeda	02h 04m	+42° 20'	Multiple Star	★★★	★★★	187
Al Giedi	Capricornus	20h 18m	-12° 33'	Multiple Star	★★★	★★	245
Achird	Cassiopeia	00h 49m	+57° 49'	Multiple Star	★★★	★★★	188
Owl Cluster	Cassiopeia	01h 20m	+58° 17'	Open Cluster	★	★★★	189
Messier 103	Cassiopeia	01h 33m	+60° 39'	Open Cluster	★	★	190
NGC 663	Cassiopeia	01h 46m	+61° 13'	Open Cluster	★★	★★★	191
Albireo	Cygnus	19h 31m	+27° 58'	Multiple Star	★★★	★★★	241
Messier 29	Cygnus	20h 24m	+38° 30'	Open Cluster	★	★★	242
Gamma Delphini	Delphinus	20h 47m	+16° 08'	Multiple Star	★★★	★★	246
Kuma	Draco	17h 32m	+55° 11'	Multiple Star	★★★	★★★	233
Keystone Cluster	Hercules	16h 42m	+36° 28'	Globular Cluster	★★	★★★	231
Rho Herculis	Hercules	17h 24m	+37° 08'	Multiple Star	★★	★★	232
Double Double	Lyra	18h 44m	+39° 40'	Multiple Star	★★★	★★★	235
Sheliak	Lyra	18h 50m	+33° 22'	Multiple Star	★★★	★★	236
Ring Nebula	Lyra	18h 54m	+33° 02'	Multiple Star	★	★★★	237
Messier 15	Pegasus	21h 30m	+12° 10'	Globular Cluster	★	★★	247
Double Cluster	Perseus	02h 22m	+57° 09'	Open Clusters	★★★	★★★	194
Messier 34	Perseus	02h 42m	+42° 45'	Open Cluster	★★	★★	195
Zeta Sagittae	Sagitta	19h 49m	+19° 09'	Multiple Star	★★	★★	243
Messier 71	Sagitta	19h 54m	+18° 47'	Globular Cluster	★	★	244
Messier 22	Sagittarius	18h 36m	-23° 54'	Globular Cluster	★★	★★	234
Wild Duck Cluster	Scutum	18h 51m	-06° 16'	Open Cluster	★★	★★	238
Mizar & Alcor	Ursa Major	13h 24m	+54° 56'	Multiple Star	★★★	★★★	221
Polaris	Ursa Minor	02h 32m	+89° 16'	Multiple Star	★★★	★	196
Coathanger Cluster	Vulpecula	19h 26m	+20° 13'	Open Cluster	★★	★★	239
Dumbbell Nebula	Vulpecula	20h 00m	+22° 43'	Planetary Nebula	★	★★	240

Chart 22

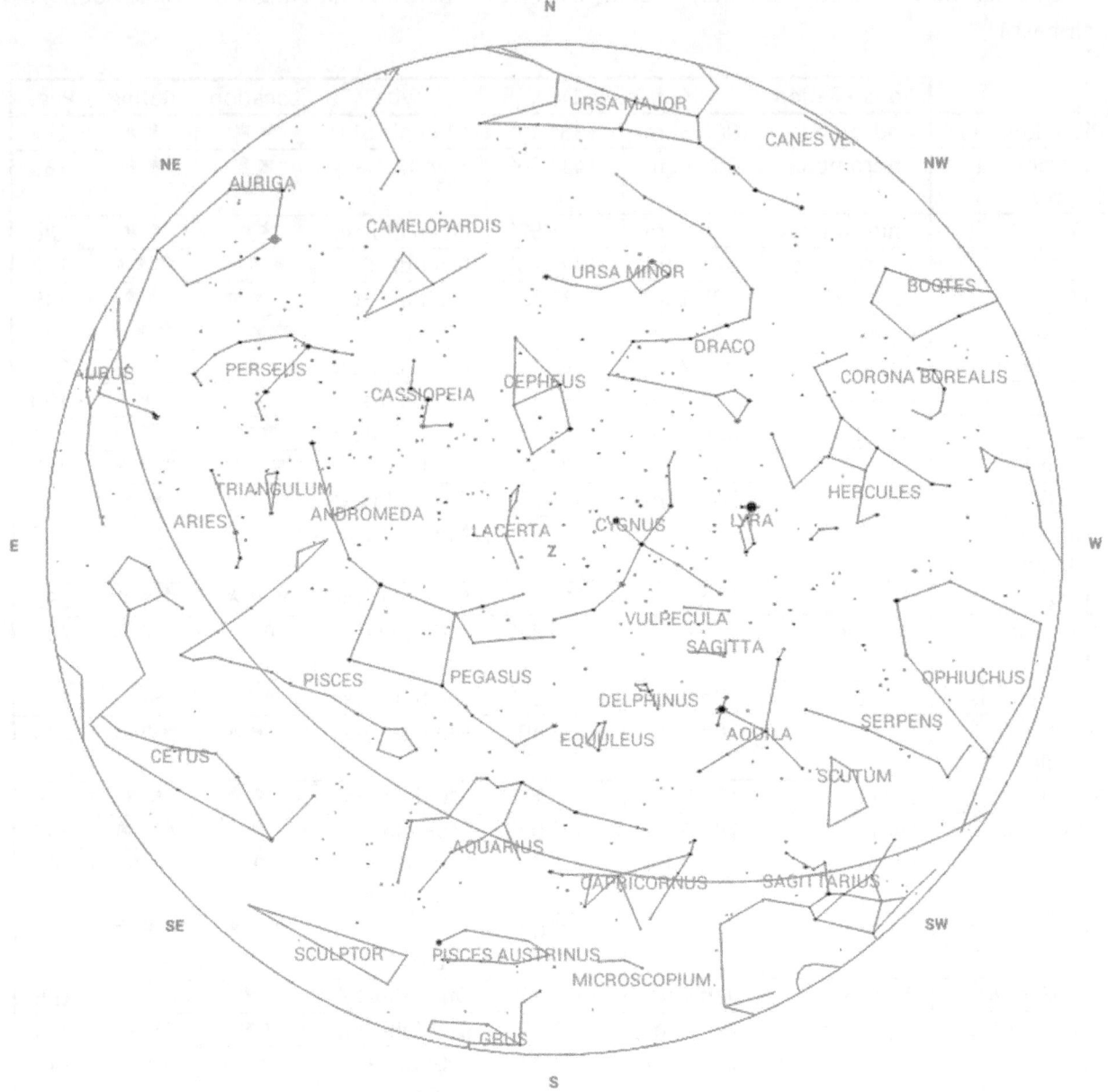

The following table shows the constellations at their best visibility at this time.

Object	Page
Andromeda	104
Aquarius	106
Aquila	108
Aries	110
Capricornus	124
Cassiopeia	126
Cepheus	128

Object	Page
Corona Borealis	134
Cygnus	138
Delphinus	140
Draco	142
Equuleus	160
Hercules	146
Lyra	154

Object	Page
Pegasus	160
Perseus	162
Pisces	164
Sagitta	166
Ursa Minor	178
Vulpecula	166

The following table shows telescopic objects at their best visibility (listed by constellation. NB: Not all visible objects may be listed; there may be some close to the horizon which might be visible but not at their best.)

	Constellation	R.A.	Dec.	Type	Location	Rating	Page
Pi Andomedae	Andromeda	00h 37m	+33° 43'	Multiple Star	★★	★★	184
Andromeda Galaxy	Andromeda	00h 43m	+41° 16'	Spiral Galaxy	★★	★★	185
NGC 752	Andromeda	01h 58m	+37° 52'	Open Cluster	★★	★★	186
Almach	Andromeda	02h 04m	+42° 20'	Multiple Star	★★★	★★★	187
Mesarthim	Aries	01h 53m	+19° 18'	Multiple Star	★★★	★★★	192
Lambda Arietis	Aries	01h 58m	+23° 36'	Multiple Star	★★★	★★★	193
Al Giedi	Capricornus	20h 18m	-12° 33'	Multiple Star	★★★	★★	245
Achird	Cassiopeia	00h 49m	+57° 49'	Multiple Star	★★★	★★★	188
Owl Cluster	Cassiopeia	01h 20m	+58° 17'	Open Cluster	★	★★★	189
Messier 103	Cassiopeia	01h 33m	+60° 39'	Open Cluster	★	★	190
NGC 663	Cassiopeia	01h 46m	+61° 13'	Open Cluster	★★	★★★	191
Albireo	Cygnus	19h 31m	+27° 58'	Multiple Star	★★★	★★★	241
Messier 29	Cygnus	20h 24m	+38° 30'	Open Cluster	★	★★	242
Gamma Delphini	Delphinus	20h 47m	+16° 08'	Multiple Star	★★★	★★	246
Kuma	Draco	17h 32m	+55° 11'	Multiple Star	★★★	★★★	233
Keystone Cluster	Hercules	16h 42m	+36° 28'	Globular Cluster	★★	★★★	231
Rho Herculis	Hercules	17h 24m	+37° 08'	Multiple Star	★★	★★	232
Double Double	Lyra	18h 44m	+39° 40'	Multiple Star	★★★	★★★	235
Sheliak	Lyra	18h 50m	+33° 22'	Multiple Star	★★★	★★	236
Ring Nebula	Lyra	18h 54m	+33° 02'	Multiple Star	★	★★★	237
Messier 15	Pegasus	21h 30m	+12° 10'	Globular Cluster	★	★★	247
Double Cluster	Perseus	02h 22m	+57° 09'	Open Clusters	★★★	★★★	194
Messier 34	Perseus	02h 42m	+42° 45'	Open Cluster	★★	★★	195
Zeta Sagittae	Sagitta	19h 49m	+19° 09'	Multiple Star	★★	★★	243
Messier 71	Sagitta	19h 54m	+18° 47'	Globular Cluster	★	★	244
Wild Duck Cluster	Scutum	18h 51m	-06° 16'	Open Cluster	★★	★★	238
Polaris	Ursa Minor	02h 32m	+89° 16'	Multiple Star	★★★	★	196
Coathanger Cluster	Vulpecula	19h 26m	+20° 13'	Open Cluster	★★	★★	239
Dumbbell Nebula	Vulpecula	20h 00m	+22° 43'	Planetary Nebula	★	★★	240

Chart 23

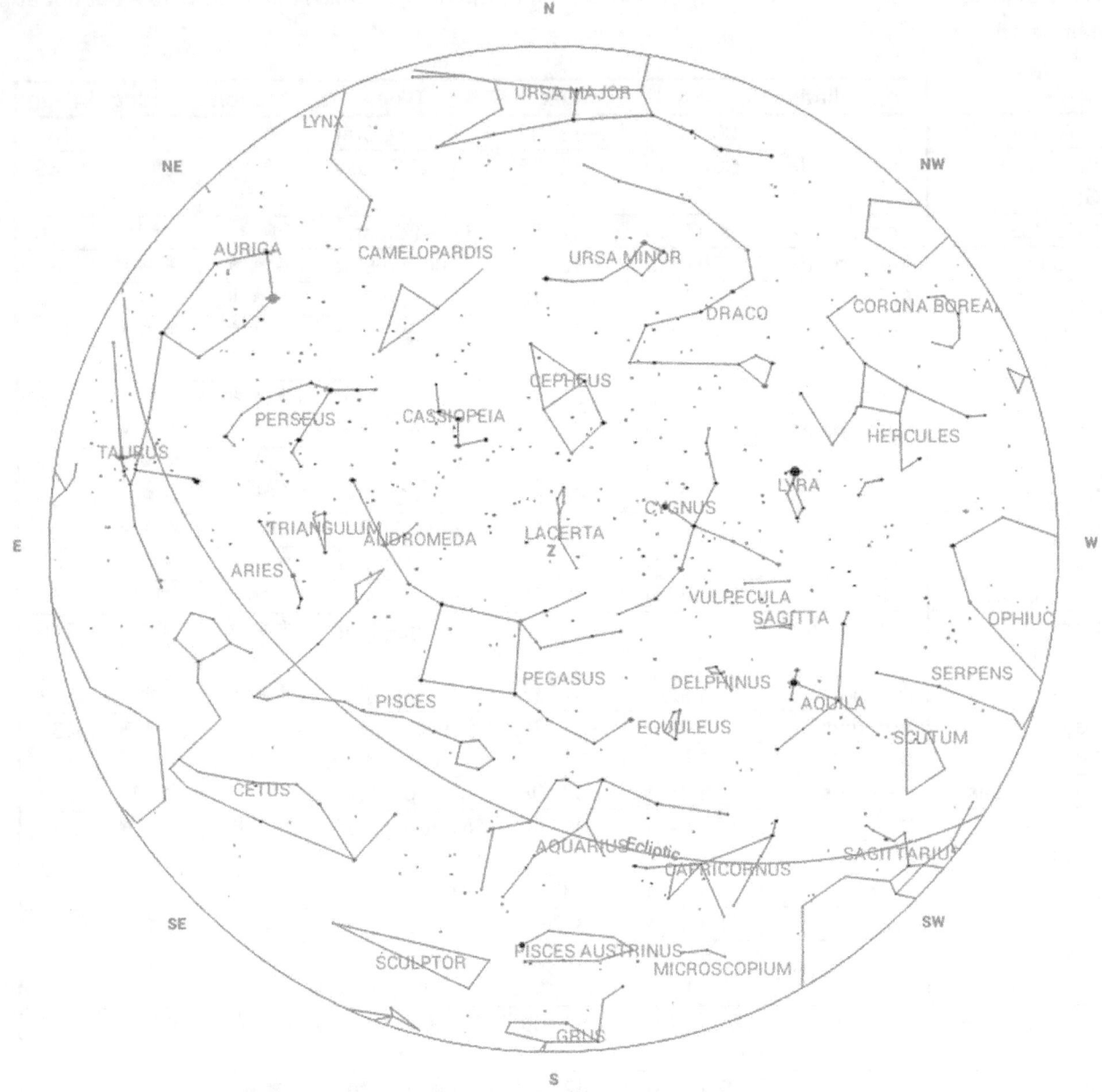

The following table shows the constellations at their best visibility at this time.

Object	Page
Andromeda	104
Aquarius	106
Aquila	108
Aries	110
Auriga	112
Capricornus	124
Cassiopeia	126

Object	Page
Cepheus	128
Cetus	130
Cygnus	138
Delphinus	140
Draco	142
Equuleus	160
Hercules	146

Object	Page
Lyra	154
Pegasus	160
Perseus	162
Pisces	164
Sagitta	166
Ursa Minor	178
Vulpecula	166

The following table shows telescopic objects at their best visibility (listed by constellation. NB: Not all visible objects may be listed; there may be some close to the horizon which might be visible but not at their best.)

	Constellation	R.A.	Dec.	Type	Location	Rating	Page
Pi Andomedae	Andromeda	00h 37m	+33° 43'	Multiple Star	★★	★★	184
Andromeda Galaxy	Andromeda	00h 43m	+41° 16'	Spiral Galaxy	★★	★★	185
NGC 752	Andromeda	01h 58m	+37° 52'	Open Cluster	★★	★★	186
Almach	Andromeda	02h 04m	+42° 20'	Multiple Star	★★★	★★★	187
Mesarthim	Aries	01h 53m	+19° 18'	Multiple Star	★★★	★★★	192
Lambda Arietis	Aries	01h 58m	+23° 36'	Multiple Star	★★★	★★★	193
Messier 37	Auriga	05h 52m	+32° 33'	Open Cluster	★	★★★	204
Al Giedi	Capricornus	20h 18m	-12° 33'	Multiple Star	★★★	★★	245
Achird	Cassiopeia	00h 49m	+57° 49'	Multiple Star	★★★	★★★	188
Owl Cluster	Cassiopeia	01h 20m	+58° 17'	Open Cluster	★	★★★	189
Messier 103	Cassiopeia	01h 33m	+60° 39'	Open Cluster	★	★	190
NGC 663	Cassiopeia	01h 46m	+61° 13'	Open Cluster	★★	★★★	191
Albireo	Cygnus	19h 31m	+27° 58'	Multiple Star	★★★	★★★	241
Messier 29	Cygnus	20h 24m	+38° 30'	Open Cluster	★	★★	242
Gamma Delphini	Delphinus	20h 47m	+16° 08'	Multiple Star	★★★	★★	246
Kuma	Draco	17h 32m	+55° 11'	Multiple Star	★★★	★★★	233
Keystone Cluster	Hercules	16h 42m	+36° 28'	Globular Cluster	★★	★★★	231
Rho Herculis	Hercules	17h 24m	+37° 08'	Multiple Star	★★	★★	232
Double Double	Lyra	18h 44m	+39° 40'	Multiple Star	★★★	★★★	235
Sheliak	Lyra	18h 50m	+33° 22'	Multiple Star	★★★	★★	236
Ring Nebula	Lyra	18h 54m	+33° 02'	Multiple Star	★	★★★	237
Messier 15	Pegasus	21h 30m	+12° 10'	Globular Cluster	★	★★	247
Double Cluster	Perseus	02h 22m	+57° 09'	Open Clusters	★★★	★★★	194
Messier 34	Perseus	02h 42m	+42° 45'	Open Cluster	★★	★★	195
Zeta Sagittae	Sagitta	19h 49m	+19° 09'	Multiple Star	★★	★★	243
Messier 71	Sagitta	19h 54m	+18° 47'	Globular Cluster	★	★	244
Polaris	Ursa Minor	02h 32m	+89° 16'	Multiple Star	★★★	★	196
Coathanger Cluster	Vulpecula	19h 26m	+20° 13'	Open Cluster	★★	★★	239
Dumbbell Nebula	Vulpecula	20h 00m	+22° 43'	Planetary Nebula	★	★★	240

Chart 24

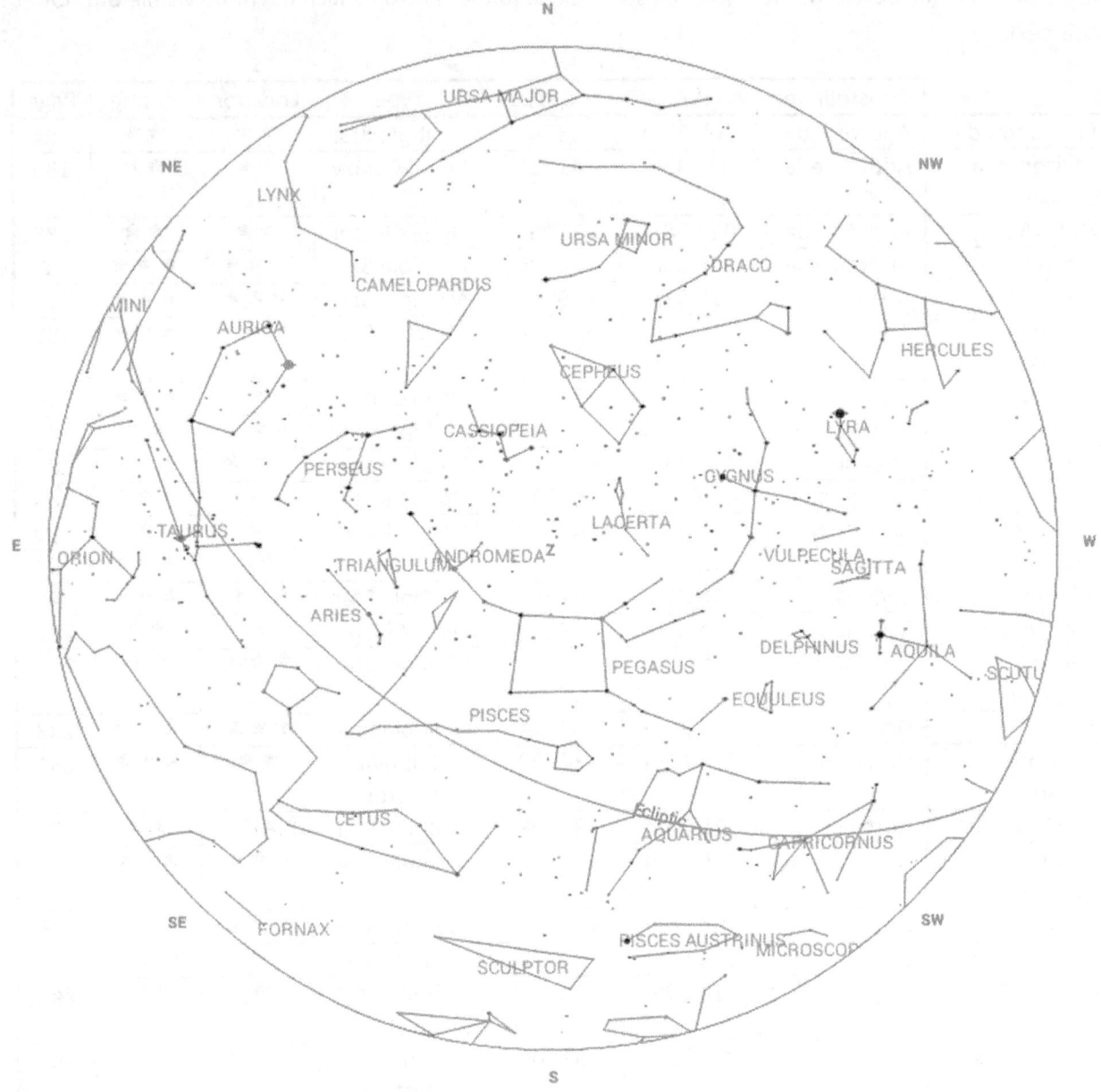

The following table shows the constellations at their best visibility at this time.

Object	Page
Andromeda	104
Aquarius	106
Aquila	108
Aries	110
Auriga	112
Capricornus	124
Cassiopeia	126

Object	Page
Cepheus	128
Cetus	130
Cygnus	138
Delphinus	140
Draco	142
Equuleus	160
Lyra	154

Object	Page
Pegasus	160
Perseus	162
Pisces	164
Sagitta	166
Taurus	174
Ursa Minor	178
Vulpecula	166

The following table shows telescopic objects at their best visibility (listed by constellation. NB: Not all visible objects may be listed; there may be some close to the horizon which might be visible but not at their best.)

	Constellation	R.A.	Dec.	Type	Location	Rating	Page
Pi Andomedae	Andromeda	00h 37m	+33° 43'	Multiple Star	★★	★★	184
Andromeda Galaxy	Andromeda	00h 43m	+41° 16'	Spiral Galaxy	★★	★★	185
NGC 752	Andromeda	01h 58m	+37° 52'	Open Cluster	★★	★★	186
Almach	Andromeda	02h 04m	+42° 20'	Multiple Star	★★★	★★★	187
Mesarthim	Aries	01h 53m	+19° 18'	Multiple Star	★★★	★★★	192
Lambda Arietis	Aries	01h 58m	+23° 36'	Multiple Star	★★★	★★★	193
Messier 37	Auriga	05h 52m	+32° 33'	Open Cluster	★	★★★	204
Al Giedi	Capricornus	20h 18m	-12° 33'	Multiple Star	★★★	★★	245
Achird	Cassiopeia	00h 49m	+57° 49'	Multiple Star	★★★	★★★	188
Owl Cluster	Cassiopeia	01h 20m	+58° 17'	Open Cluster	★	★★★	189
Messier 103	Cassiopeia	01h 33m	+60° 39'	Open Cluster	★	★	190
NGC 663	Cassiopeia	01h 46m	+61° 13'	Open Cluster	★★	★★★	191
Albireo	Cygnus	19h 31m	+27° 58'	Multiple Star	★★★	★★★	241
Messier 29	Cygnus	20h 24m	+38° 30'	Open Cluster	★	★★	242
Gamma Delphini	Delphinus	20h 47m	+16° 08'	Multiple Star	★★★	★★	246
Kuma	Draco	17h 32m	+55° 11'	Multiple Star	★★★	★★★	233
Keystone Cluster	Hercules	16h 42m	+36° 28'	Globular Cluster	★★	★★★	231
Rho Herculis	Hercules	17h 24m	+37° 08'	Multiple Star	★★	★★	232
Double Double	Lyra	18h 44m	+39° 40'	Multiple Star	★★★	★★★	235
Sheliak	Lyra	18h 50m	+33° 22'	Multiple Star	★★★	★★	236
Ring Nebula	Lyra	18h 54m	+33° 02'	Multiple Star	★	★★★	237
Messier 15	Pegasus	21h 30m	+12° 10'	Globular Cluster	★	★★	247
Double Cluster	Perseus	02h 22m	+57° 09'	Open Clusters	★★★	★★★	194
Messier 34	Perseus	02h 42m	+42° 45'	Open Cluster	★★	★★	195
Zeta Sagittae	Sagitta	19h 49m	+19° 09'	Multiple Star	★★	★★	243
Messier 71	Sagitta	19h 54m	+18° 47'	Globular Cluster	★	★	244
The Pleiades	Taurus	03h 48m	+24° 10'	Open Cluster	★★★	★★★	197
Polaris	Ursa Minor	02h 32m	+89° 16'	Multiple Star	★★★	★	196
Coathanger Cluster	Vulpecula	19h 26m	+20° 13'	Open Cluster	★★	★★	239
Dumbbell Nebula	Vulpecula	20h 00m	+22° 43'	Planetary Nebula	★	★★	240

The Constellations

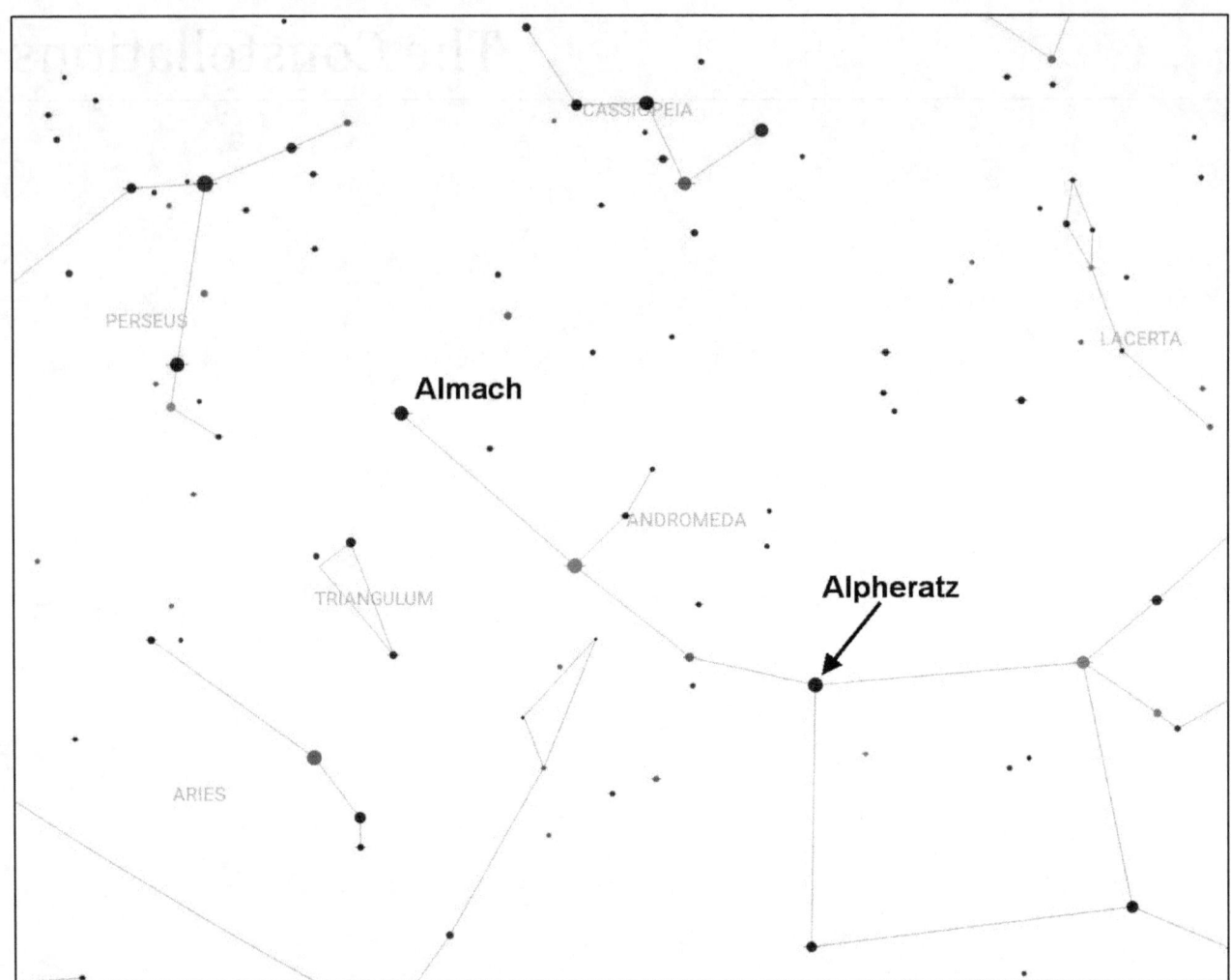

Andromeda

Andromeda is one of the more prominent constellations of the autumn sky and is conveniently located close to Cassiopeia and Pegasus. In Greek mythology, Andromeda was a princess and her story was made famous by the movie *Clash of the Titans*.

The legend involves a number of other constellations, all found nearby, most notably Cassiopeia, Cepheus, Pegasus and Perseus.

Andromeda was the daughter of Cassiopeia, the Queen and Cepheus, the King. Cassiopeia was a vain woman who boasted that her daughter was more beautiful than the sea-nymphs, thereby angering the god Poseidon.

In retribution, Poseidon sent a monster to terrorize the kingdom. Cepheus consulted an oracle who advised him to sacrifice his daughter to the monster. This would appease the god and save his kingdom.

Poor Andromeda was chained to a rock and was about to be devoured when Perseus came to her rescue. Perseus was returning from slaying the Medusa, a half-woman/half-serpent whose gaze could turn living creatures to stone.

Swooping down on his trusty steed, Pegasus the Flying Horse, Perseus pulled the head of Medusa from his bag and turned the monster to stone, thereby saving the princess. Of course, in true fairy tale fashion, the pair fell in love, married and lived happily ever after.

The constellation has a distinctive curve that outlines the princess's body, but in a game of stellar join-the-dots, if you look closely you can see a second curved line of fainter stars just to the north.

It's one of only two pairs of constellations that share a star. Alpha Andromedae actually marks the north-eastern corner of the square of Pegasus, the Flying Horse and is more commonly known as Alpheratz, from the Arabic for "the navel of the mare" – a reminder of its role in that constellation.

Although it appears to be a single magnitude 2.1 star to your eyes, it's actually a binary system about 100 light years away.

Besides Alpheratz, Andromeda is notable for containing the most distant object easily seen with just your eyes. Messier 31, the Andromeda Galaxy, is a sister galaxy to our own Milky Way and is thought to be about 2 ¼ million light years away.

If you look carefully, you can barely glimpse it as a very faint, misty patch but you'll need to be away from the lights of your town or city to see it. If you can't get away, try scanning the area with binoculars or a small telescope.

Further to the east is Gamma, also known as Almach, a magnitude 2.3 star that's a fine double in small telescopes. Close to Almach is NGC 752, a binocular star cluster that appears large but faint through a regular pair of 10x50's. Like Almach, this cluster is also a fine target for small telescopes with even a low power eyepiece providing an attractive view.

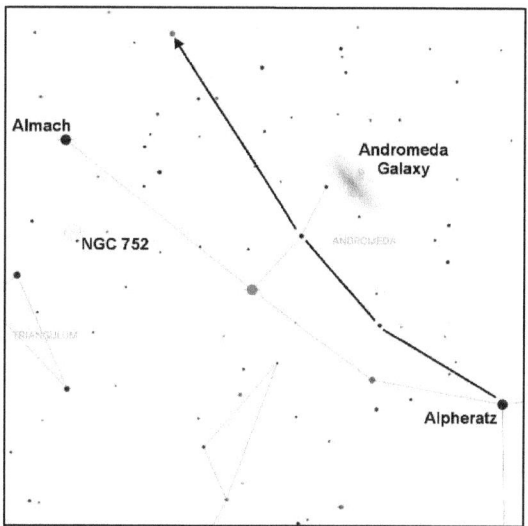
Features of Andromeda

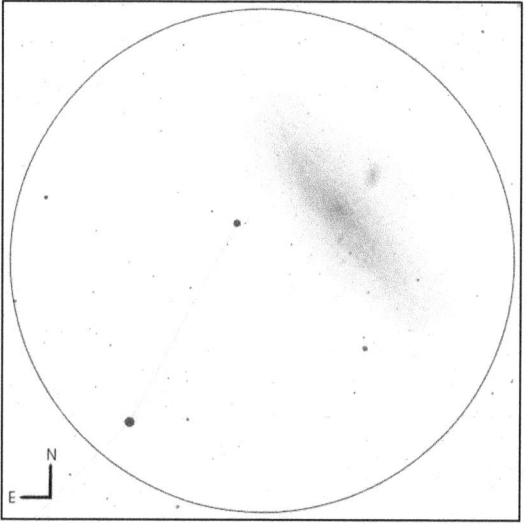
Messier 31 binocular view

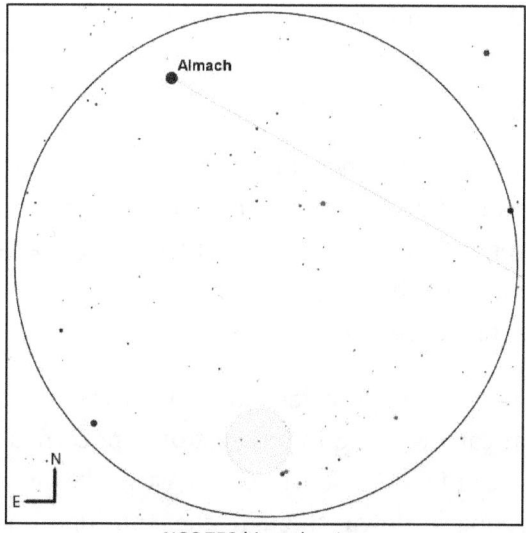
NGC 752 binocular view

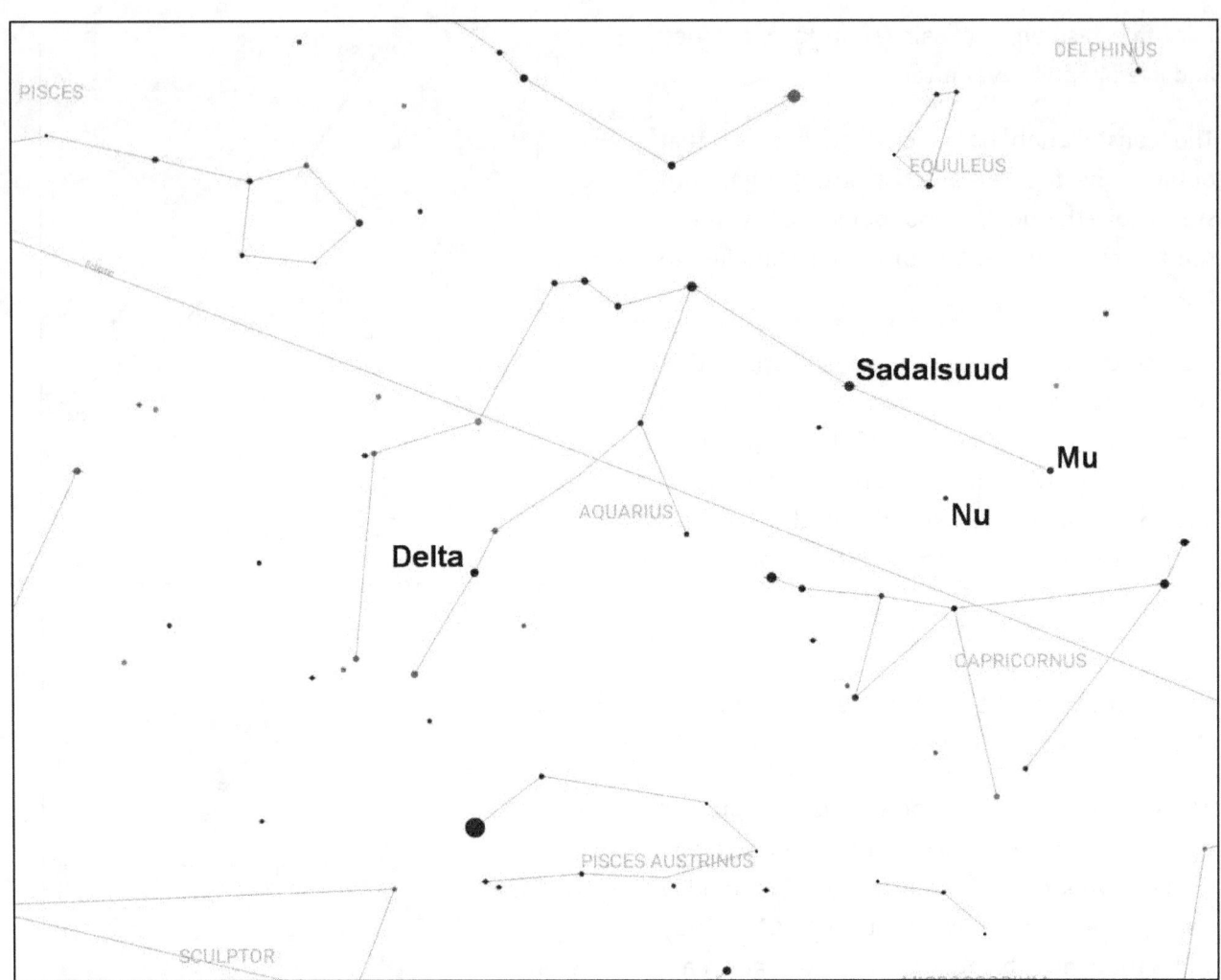

Aquarius

Aquarius is also known as "the water carrier" and, in Greek mythology, it's often associated with Ganymede, a Trojan boy who was kidnapped by Zeus to be the cup bearer of the gods.

Aquarius is one of the faint autumnal constellations and is not easily seen or recognized. You'll probably need to get away from the lights of your nearest town or city to identify its outline.

It's a sprawling constellation, 10th in size overall, that can be found to the east of Capricornus. If you can find the square of Pegasus, draw a line diagonally down through the top right and bottom right stars of the square. Keep going down until you come to a bright star. This is Fomalhaut, the brightest star in Pisces Austrinus, the Southern Fish. Midway between the two is Aquarius.

The brightest star in Aquarius is Beta, called Sadalsuud, an Arabic name that means "luck of lucks." Located some 550 light years away, it's a yellow star some fifty times the size of the Sun that's over 2,000 times brighter than our own star.

It's also pretty young – only about 60 million years old – which means if you could travel back in time and walk the Earth with the dinosaurs, you wouldn't see this star in the sky!

To our eyes today, it appears as a pretty ordinary star, just under magnitude 3 in brightness.

Aquarius is home to a number of deep sky objects but you'll need at least a pair of binoculars or a telescope to see them.

Brightest of these is Messier 2, a large globular cluster that should be within the range of a decent pair of binoculars. It's located about 3/4 of the way between Enif, in Pegasus and Sadalsuud. You can also try scanning westward from Sadalmelik.

Through binoculars I've noted that it appears as a small, grey misty patch while a small telescope can reveal a little more detail.

Of the other objects, two in particular stand out. The Helix Nebula is very large (almost the size of the full Moon) but is also very faint. You'll certainly need binoculars and you'll definitely need to be as far away from any lights as possible. Try to catch it close to Delta Aquarii with 66 Aquarii midway between the two.

This is a planetary nebula, which means it's the shell of a dying star that's slowly expanding outwards into space. This shell is now thought to be nearly three light years in diameter!

Another planetary is called the Saturn nebula. It's much smaller than the Helix, but has the appearance of that famous planet when observed with a telescope.

Through binoculars it can appear as a small, fairly bright star-like point, but it has the distinction of having a yellow-greenish hue. Like the Helix, this is the shell of a dying star but this particular stellar remnant is only a half light year in diameter. Look for it close to Mu and Nu Aquarii, but again, it's best to get away from any lights if you want to try your luck!

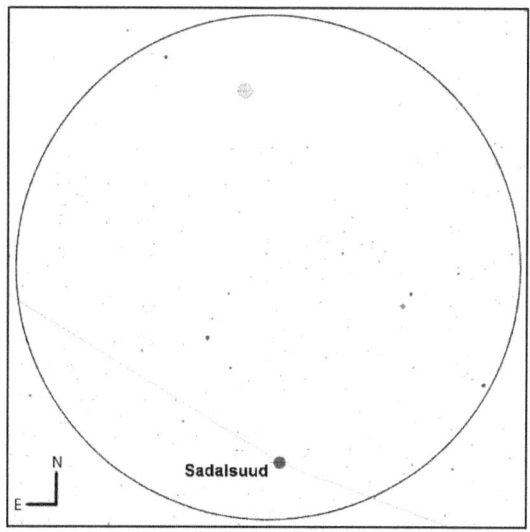

Messier2 binocular view

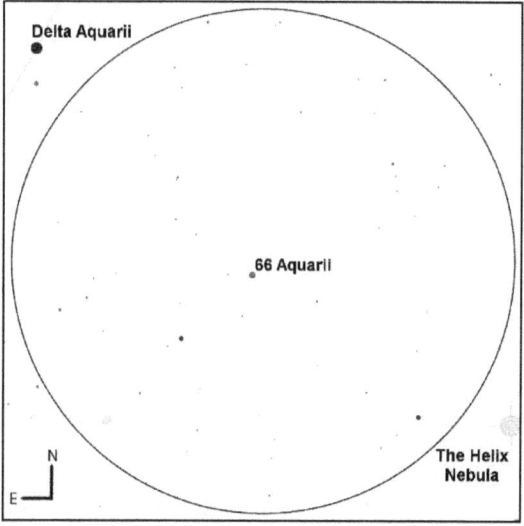

Locating the Helix Nebula, binocular view

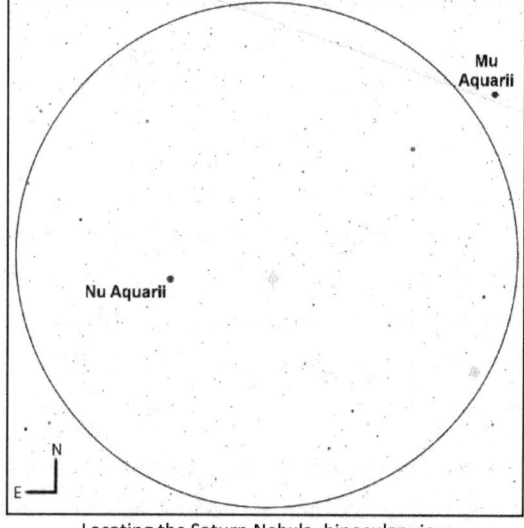

Locating the Saturn Nebula, binocular view

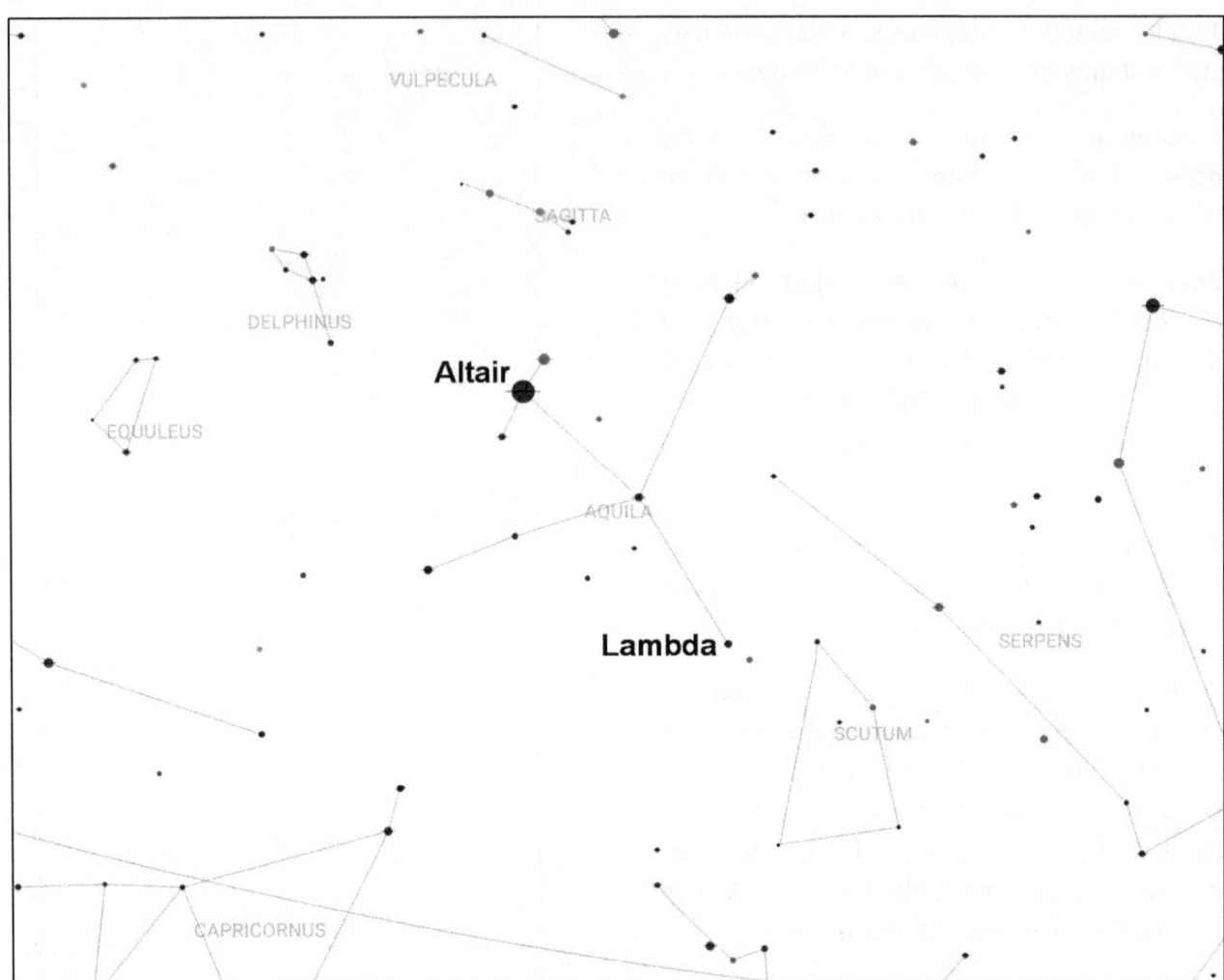

Aquila

In Greek mythology, Aquila represents the eagle that carried the thunderbolts of Zeus. This was the same eagle that carried the boy Ganymede away from his home to be the cup bearer for the gods (and was consequently immortalized in the constellation of Aquarius.)

Aquila is one of the main constellations of summer and is easily visible until the autumn. Its brightest star, Altair, marks one corner of the Summer Triangle (see page 182) with Vega (in Lyra) and Deneb (in Cygnus) completing the pattern.

Altair itself is associated with its own legend, but not one of Greek origin. In several Asian cultures, Altair represents a cow herder who fell in love with a weaver girl (or in some versions, a princess), as represented by the star Vega.

There are a number of variations, but the tale goes something like this: the girl spent all her time weaving and despaired of finding true love. Her father, the God of Heaven, wishing her to be happy, arranged for her to meet the cow herder but he didn't anticipate what happened next.

Falling deeply in love, the pair married and the girl stopped weaving and the cows wandered all over Heaven. This angered her father and so he separated them by placing the cow herder and the weaver girl on opposite sides of a river, represented by the Milky Way.

Consequently, the pair can only meet once a year, on the 7th day of the 7th month. On that day, all the magpies come together to form a bridge across the river. If it rains on that day, it's said to be the tears of the lovers, unable to meet.

In reality, Altair is the 12th brightest star in the sky and, at just under 17 light years away, one of the closest to our own. It's a pale yellow magnitude 0.8 star, nearly twice the mass of the Sun, that spins on its axis once every nine hours. This has the curious effect of flattening the star at its poles!

Altair is accompanied on either side by two stars: Alshaid (Beta Aquilae) to the east and Tarazed (Gamma Aquilae) to the west. In the legend recalled earlier, these two stars sometimes represent the two children of the forbidden lovers.

Keeping with the avian theme, Alshaid's name is derived from the Arabic for "the raven's neck" while all three stars are collectively known as the "family of Aquila."

There isn't a conspicuous deep sky object that's easily seen in the constellation; however, if you scan with binoculars just to the south-west, you'll encounter the small constellation of Scutum the Shield.

Here, quite close to Lambda Aquilae and Beta Scuti, is the Wild Duck Cluster. Discovered in 1681 by the German astronomer Gottfried Kirch, it has an estimated 2,900 stars and is over 6,000 light years away.

Through binoculars it appears as a small, faint globular patch but it's better observed with a small telescope under low or moderate power.

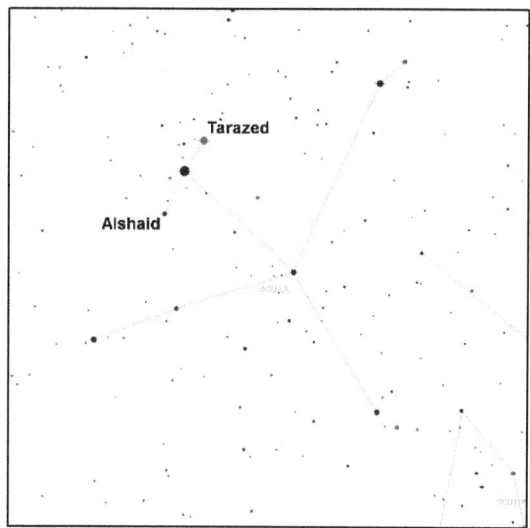

Alshaid and Tarazed

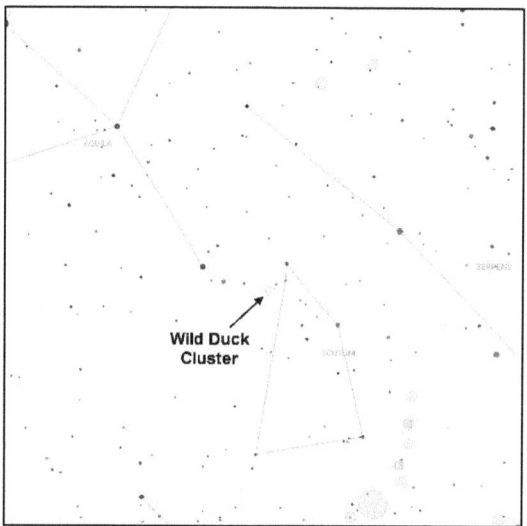

The location of the Wild Duck Cluster

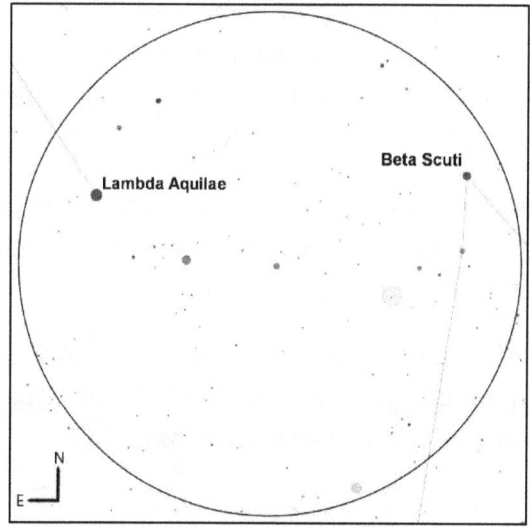

The Wild Duck Cluster, binocular view

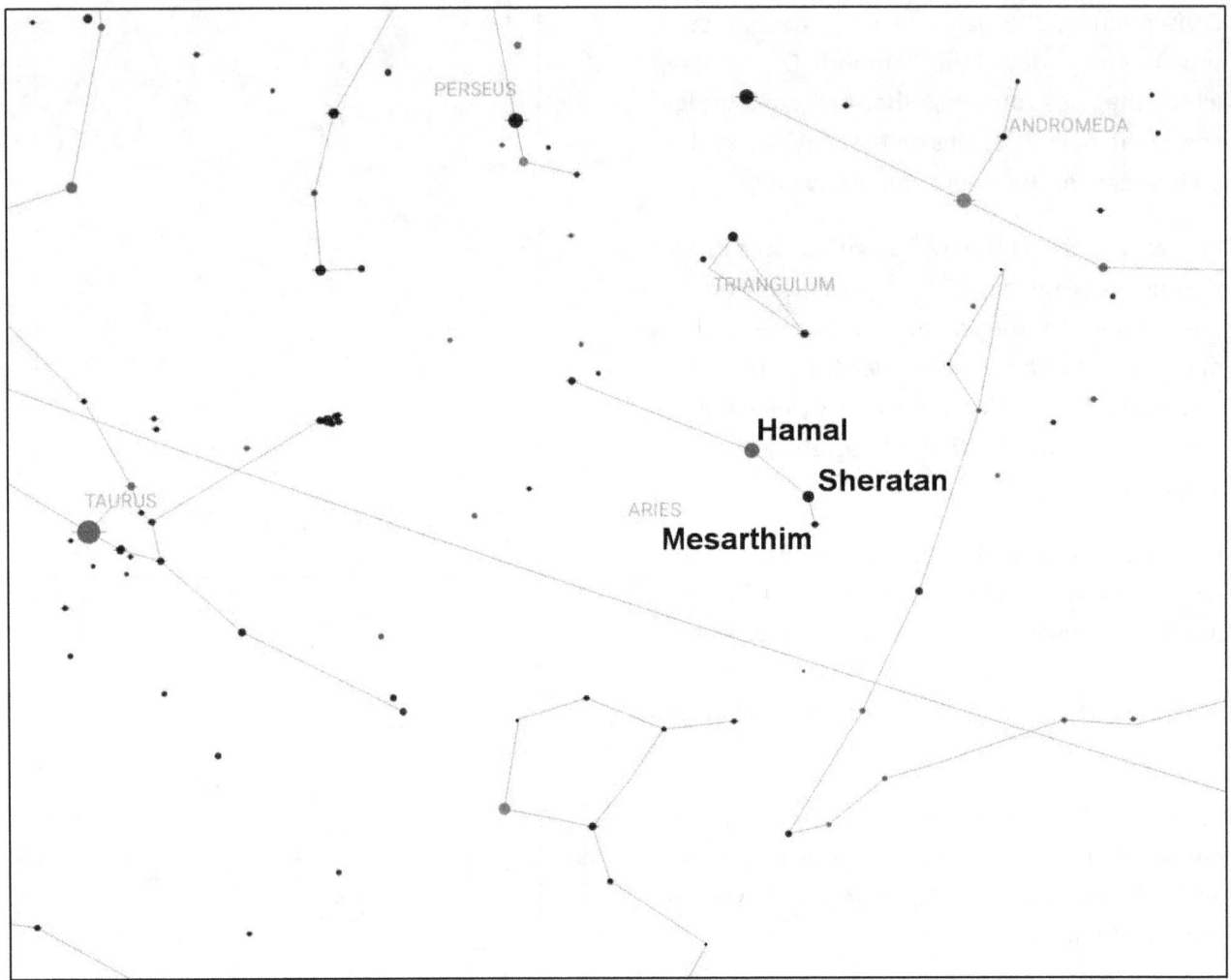

Aries

Aries is a small, autumnal constellation that has represented a ram across numerous cultures and civilizations since antiquity. Nowadays it's commonly associated with the ram that saved the lives of the twins Helle and Phrixus in Greek mythology.

The fleece of the ram became known as the Golden Fleece and features prominently in the legend of Jason and the Argonauts.

Aries is the first of twelve zodiac constellations and, as such, the Sun, Moon and planets all pass through it as they traverse across the sky.

Aries is particularly noteworthy as it once contained the First Point of Aries. This is where the Sun would cross from the southern celestial hemisphere to the north during the Spring equinox.

First noted over two thousand years ago, this point has since moved backwards into the constellation of Pisces but it still retains the name.

The constellation predominantly consists of three main stars that form a curved line between Taurus and Pegasus. From west to east, they are Mesarthim (the faintest of the three, magnitude 3.9), Sheratan (magnitude 2.7) and Hamal (the brightest, at magnitude 2.0)

Hamal means "head of the ram" and is an orange giant star some 66 light years away. It's thought to have at least one planet orbiting it once every 380 days. This planet is also about the same distance from its parent star as the Earth is from the Sun, but unfortunately, that's where the similarities end.

This planet is thought to be a giant about the size of Saturn and with nearly twice the mass of Jupiter. It also orbits the star far outside the habitable zone, so alien life is not likely to be there.

Sheratan, the second brightest star in the constellation, is slightly closer at about 60 light years away. It consists of a close pair of stars that orbit one another once every 107 days.

Unfortunately, you won't be able to see them both with only binoculars or a small telescope, but Aries is home to a particularly fine gem.

If you have a telescope, take a look at Mesarthim, the faintest of the three stars. With medium power (about 50x), you can easily split the star into two bright white components of equal brightness.

This is a genuine double star system that lies about 165 light years away with the two stars taking more than 5,000 years to orbit one another.

As for the name, its origins are not definitively known with links to the Arabic and Sanskrit words for ram.

One other star deserves a mention: Teegarden's star is a very faint red dwarf star, far beyond the limits of your eyes or binoculars. It wasn't discovered until 2003 but, at 12 light years away, is one of our closest neighbors in space.

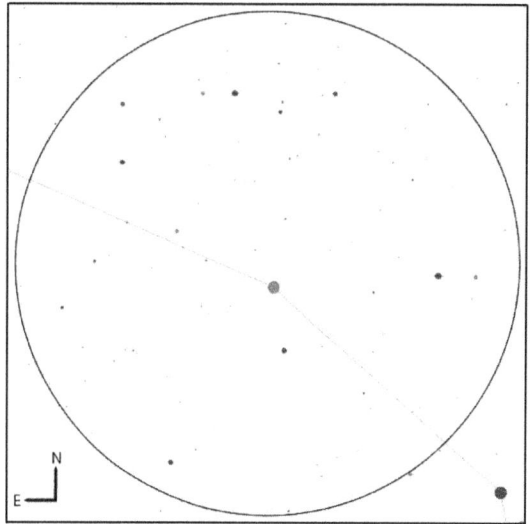

Hamal, binocular view

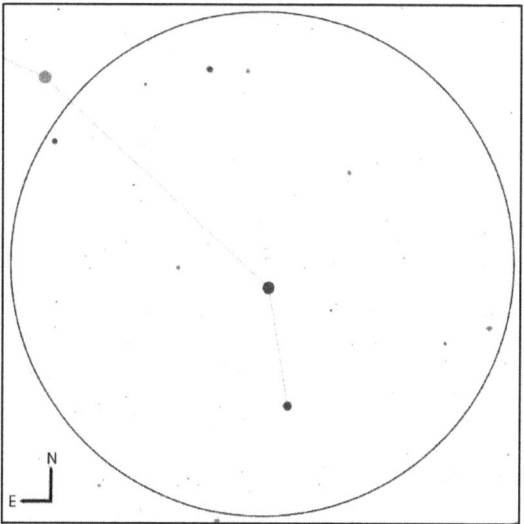

Sheratan, binocular view

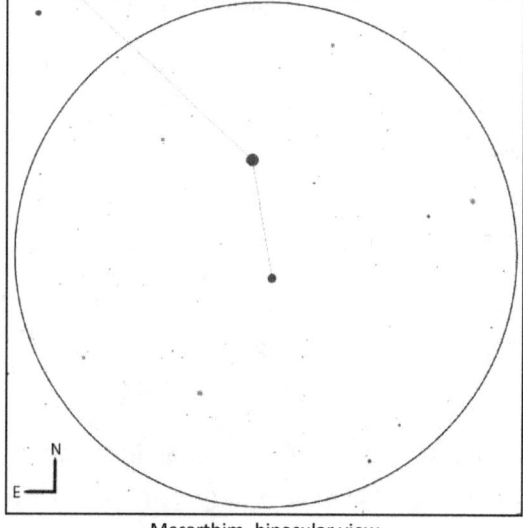

Mesarthim, binocular view

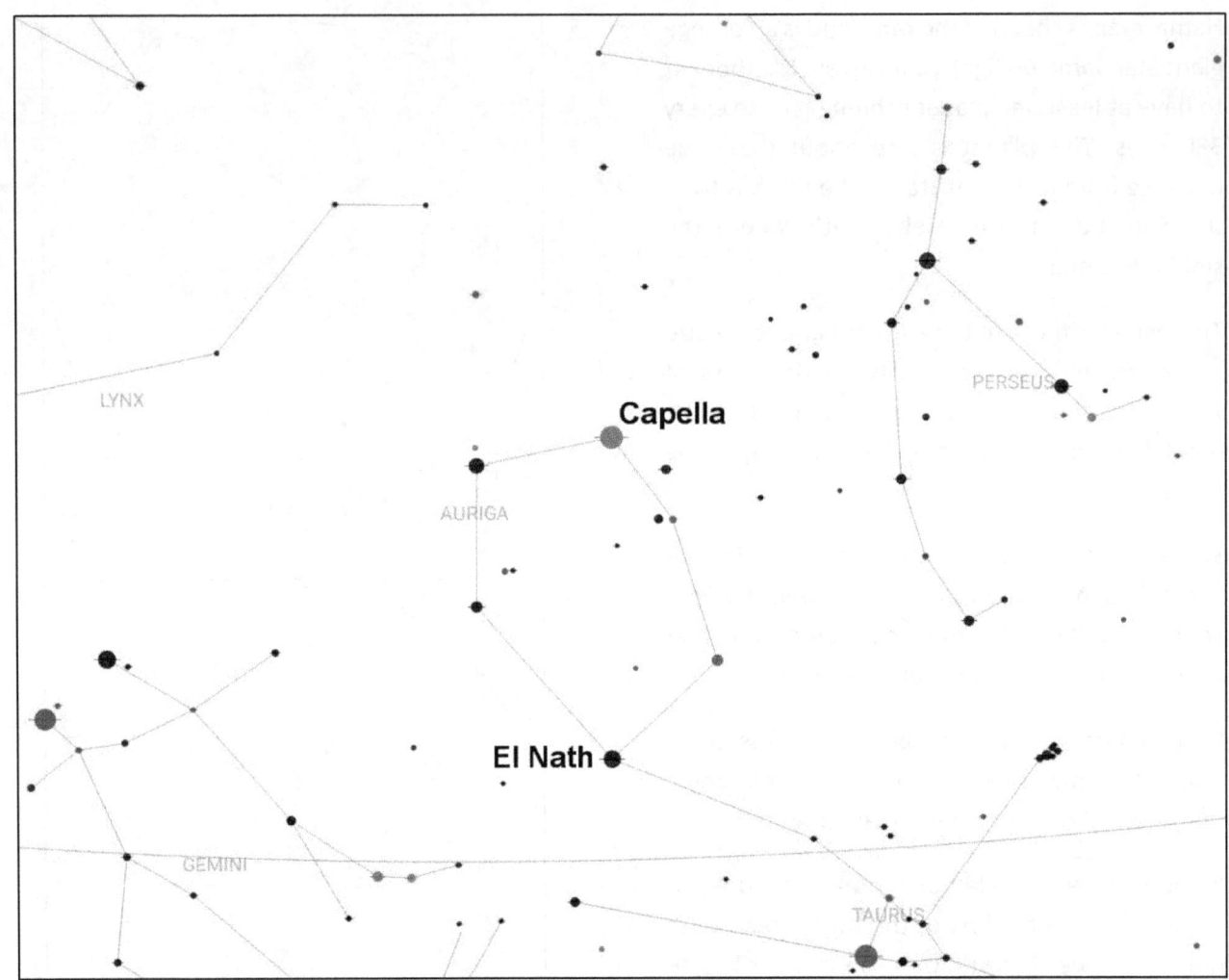

Auriga

Auriga, the Charioteer, is thought to represent Erichthonius, the lame footed king of Athens who invented the chariot as a means of transport.

It's a mis-shapen hexagon halfway between Perseus and Gemini, easily visible throughout the entire winter and made distinctive by its brightest star, Capella.

Not only is Capella the brightest star in the constellation, but at magnitude 0.1, it's also the sixth brightest in the entire night sky.

Like many other bright stars, at just under 43 light years, Capella is one of our closest neighbors but its most fascinating feature is invisible to us. To the naked eye it appears as a single star but it's actually a quadruple system made up of two pairs of stars.

The first pair are two yellow stars, similar to the Sun but each about 2-3 times more massive. They orbit one another once every 100 days or so with a gap of about ¾ the distance of the Earth to the Sun between them.

The other pair consists of two red dwarfs, thousands of times further out from the first pair.

Capella is also one of the few stars that has its own mythological associations.

According to Greek myth, the star represents the she-goat Amalthea who nursed the baby Zeus.

If you look closely at Capella you'll see a small elongated triangle of three stars, just to the west and on the Perseus side of the constellation.

Known as the *Haedi* (or "the kids") the bottom two stars represent the young goats sometimes depicted in the arms of the charioteer himself.

The northernmost star in the triangle may also be worth a look. Epsilon Aurigae is a variable star; in other words, it appears to grow dim and then brighten again. Unfortunately, it doesn't happen very often – in fact, it only happens to Epsilon once every 27 years and it won't happen again until approximately 2036.

At that time, the star will appear to fade from magnitude 3.0 to 3.8 and will stay this way for about a year or so. This is because the star is actually a binary system. Once every 27 years the fainter companion passes in front of the brighter primary star and eclipses it, causing the star's magnitude to drop.

Auriga, like Andromeda, actually shares one of its stars with another constellation. If you look carefully at the middle image on this page, you'll notice that the southernmost star, El Nath, is also a part of Taurus the Bull.

Once known as Gamma Aurigae, it now officially belongs to only Taurus and is known as Beta Tauri.

Lastly, if you have binoculars be sure to track down Auriga's three open star clusters – M36, M37 and M38. Of the three, M37 is the brightest and densest but M36 is smaller and, hence, its light is more condensed and the cluster can be easier to spot. M38 may only appear as a faint glow. All are best seen with a small telescope.

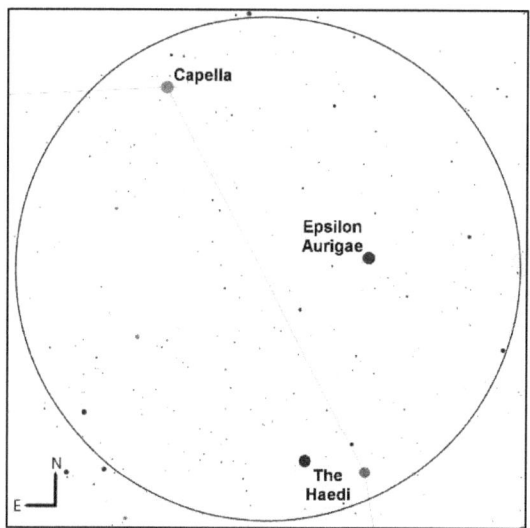

Capella, Epsilon and the Haedi, binocular view

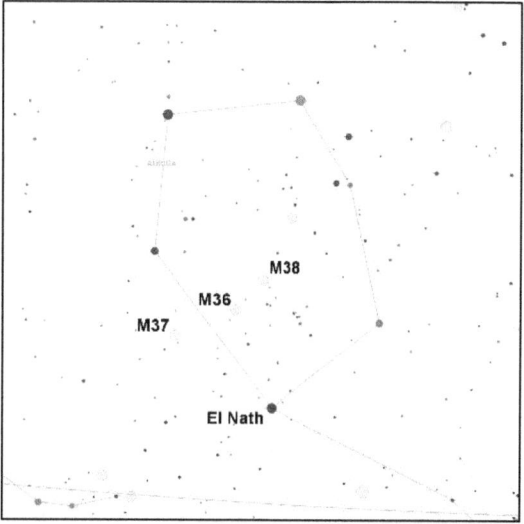

Location of El Nath, M36, M37 and M38

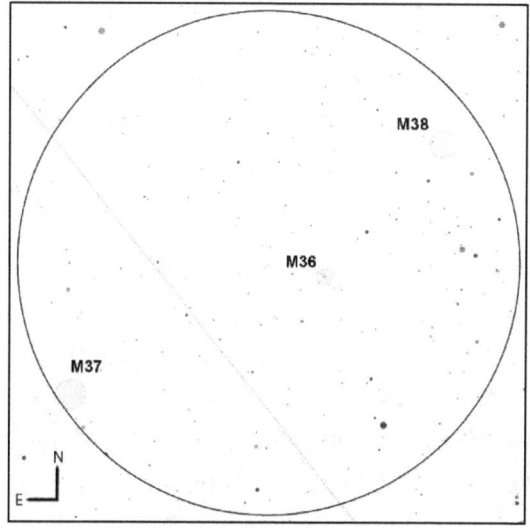

M36, M37 and M38, binocular view

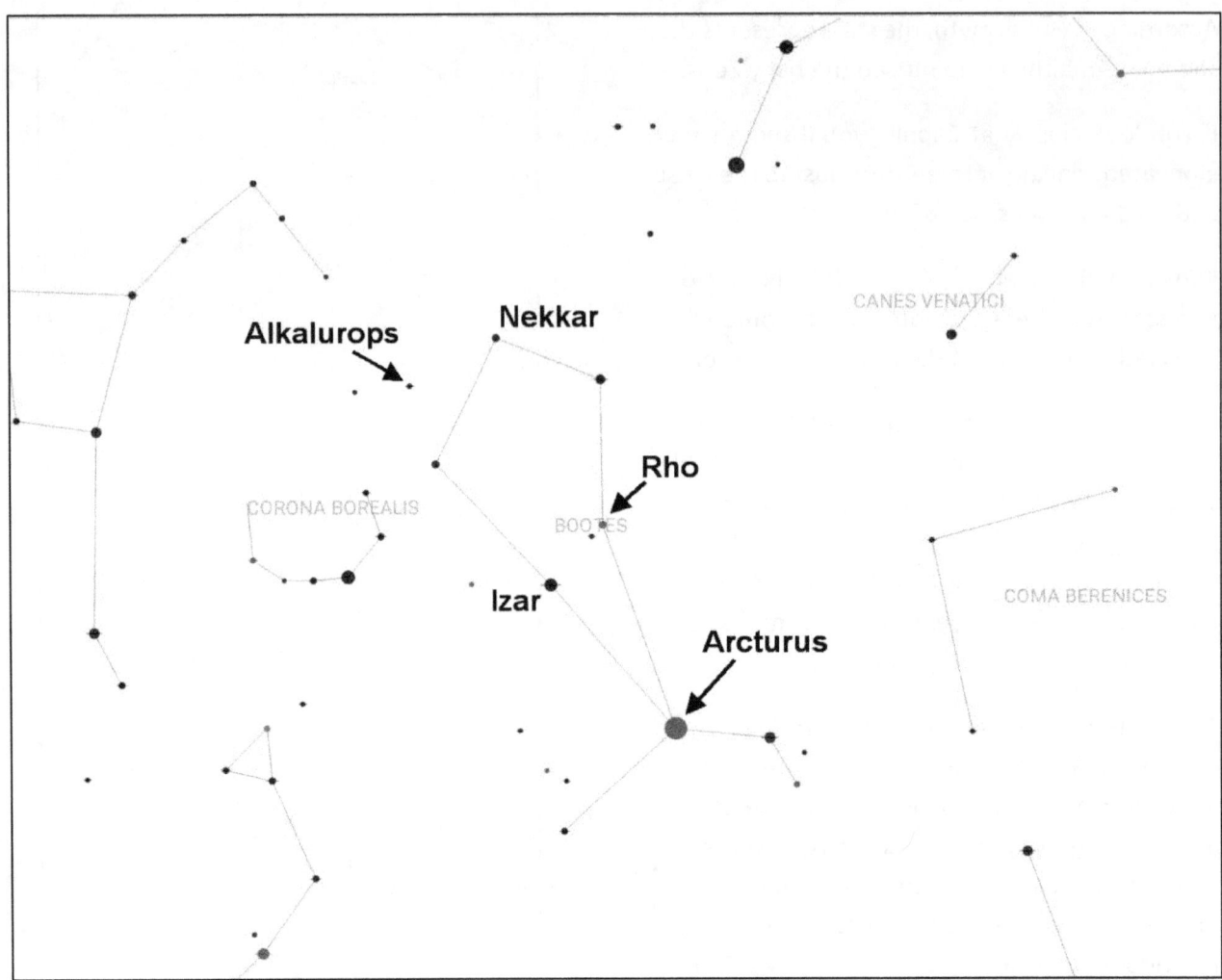

Boötes

Boötes is a large, kit-shaped constellation that can be quite easily seen overhead throughout the Spring and Summer.

Collectively the stars represent a herdsman, an association that dates back to Babylonian times when astronomers associated the constellation with farmers.

To the Greeks the constellation was often tied to Ursa Major, which was sometimes represented as oxen. Boötes was seen to be driving the oxen or, sometimes a plough – which would explain why the seven brightest stars in Ursa Major are known as the Plough in Europe today.

In fact, to find Boötes you only have to first find the Plough and you can use the three stars of its handle to curve down to Arcturus, the brightest star in Boötes. (If it helps, remember to "arc down to Arcturus." Also see page 181.)

Arcturus itself is the fourth brightest star in the entire night sky and is hard to miss on a Spring or Summer evening.

An orange giant star only 37 light years away, it has a distinctive hue when observed with the naked eye. The name is derived from an ancient Greek word meaning "guardian of the bear" – an obvious reference to nearby Ursa Major.

Many other cultures across the world are also familiar with the star and it's one of the few stars mentioned in some versions of the Bible (Job 38:32, "or canst thou guide Arcturus with his sons?")

Perhaps most curiously, light from Arcturus was actually used to open the 1933 World's Fair in Chicago. It was thought the light from the star had started on its journey during the last World Fair in 1893.

Unfortunately, this was wrong as the star is only 37 light years away – not the required 40 – although it's possible that contemporary astronomers believed it to be that distance.

Of the other stars there are a number of multiples that might be worth seeking out in binoculars or a small telescope. For starters, take a look near the top of the kite for two easy-to-see binocular doubles.

Firstly, there's Nu¹ and Nu² Boötis, a relatively wide pair of stars of equal brightness. Both have appeared white to me but others have reported hints of blue and orange.

Within the same field of view is Alkalurops (Mu Boötis), another easy binocular double with the primary appearing about twice as bright as the secondary.

While you're in the area, take a look at Beta, also known as Nekkar. This slightly variable star has a curved trail of fainter stars to the south and a close pair of magnitude 8 stars to the south-west.

Lastly, move south to Epsilon (Izar) and Rho. Izar is a white star with an unrelated companion to the south. A magnitude 4.5 star appears close to Rho and a much closer magnitude 7.8 companion may also be within binocular reach.

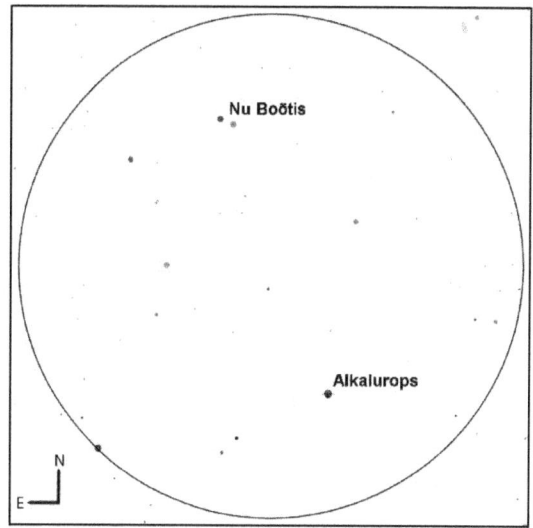
Alkalurops and Nu Boötis, binocular view

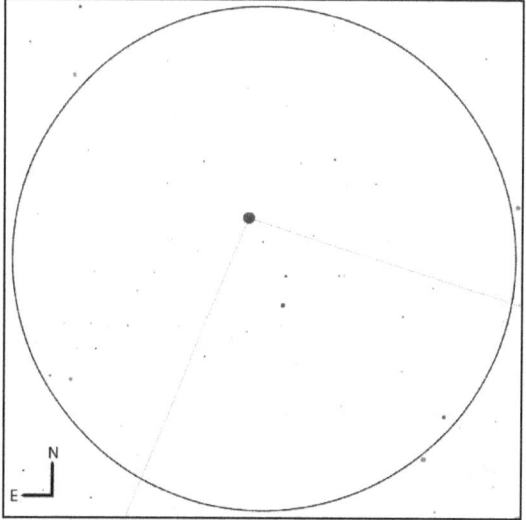
Nekkar, binocular view

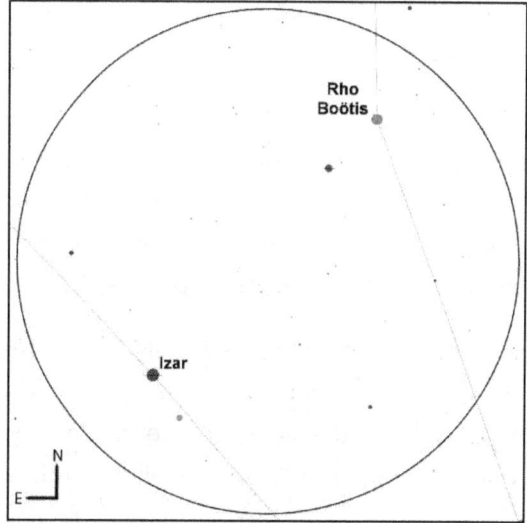
Izar and Rho Boötis, binocular view

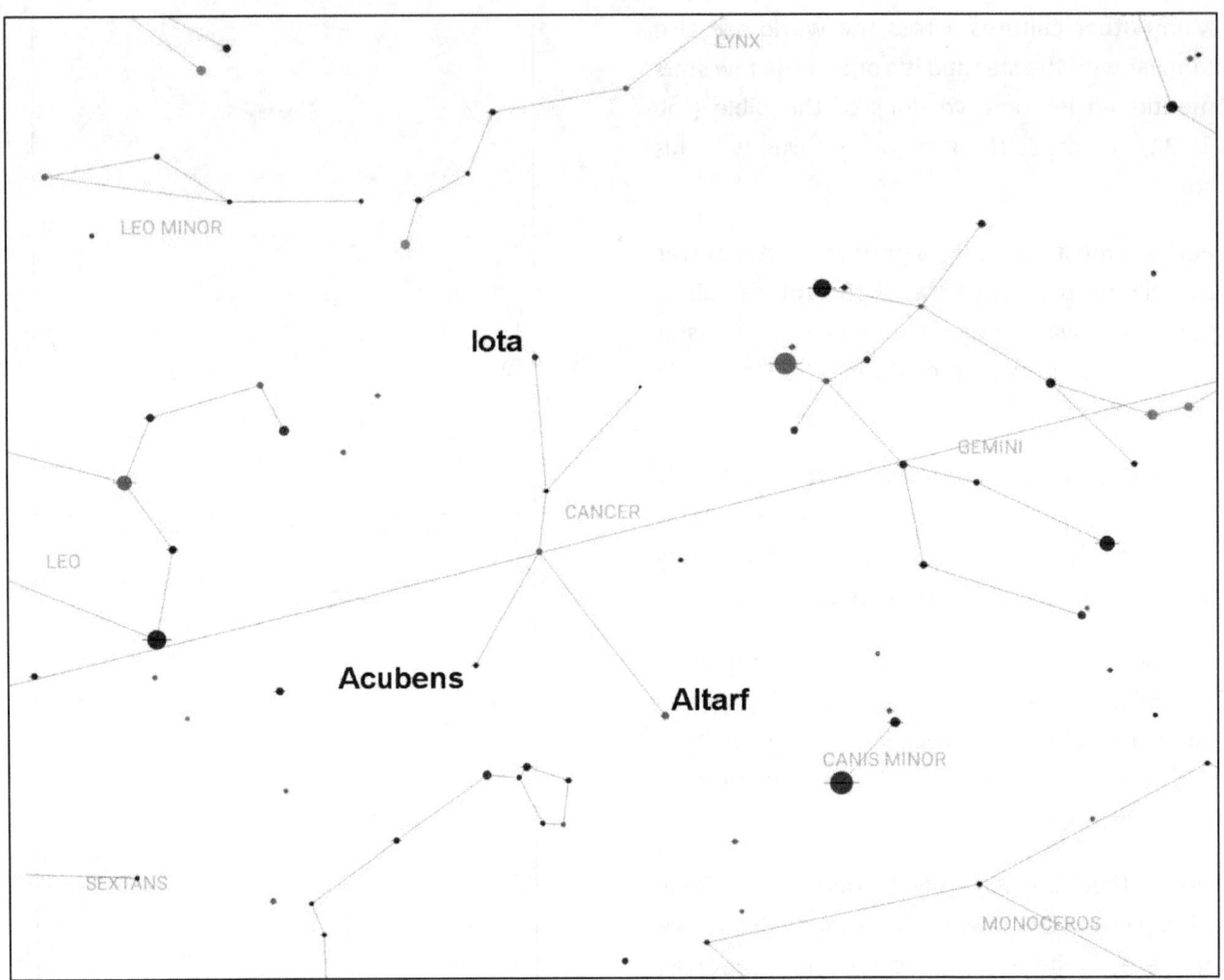

Cancer

Cancer, as many people know, represents a Crab and is one of the twelve signs of the zodiac. It's also the faintest, with no stars brighter than magnitude 3.5.

Poor Cancer suffers mythologically too as there are no grand stories associated with it. It's said to be the crab that Hercules crushed with his foot after the poor creature nipped him with its claws as he battled the sea-serpent Hydra.

Unlike many of the other constellations, it does at least partly resemble the crab it represents. Get away from the lights of the city and you should see a faint K shaped constellation midway between the bright stars of Pollux in Gemini and Regulus in Leo.

Its own brightest star is Altarf, also known as Beta Cancri, a magnitude 3.5 orange star some 290 light years away that's orbited by a red dwarf companion.

The companion is too faint to be seen by most amateurs, but there's at least one double that's an easy target for small telescopes – and it comes with a binocular double nearby.

Iota is the northernmost star of the K and can easily be found, even without optical aid. Turn a telescope toward it and low power (about 27x) will reveal the star's two components.

The brighter of the two appears to be white-gold and about three times brighter than its pale blue companion.

If you want a challenge, try splitting the pair with just your binoculars. In theory, a pair of 10x50's might be up to the job but it's tricky under the best of circumstances.

While you're in the area, look just a little to the east of Iota for 53 and 55 Cancri. The pair is often listed as Rho Cancri in books and makes for an easy binocular target. Through a regular pair of 10x50's, you'll see two coppery stars of almost equal magnitude.

53 Cancri is itself a double but you'll need a telescope to split it and the companion is pretty faint. 55 Cancri, also a binary, was one of the first stars discovered to have its own system of planets. At only 41 light years away, this star is currently known to have five planets orbiting it but, unfortunately, none of them are good candidates for extra-terrestrial life.

Double stars aside, the real gem of Cancer is the Praesepe open star cluster, also known as the Beehive or, occasionally, the Manger. Seen with just the unaided eye under clear dark skies, it was once used to predict the weather. If the cluster could not be seen on a clear night, it was said that rain was on its way.

It's easily found in the heart of the crab and is an attractive sight in binoculars as its Beehive shape is becomes apparent. However, it's best seen in a small telescope at low power when many of its thousand stars will readily come into view.

Lastly, be sure to also look out for Messier 67, a tightly packed star cluster close to Acubens at the southern-eastern edge of the constellation. A faint cluster, it's also one of the oldest clusters known with an estimated age of 10 billion years!

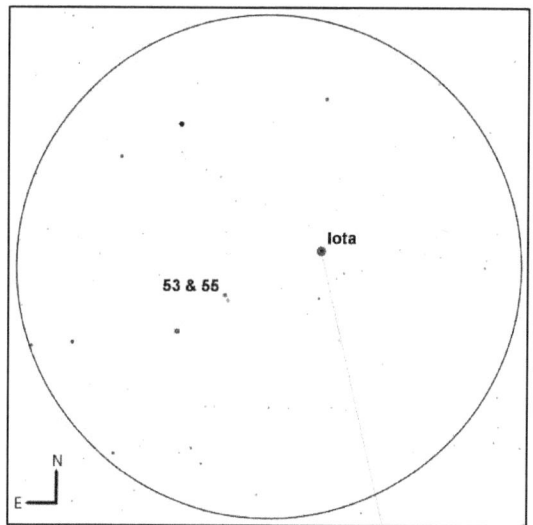

Iota, 53 and 55 Cancri, binocular view

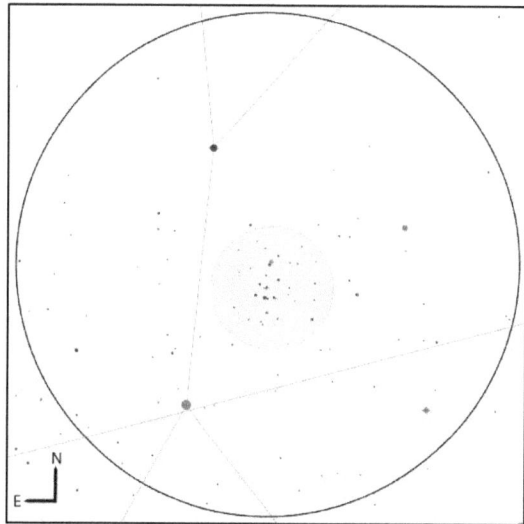

The Praesepe, binocular view

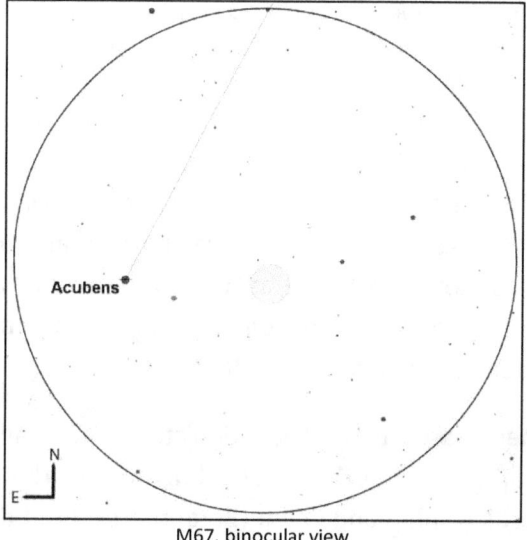

M67, binocular view

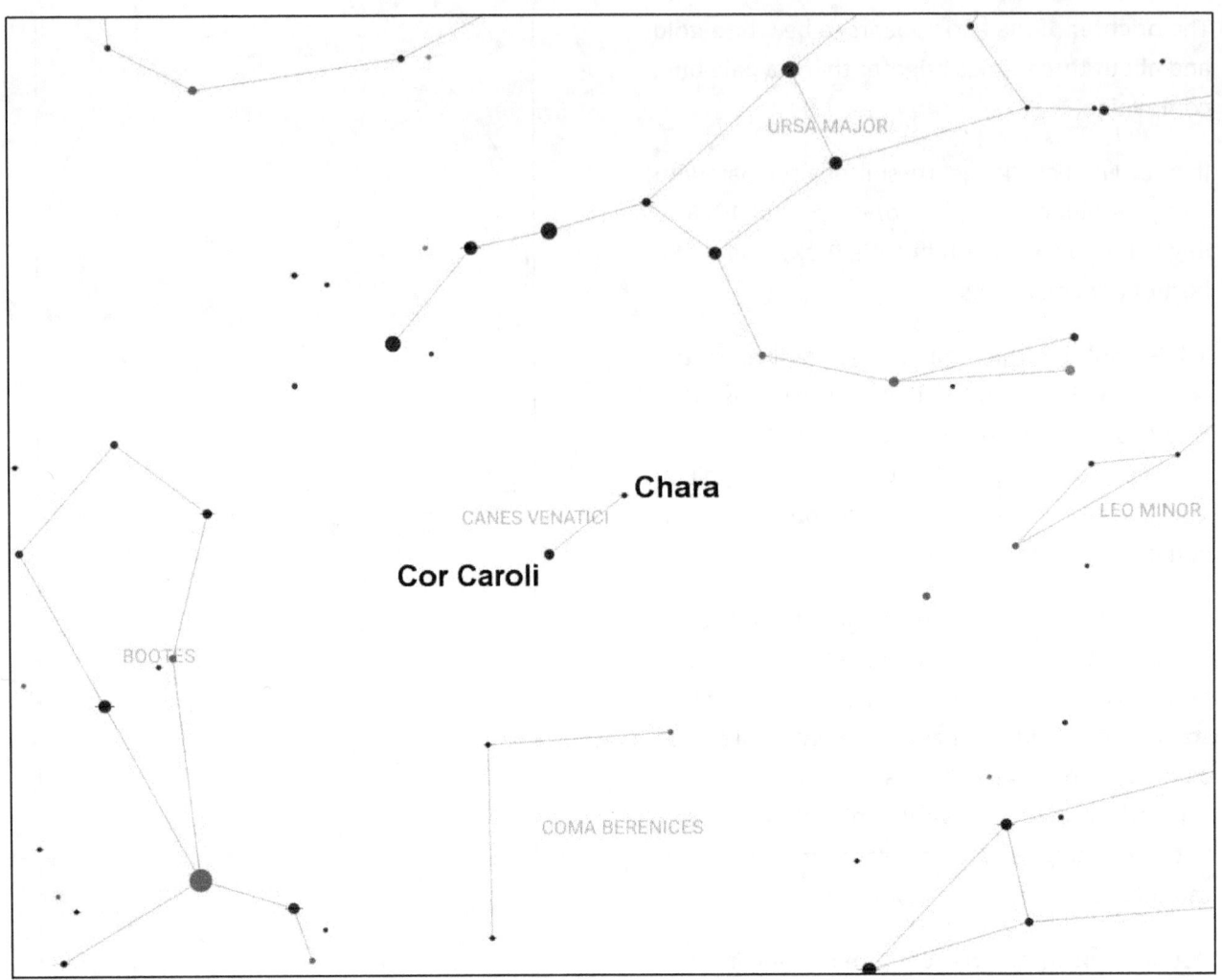

Canes Venatici

A relatively small constellation with only two prominent stars, Canes Venatici represents the Hunting Dogs associated with Boötes, the Herdsman. However, it wasn't always this way.

Originally, the Greeks recognized the stars as the club of Boötes, but due to multiple mistranslations the stars became associated with dogs. Canes Venatici was then created by the astronomer Johannes Hevelius in 1687 when he set the stars apart from Boötes and made the stars into their own constellation.

There are two dogs depicted here and, traditionally, both the two brightest stars Cor Caroli (Alpha) and Chara (Beta) represented the southern hound. In fact, the name Chara refers to the name of the dog itself as Beta originally had no name. Hevelius named the northern dog Asterion, which means "little star" – an appropriate name since no bright stars mark its position!

The star Chara is relatively unremarkable. At magnitude 4.3 it might be tricky to spot from the light polluted skies of a town or city and there's not much else to set it apart from the surrounding stars.

It's a yellow star, similar to the Sun, that's only 27 light years away and, as such, it's a potential candidate for nearby alien civilisations.

Unfortunately, so far there's been no indication of any planets, much less little green men.

Now move a little toward the north-west and try looking for the appropriately named La Superba. It's a variable star that changes magnitude from 4.9 to 7.3 over a period of roughly 160 days. Try observing the star's brightness and comparing it with the surrounding stars in the area. Then come back a few weeks later – does it seem brighter or fainter than before?

If you're looking for the star, you might be wondering how to recognize it. Fortunately, La Superba makes it easy, thanks to its deep red color. It's a rare type of carbon-rich star that, despite its low surface temperature, shines with a luminosity several thousand times that of the Sun.

Alpha, commonly known as Cor Caroli, the brightest star in the constellation, is a stellar gem for observers with a small telescope. Its name means "Charles's Heart" and was given to the star in 1660 when it's said the star brightened upon the return of King Charles to England. To the naked eye, it's a moderately bright star of magnitude 2.8 and although it *does* change a little in brightness, it's too slight to be seen with just your eyes.

At about 110 light years, it's another nearby star but unlike Chara, this one's a beautiful double, easily split with a small telescope. If you have binoculars, look out for 15 & 17 Canum Venaticorum, a close pair of stars to the east.

The constellation is also home to a number of other deep sky objects, but unfortunately none are easily seen with just your eyes. However, you may be able to glimpse the bright globular cluster, Messier 3, with a pair of binoculars.

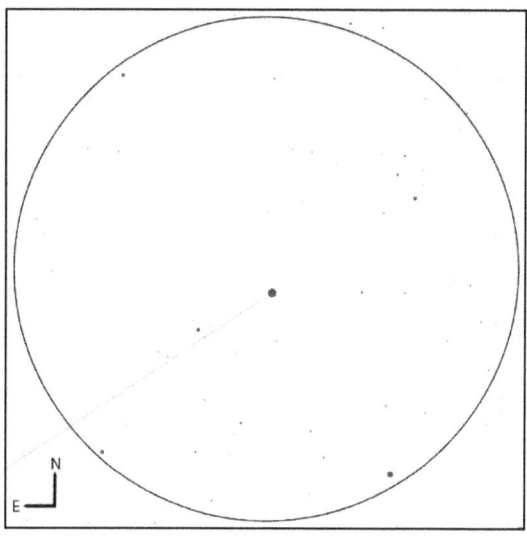
Chara, binocular view

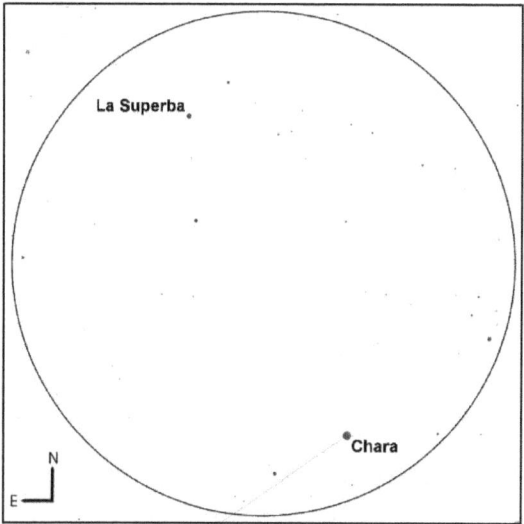
La Superba, binocular view

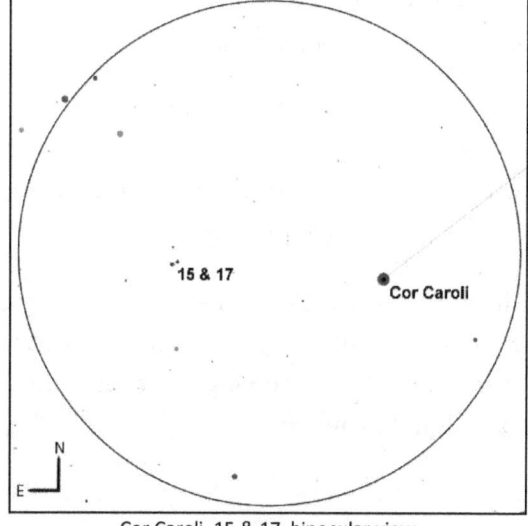
Cor Caroli, 15 & 17, binocular view

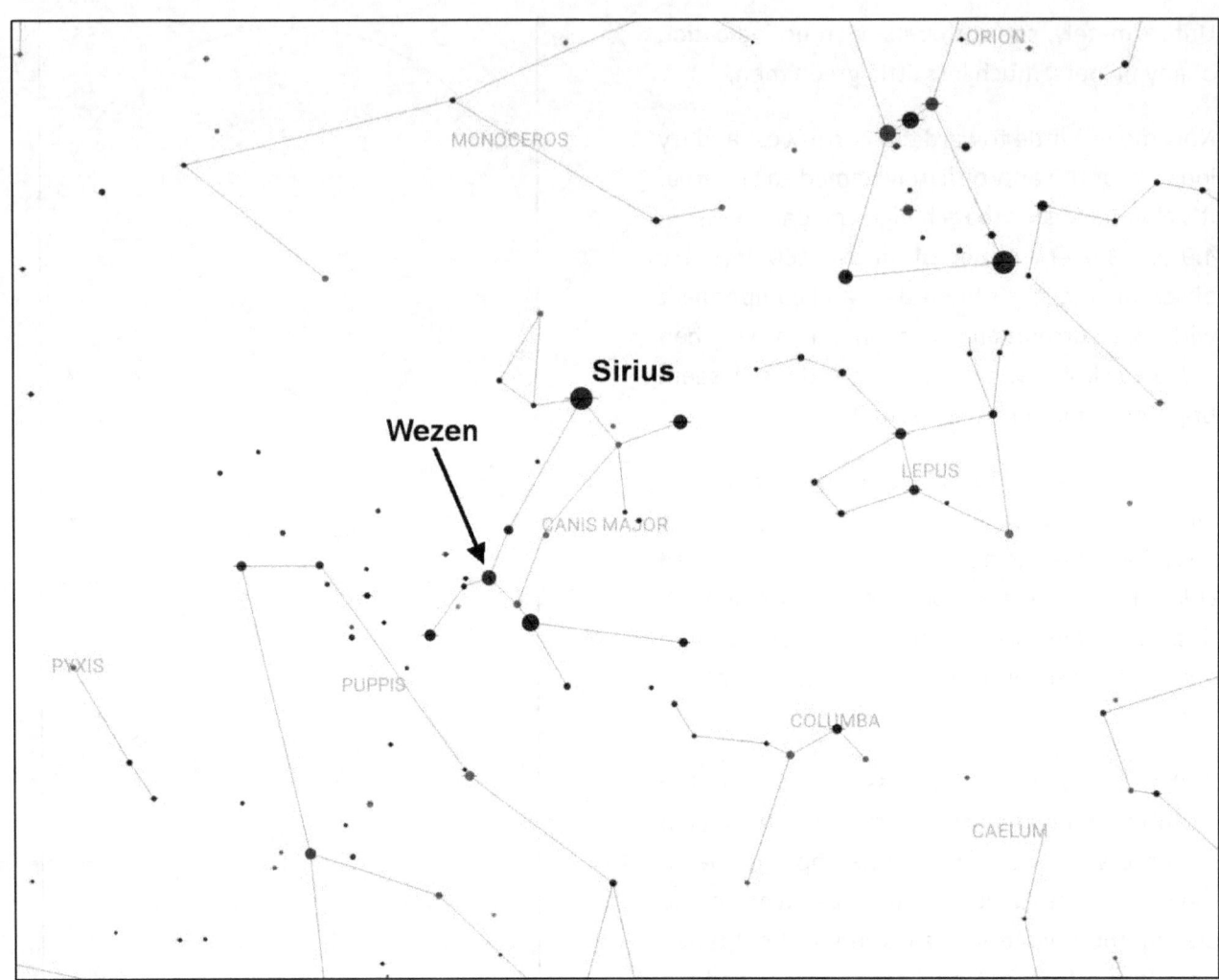

Canis Major

Canis Major is the most prominent of three canine constellations (the other two being Canis Minor and Canes Venatici) and represents the greater of Orion's two hunting dogs.

Arguably, the constellation would not be so prominent if it weren't for two things: its location and its brightest star – the brilliant Sirius, also known as "the dog star."

With the constellation being conveniently located to the south-east of his master, Sirius can be easily found by following the three stars of Orion's belt downward.

Not only is the star the brightest in the constellation, but as many folks know, it's also the brightest star in the sky and can be a dazzling sight. In fact, its very name is derived from the Greek word for "searing."

This can be especially true when the star is close to the horizon. At this time, it appears to flash a myriad of colors, including red, white and blue, and has even been mistaken for a UFO.

Given its prominence, it's not surprising to learn that the star was known to ancient civilizations around the world. It was particularly revered by the Egyptians, who associated it with their goddess Isis and used its pre-dawn rising to predict the flooding of the Nile.

What they couldn't know is that Sirius actually has a small, white dwarf companion, affectionately known as "the pup" and all but invisible to the vast majority of amateur telescopes.

They were also blissfully unaware that the system is a mere 8 light years away, making it one of the closest to the Sun.

Canis Major contains a number of deep sky objects. The easiest to observe is Messier 41, an open star cluster found almost directly due south of Sirius and midway between that star and Wezen.

This cluster is a highlight of winter skies but is often overlooked in favor of the larger and brighter Pleiades. Even from the suburbs it's visible as a small, conspicuous patch in binoculars and has even reminded me of a lobster on occasion.

That being said, the cluster truly shines when observed with a small telescope. Low power will reveal a large cluster, predominantly made up of uniformly bright blue-white stars with a couple of older orange stars thrown in for good measure.

There are a number of other highlights for binocular observers. If Sirius is one of our closest neighbors, then VY Canis Majoris – at a distance of 3,900 light years - is one of our most distant and most fascinating.

Shining with the light of over 60,000 suns and with a diameter of over 600 suns, it's one of the largest and most luminous stars known. It's also variable, fading from magnitude 6.5 to 9.6 roughly every 2,000 days. Tricky to find, look for it within the same binocular field of view as Delta, also known as Wezen.

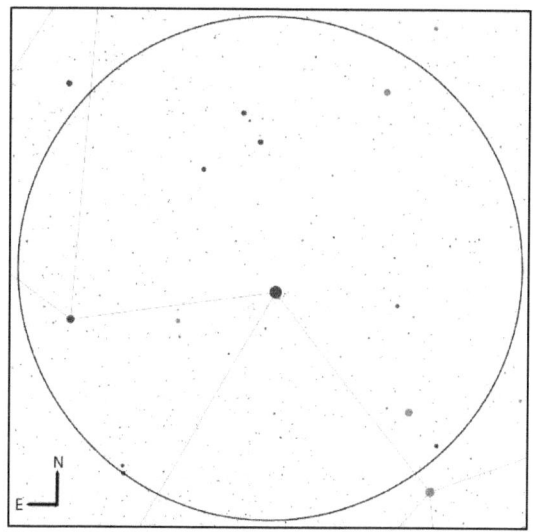

Sirius, binocular view

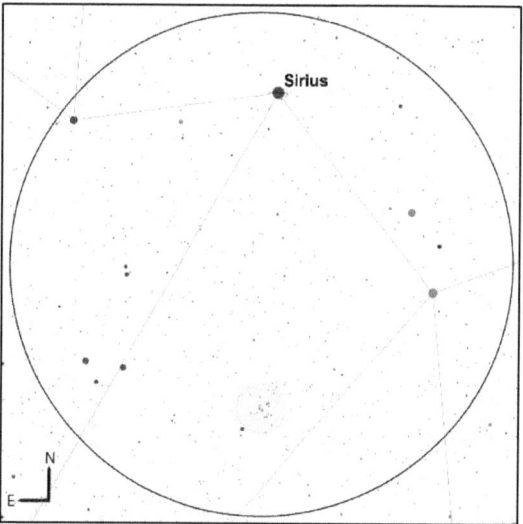

M41, binocular view

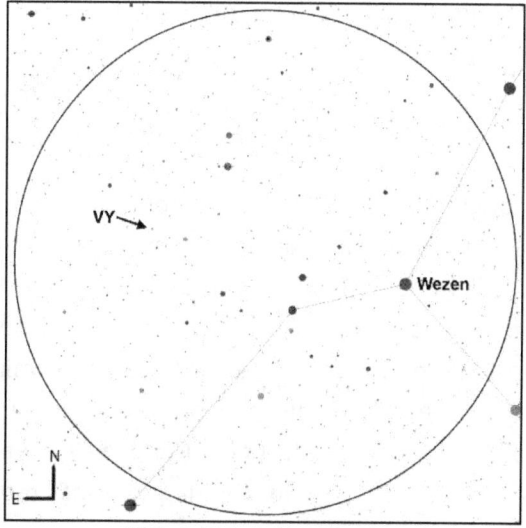

VY Canis Majoris, binocular view

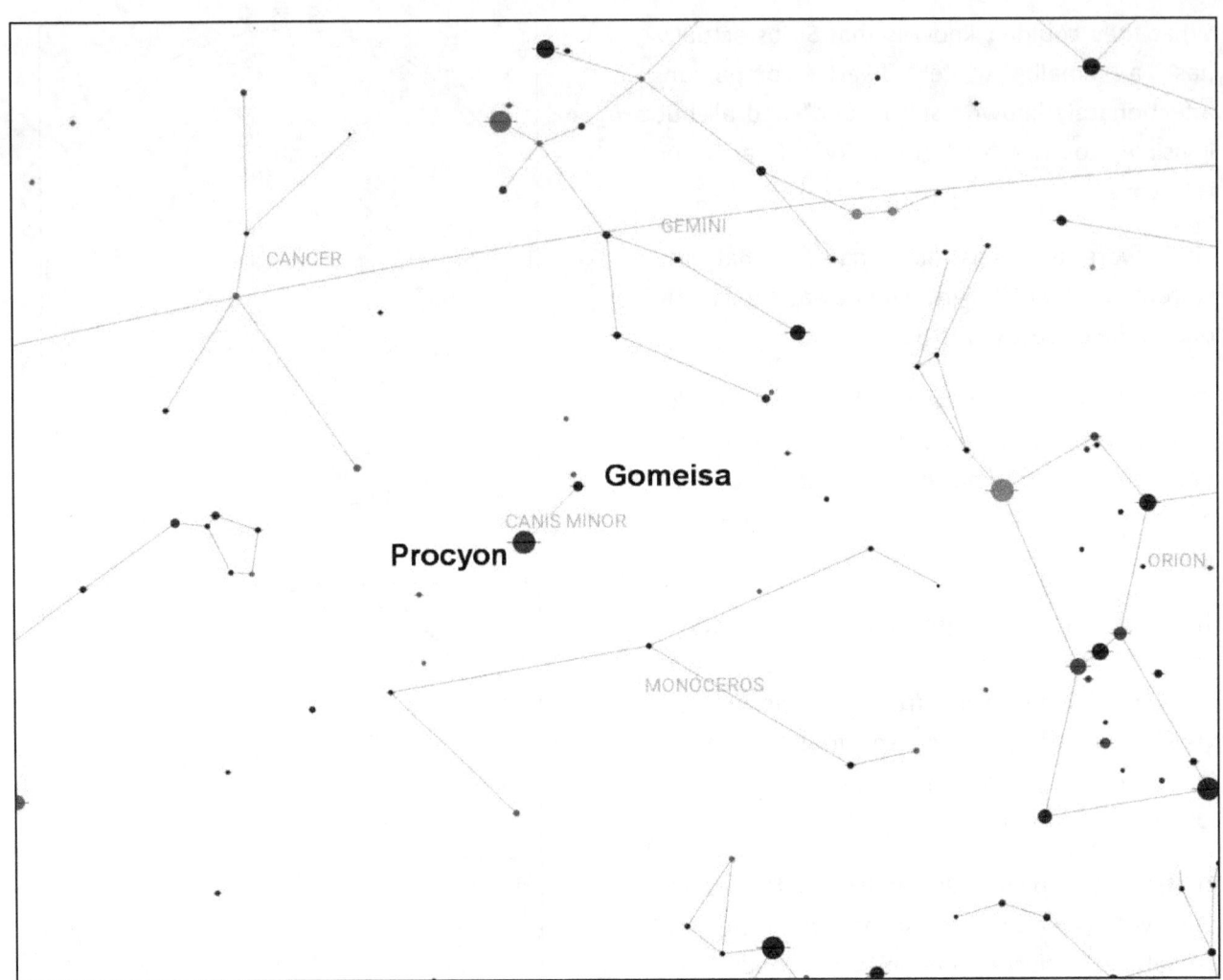

Canis Minor

Representing the second of Orion's hunting dogs, this constellation depicts the lesser and presumably smaller canine. Out of the 88 constellations in the sky, it ranks 71st in size and, like Canes Venatici, another hunting canine constellation, it consists of only two bright stars.

Its brightest is Procyon, the 8th brightest star in the sky and, like its neighbor Sirius, at just under 11 ½ light years it's also one of the nearest. Its name was originally given to the whole constellation by the Greeks and means "coming before the dog." This is a reference to Canis Major as Procyon and Canis Minor will always rise before Sirius and the larger (and more southerly) constellation.

A white star just moving into the subgiant stage, Procyon also has a white dwarf companion (like Sirius) that's invisible to amateur telescopes. The pair orbit one another once every 41 years at about the same distance as that between the Sun and Uranus. Binoculars reveal a faint white companion that's unrelated to Procyon itself.

Procyon also marks one corner of what's commonly known as the Winter Triangle with the other corners marked by Betelgeuse in Orion and Sirius, the brightest star in the sky.

The other bright star is Gomeisa, an unremarkable blue-white star of magnitude 2.9 that lies some 162 light years away. Its name means "the bleary-eyed woman."

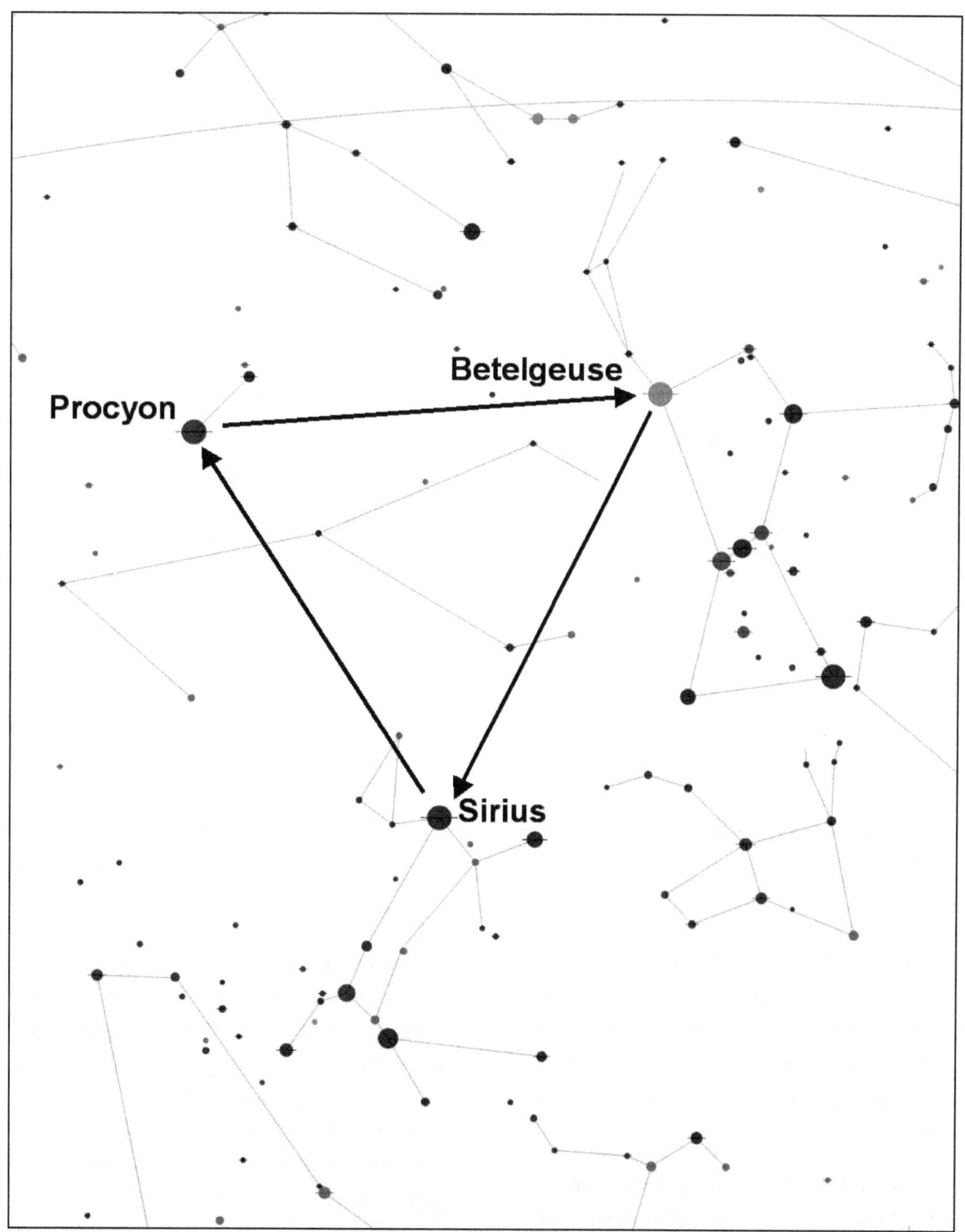

The Winter Triangle is a prominent feature of the winter sky and can be easily seen right through to early Spring. It consists of Procyon in Canis Minor, Betelgeuse in Orion and Sirius in Canis Major, three of the brightest stars in the entire sky. This large asterism is best seen in the evening hours throughout January and February. It has its counterpart in the Summer Triangle (see page 182.)

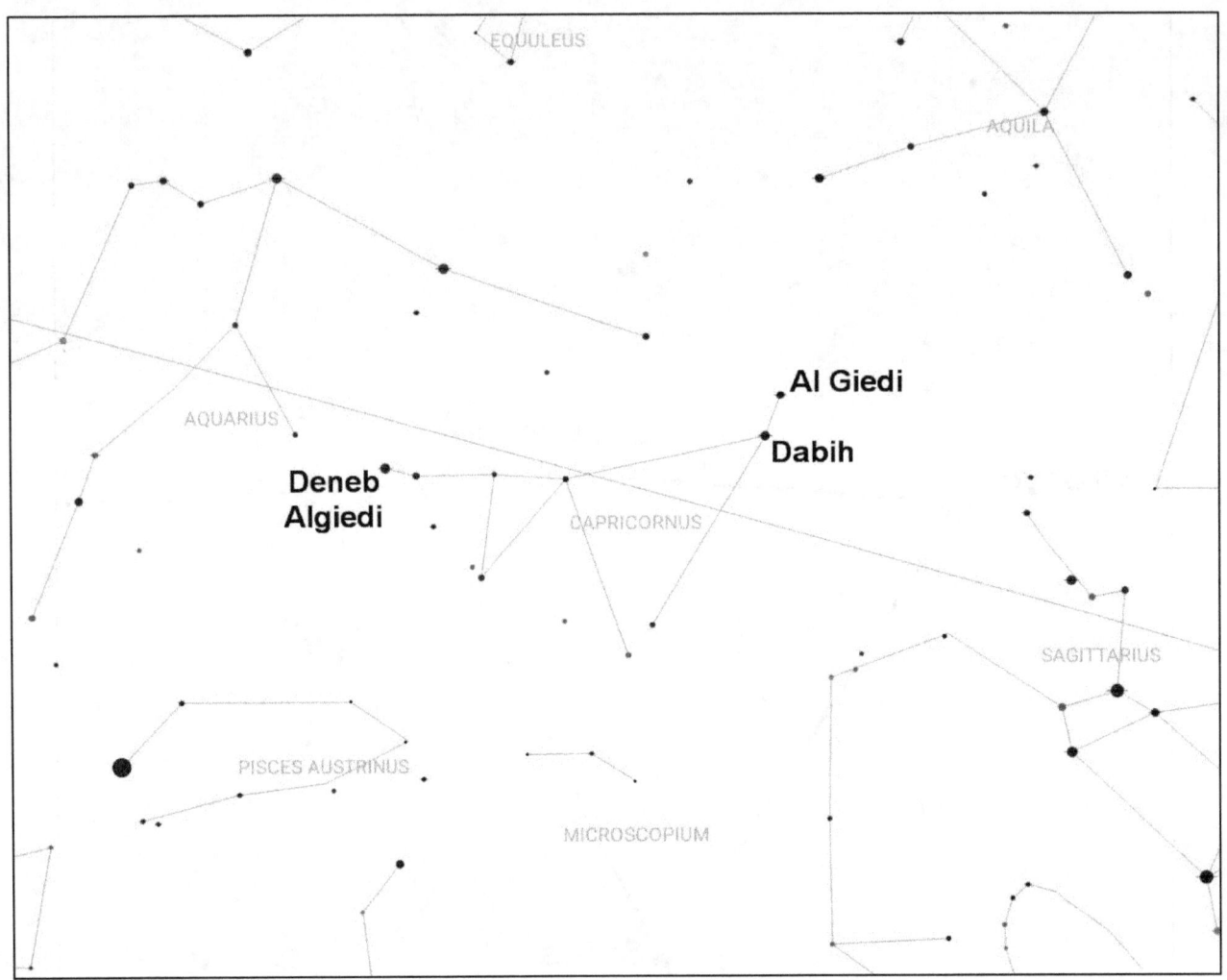

Capricornus

Capricornus, one of the twelve signs of the zodiac, represents quite a bizarre creature. Half goat, half fish, it's said to be the goat-like god Pan who jumped into a river to save himself from the monster Typhon. In doing so, he panicked and didn't give himself enough time to properly change into a fish; hence the half-goat, half-fish combination.

Sometimes affectionately known as "the smile in the sky," Capricornus is a mid-sized constellation of fairly inconspicuous stars.

It's often thought of as being an autumnal constellation, but in fact, it can be easily seen in the evening sky from mid to late Summer onwards and is found to the south-east of Aquila.

Its brightest star is Delta Capricorni. Commonly known as Deneb Algiedi, its name is Arabic for "the tail of the goat" and it is appropriately located on the eastern edge of the constellation. Like many stars, it's actually a double star system with both components taking just one day to orbit one another. The stars lie about 38 light years away.

Of greater interest is Alpha Capricorni, also known as Al Giedi, an Arabic name that translates to "the kid." Al Giedi is actually a wide pair of stars that, like Mizar & Alcor in Ursa

Major, can be seen without optical aid by sharp-eyed observers.

The most westerly is known as Alpha[1] Capricorni (or Prima Giedi) while the easterly star is Alpha[2] Capricorni (Secunda Giedi.)

The pair make for a good binocular target with both stars appearing creamy-white and of almost equal brightness.

In reality, these stars are not a true double star system and only appear close together due to a chance alignment. Alpha[1] is nearly 700 light years away while Alpha[2] is a lot closer at 109 light years. Alpha[1] must therefore be the much brighter of the two and, in fact, is about 1,000 times more luminous than the Sun.

While you're in the area, be sure to look for another double. Beta Capricorni (aka Dabih) has two components, separated by a third of a light year and taking about 700,000 years to orbit one another. An easy target for binoculars, the primary appears pale yellow and about two or three times brighter than the white secondary.

If doubles don't appeal, try your hand at a globular cluster. Messier 30 is magnitude 7.7 and well within reach of binoculars. It can be found within the same field of view as Zeta and close to the magnitude 5 star 41 Capricorni.

Discovered in 1764 by the French astronomer Charles Messier, it's about 27,000 light years away and is thought to be about 93 light years across. As with other globulars, these stars are typically much older – this cluster is thought to be nearly 13 billion years old!

Through a telescope at low power, it appears as a small, hazy patch but increasing the magnification to about 100x may help to resolve the cluster into its individual stars.

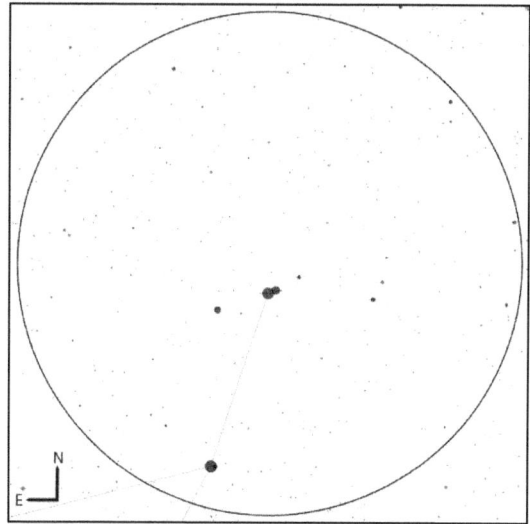

Al Giedi, binocular view

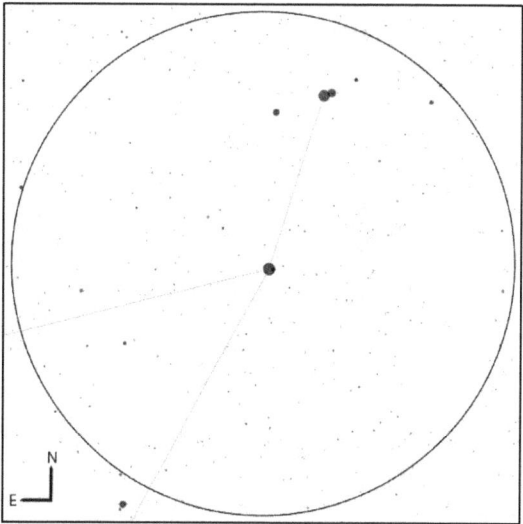

Dabih, binocular view

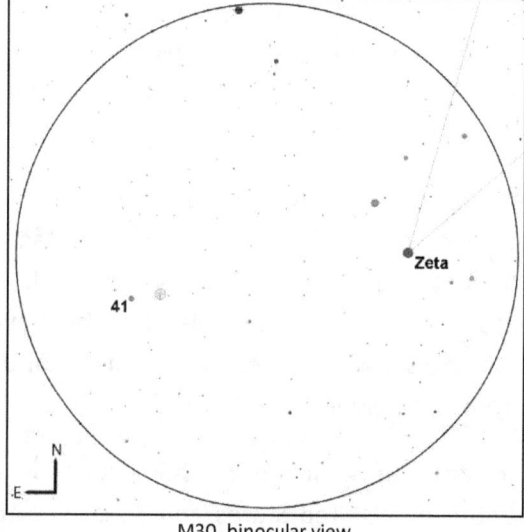

M30, binocular view

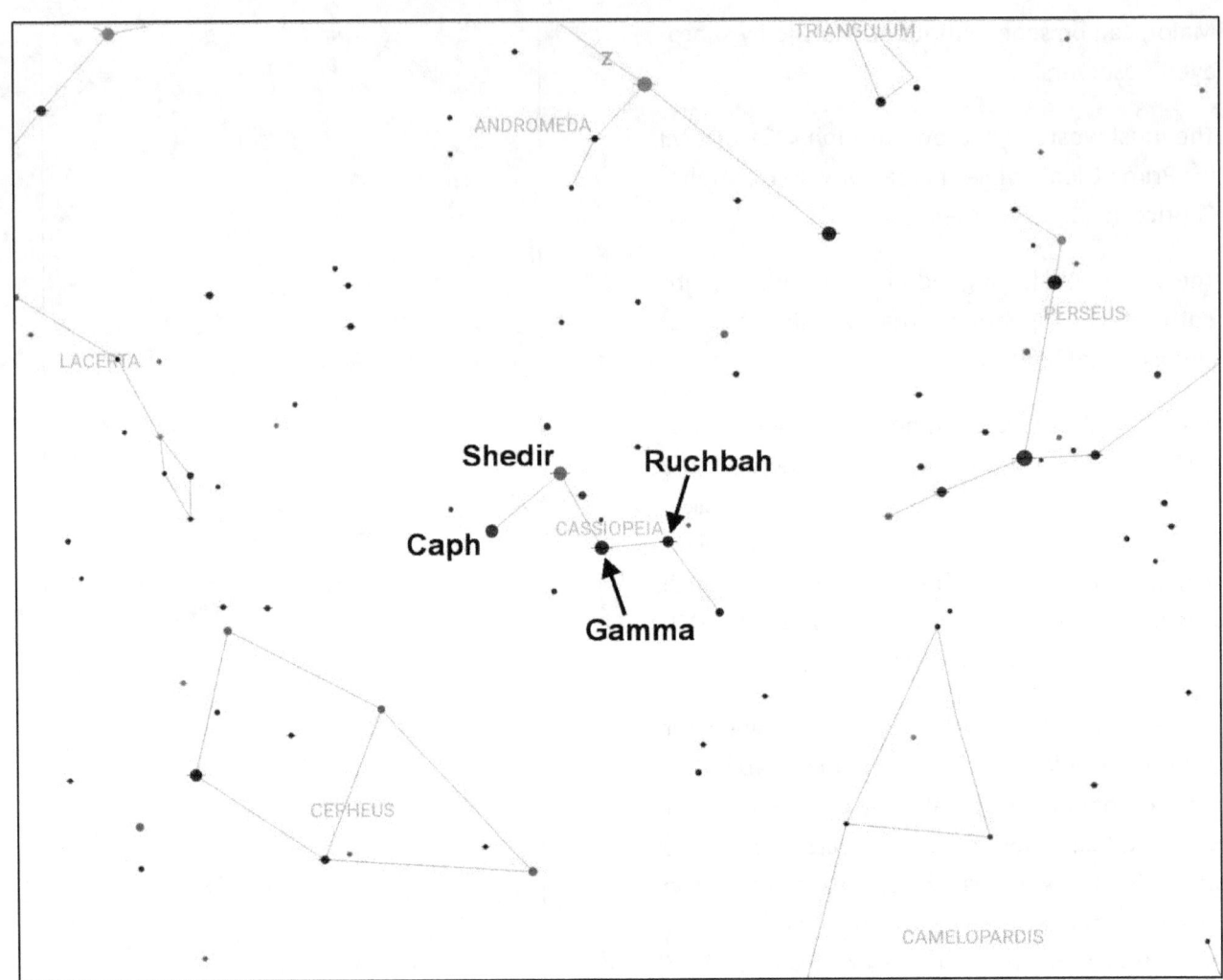

Cassiopeia

Cassiopeia represents the mother of Andromeda the Princess and a central player in that constellation's mythological story. It was Cassiopeia's vain boasts that angered Poseidon and put her daughter in mortal danger (see page 104 for the full story.)

The constellation is distinctive for a number of reasons. Firstly, it's what's known as a circumpolar constellation. In other words, it circles Polaris, the North Pole star and never appears to set when observed from much of the northern hemisphere.

For example, if you live in the United Kingdom or the northern United States and could see the stars shine through both the day and the night, you'd see it move around Polaris throughout the twenty-four hours of the day.

Which leads us to its second distinctive feature: its shape. The five brightest stars form an easily recognizable W in the sky... or M... or 3... or E... depending on its position in relation to Polaris.

For example, in the autumn it appears as an M above Polaris. In the Winter, it's an E, to the west of the star. In the Spring, it's a W below the star and just above the horizon. Finally, during the Summer months it begins its climb again and appears as a 3 to the east of the pole star.

Its last distinctive feature is the stars themselves as its five main stars are quite bright. With an average magnitude of 2.5 between them, Cassiopeia is easily found and the stars can provide convenient markers for some nice deep sky objects.

The brightest star – for now - is actually Gamma Cassiopeiae, the central star in the constellation. Curiously, unlike its neighbors, it has no name, almost as though the ancient astronomers forgot to assign it one.

I say it's brightest for now because Gamma is an unpredictable, irregular variable star. In other words, it can vary in brightness between magnitude 1.6 and 3.0 over a period of years. It's now gradually brightening and is slightly brighter than Alpha Cassiopeiae, also known as Shedir.

The Milky Way runs through Cassiopeia and we can use Shedir to find one of the constellation's open star clusters. By drawing a line through that star, moving through Beta (Caph) and continuing for about the same distance again, we come to Messier 52. This cluster appears in the same binocular field of view as Beta and may appear as a small misty patch.

Another sight for binoculars is the open cluster NGC 457. Also known as the Owl Cluster, this can be found close to Delta (Ruchbah) and Gamma Cassiopeiae. Although it can be seen with binoculars under suburban skies as a tiny, very faint smudge, it's best observed with a small telescope.

Look carefully and you'll see the double star Phi Cassiopeiae that marks the eyes of the Owl with the remaining stars forming the body of the bird. It's a large, attractive and fairly bright cluster that – somehow – Charles Messier missed as he was compiling his famous catalog.

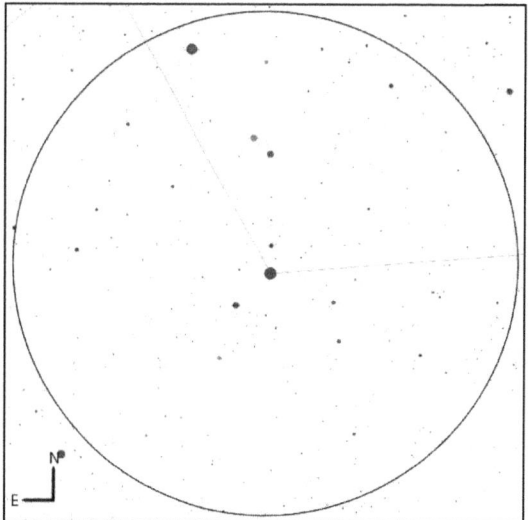

Gamma Cassiopeiae, binocular view

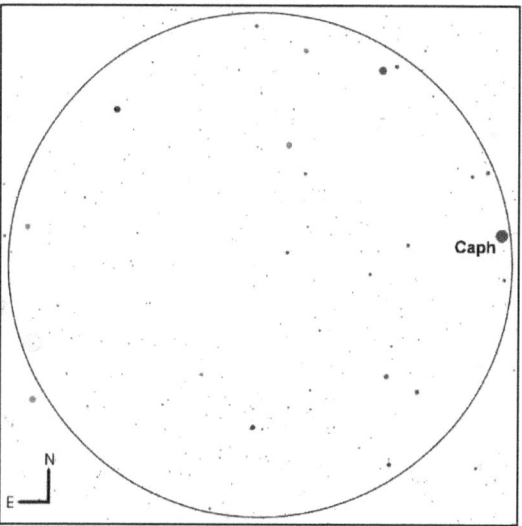

Messier 52, binocular view

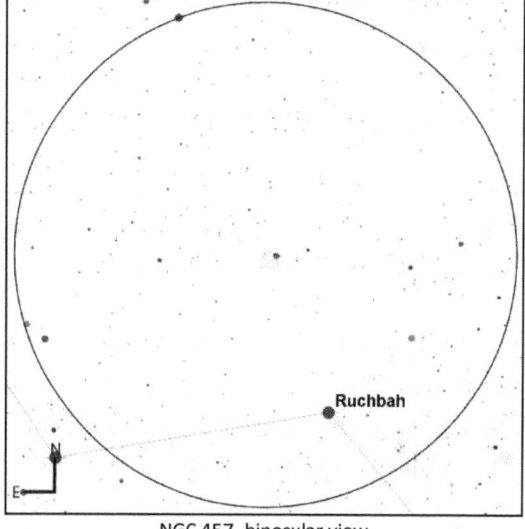

NGC 457, binocular view

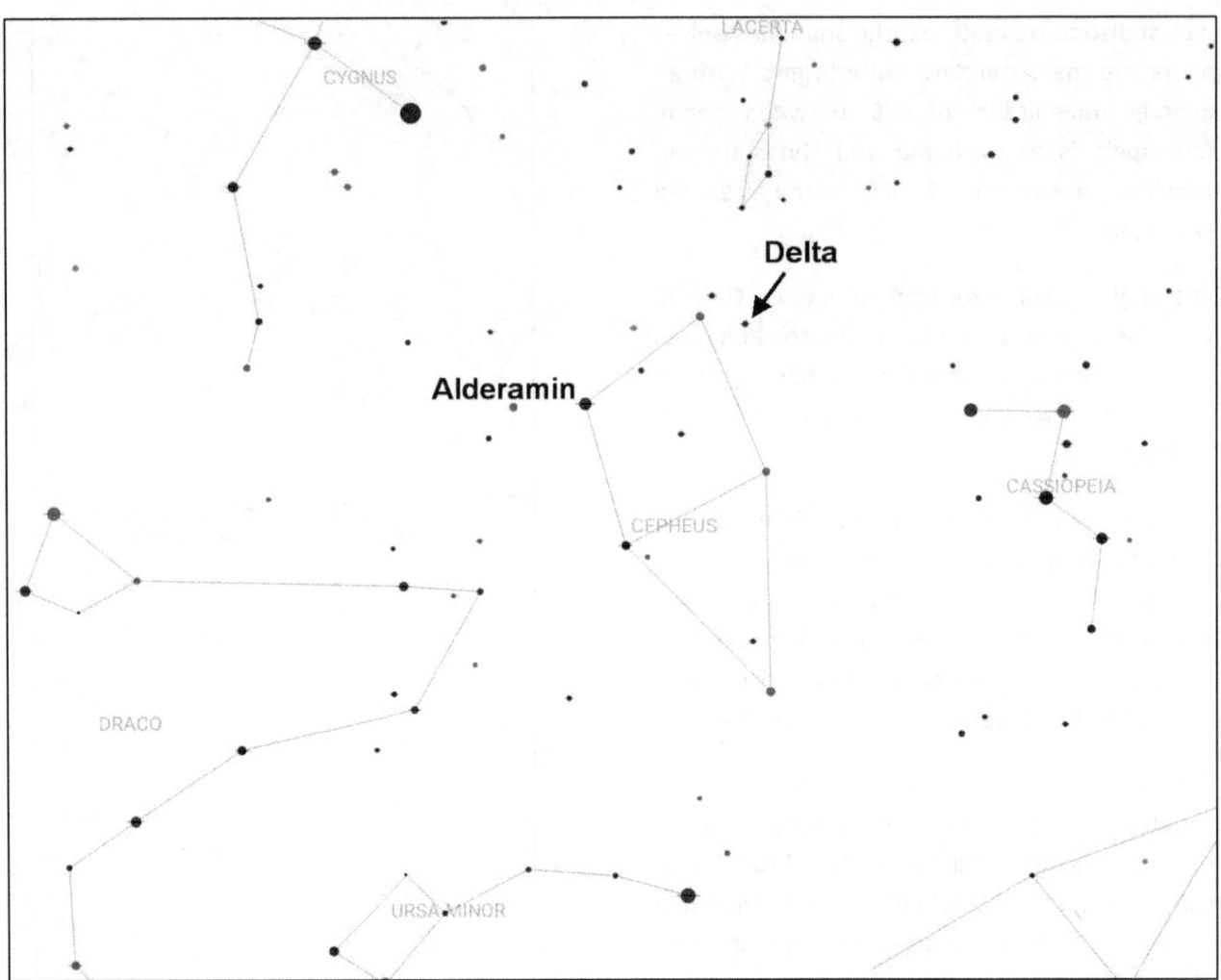

Cepheus

The constellation of Cepheus depicts the husband of vain Queen Cassiopeia and father to poor princess Andromeda. As with his wife, he plays a pivotal role in Andromeda's story as he is primarily responsible for agreeing to her sacrifice. (See page 104 for the whole story.)

It's not a particularly conspicuous constellation but it does have a fairly distinctive shape as it looks like a small child's drawing of a house.

There are no particularly bright stars to be seen here; the brightest is Alpha, a magnitude 2.5 white star only 49 light years away. Also known as Alderamin (derived from the Arabic for "the right arm") it's primary claim to fame is that it was once the celestial pole star (as Polaris is now) and, in about 5,500 years, will be once again.

Of the other stars, Delta Cepheid is certainly worthy of mention as it's the prototype of a class of variable star that bears its name. As first noted by the English astronomer John Goodricke in 1784, Delta varies in brightness from magnitude 3.48 to 4.37 over a period of 5.37 days. Today, you can track the changes with binoculars.

A quadruple star nearly 900 light years away, the brightness change is due to gas and dust smothering the light from the distant star system.

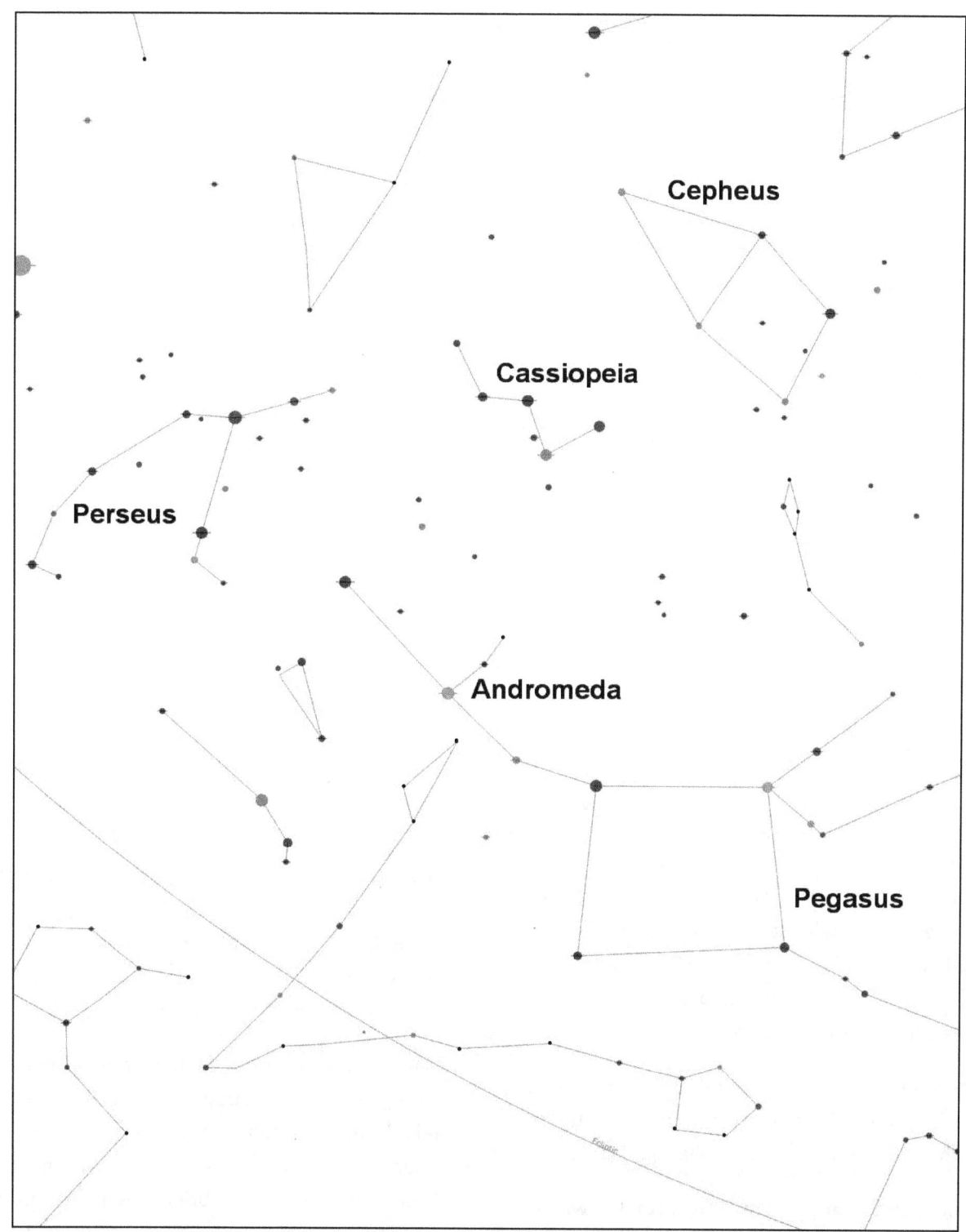

The five major players in the legend of Andromeda can be seen in the late autumn and early winter sky. Andromeda's vain mother, Cassiopeia, appears to the north while her father, Cepheus, appears to the north-west. Her hero, Perseus is to her east while his trusty steed, Pegasus, is just to the west. This image simulates the overhead view at 10pm in early November, 9pm in mid November, 8pm in early December and 7pm in mid December.

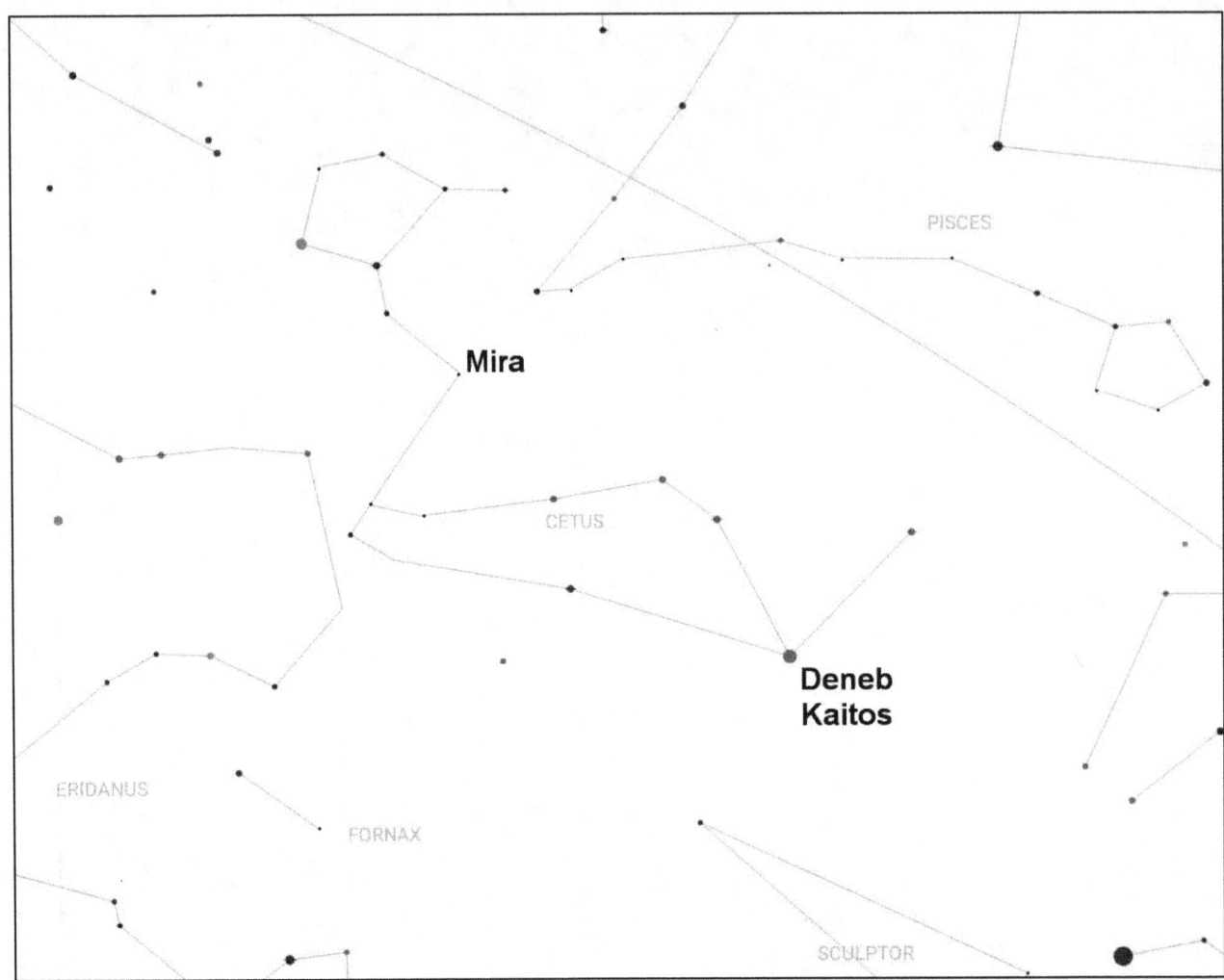

Cetus

Depending on who you ask, Cetus either represents a Whale (some associate it with the Whale that swallowed Job in the Bible) or the sea monster that nearly consumed poor Andromeda (see page 104 for the myth.)

Either way, it's a large but rather unimpressive constellation, 4th in size and found beneath another faint autumnal constellation, Pisces.

Its brightest star is Beta Ceti, also known as either Deneb Kaitos (derived from the Arabic for "the tail of the whale") or Diphda, which is taken from the Arabic word for "frog." Only 96 light years away, this magnitude 2.0 orange giant is otherwise quite unremarkable.

However, there is one star in the constellation that's famous with astronomers across the globe. Omicron Ceti, also known as Mira (which very aptly means "the wonderful") was one of the first variable stars to be discovered.

A red giant star hundreds of light years away, on average it shines at about magnitude 8 –beyond naked eye visibility but within reach of binoculars. However, every 11 months (332 days, to be precise) it brightens to naked eye visibility, usually to around magnitude 3, but sometimes as high as 2.0.

Check with the American Association of Variable Star Observers (www.aavso.org) for the most up-to-date predictions.

The Helix Nebula in the constellation Aquarius. Image by the author using Slooh. (www.slooh.com)

Messier 37 in the constellation Auriga. Image by the author using Slooh. (www.slooh.com)

Messier 30 in the constellation Capricornus. Image by the author using Slooh. (www.slooh.com)

The Owl Cluster in the constellation Cassiopeia. Image by the author using Slooh. (www.slooh.com)

The Blackeye Galaxy in the constellation Coma Berenices. Image by the author using Slooh. (www.slooh.com)

Messier 68 in the constellation Hydra. Image by the author using Slooh. (www.slooh.com)

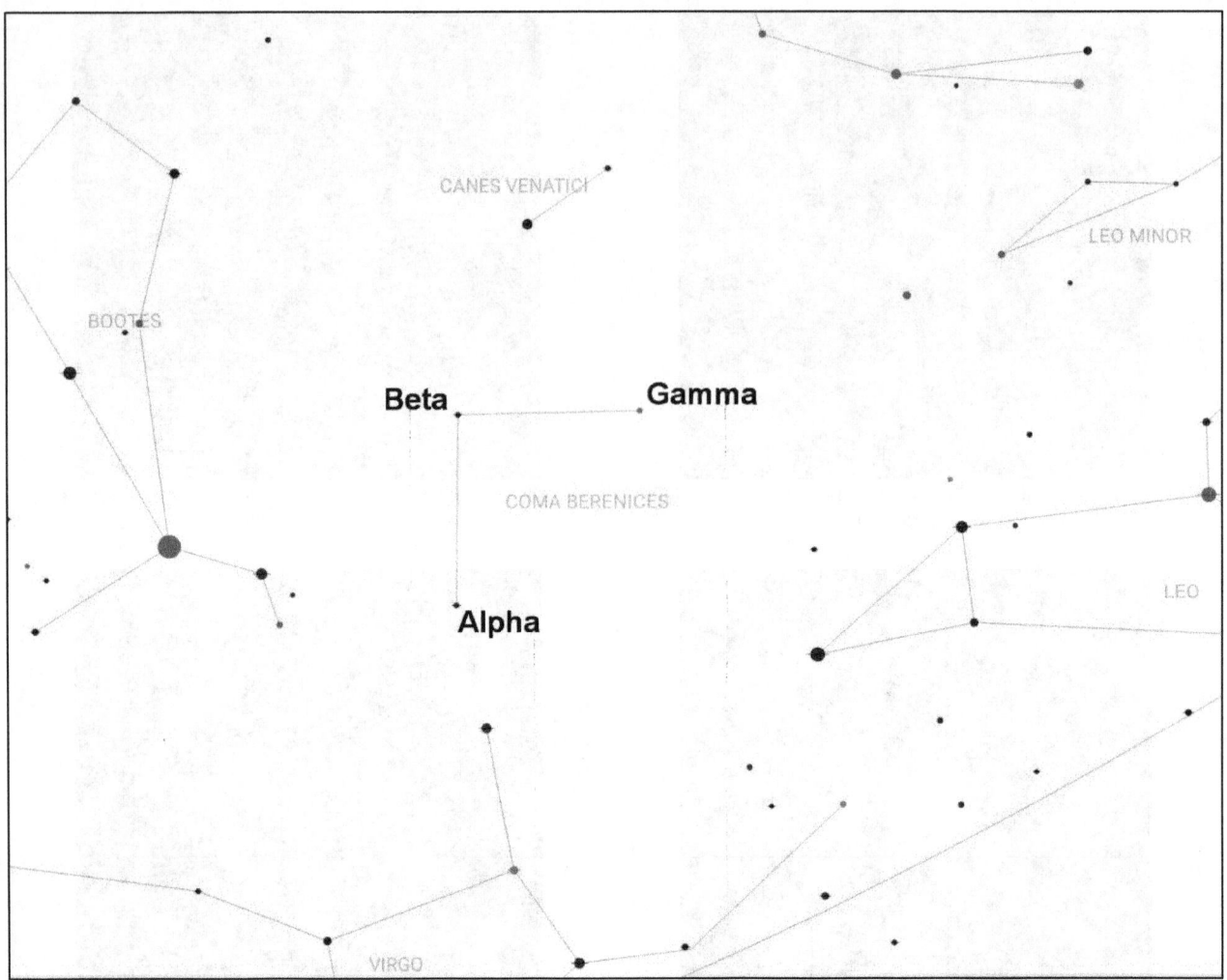

Coma Berenices

Coma Berenices (or just "Coma") is a mid-sized but very faint constellation at its best visibility during the Spring and Summer months.

Many of the constellations were created and named by the Greeks and Coma is no exception. However, what sets it apart is that it doesn't represent a mythological figure at all, but rather the hair of Queen Berenice.

Queen Berenice was an historical figure who lived in the third century B.C.E. Married to King Ptolemy of Egypt, she swore to the goddess Aphrodite that she would cut off her beautiful hair if her husband safely returned from battle.

Sure enough, the king came home and Berenice cut off her hair and left it for Aphrodite in her temple. The hair vanished overnight and was said to have been placed amongst the stars by the goddess herself.

Located to the west of Boötes, the constellation has no bright stars and will require dark skies to find it. Its brightest star is actually Beta, which shines (if you can call it that) at a dim magnitude 4.3.

That doesn't mean it's not interesting – far from it, because Beta is only 30 light years from Earth and is actually remarkably Sun-like.

There are a few minor differences. The Sun is approximately 4½ billion years old whereas Beta is only three and Beta is also slightly larger and brighter. However, liquid water could exist on a hypothetical planet at the same distance as the Earth is from the Sun. (Alas, no planets are known to orbit the star.)

Coma's second brightest star, Alpha, is only slightly fainter and is commonly known as Diadem. The name refers to the crown or headband that Berenice wore upon her head.

Very close to Alpha is Messier 53, a globular cluster that should be within range of binoculars. Look for a faint, fuzzy star within the same field-of-view as Alpha. A small telescope will help, but you'll need a mid-sized 'scope to see the individual stars within the cluster.

Coma is also well-known for containing a high concentration of galaxies, including seven from Charles Messier's famous catalog. One (M64, the Blackeye Galaxy) may be glimpsed with binoculars but as it lies in such a barren area of sky, it might be quite difficult for the inexperienced astronomer to find.

The reason for the multitude of galaxies has to do with our own Milky Way. The heart of our galaxy is best seen in the summer, but during the Spring you're actually looking directly over the north pole of our galaxy and into the deep depths of the universe itself!

While you're using your binoculars, scan to the north-west for Gamma and the Coma Star Cluster (Melotte 111). A large sprinkling of faint stars barely visible under dark skies with the unaided eye, it's worth a look with binoculars when more of its fifty members will come into view. A true cluster, it lies only 288 light years away.

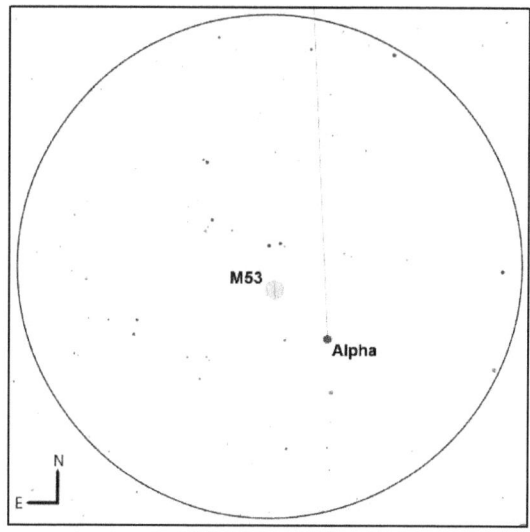

M53, binocular view

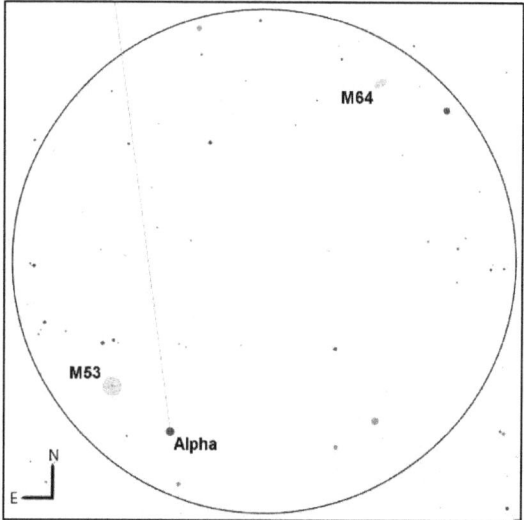

M64, binocular view

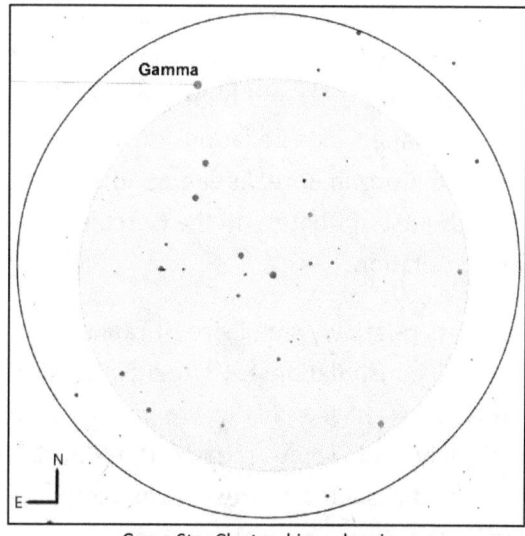

Coma Star Cluster, binocular view

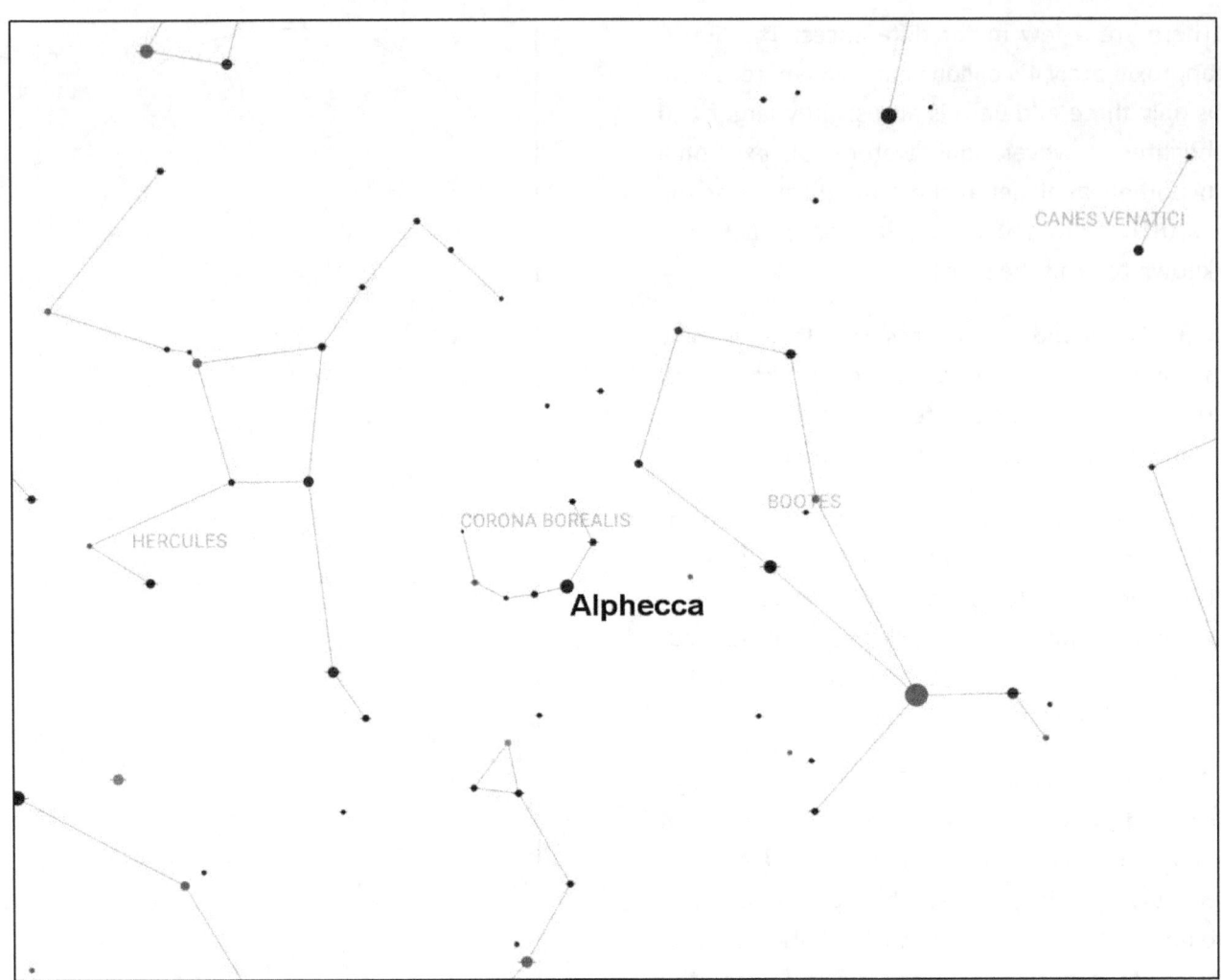

CANES VENATICI

CORONA BOREALIS

HERCULES

BOÖTES

Alphecca

Corona Borealis

Corona Borealis is a small but noticeable constellation that passes high overhead through the northern Spring and Summer skies. First find Arcturus by following the curved tail of Ursa Major (see page 181) and then use it to identify the kite-shaped constellation of Boötes the Herdsman. Corona Borealis can be found just to the north-east of Boötes, on the Hercules side of the constellation.

Despite its relatively small size (it ranks 73rd out of the 88 constellations) it's relatively easy to identify as it has a distinctive C shape. Not surprisingly, its name means the "northern crown" and is said to represent the crown once worn by the Cretan princess Aridane.

According to Greek myth, the crown was given to her by her husband, Theseus, but when he left her and she re-married, the crown was transferred into the starry heavens above.

It wasn't always identified with a crown; originally, the ancient Greeks saw a wreath here while other cultures saw a platter, sisters and, to the indigenous people of Australia, a boomerang.

Unfortunately, the constellation is a little bereft of deep sky wonders. Its brightest star, Alpha, is known as Alphecca and shines at a respectable magnitude 2.2. The name is derived from Arabic for "the bright one of the dish" but others know it more appropriately as Gemma – literally, the jewel.

A binary system some 75 light years away, the primary component is a white star about three times the size of the Sun while its companion is a yellow star of about the Sun's size.

It's thought the system might be surrounded by a disk of gas and dust, which might indicate a young solar system in the process of forming.

There are several double stars for small telescopes in the area but they lie away from the main constellation. However, Omicron Coronae Borealis lies near the eastern tip of the crown and reveals two tiny unrelated companion stars through binoculars. The star has one confirmed planet orbiting it.

There's also R Coronae Borealis. An irregular variable, it normally shines at about magnitude 6 but it will unpredictably fade to magnitude 13 or 14 over the course of several months– making it almost invisible to all but the largest of amateur telescopes. This fading is thought to be caused by a build-up of carbon dust in the star's atmosphere.

At the moment (March 2016) it's still quite faint (around magnitude 8 or so) and has been randomly brightening and fading by several magnitudes for nearly ten years now. Take a look – can you see it?

Similarly, T Coronae Borealis is known as the Blaze Star and is a recurring nova. In other words, it randomly and suddenly brightens to around magnitude 2 or 3. However it's most recent outburst was in 1946 and, at a regular magnitude of about 10, it may be tricky to spot in binoculars.

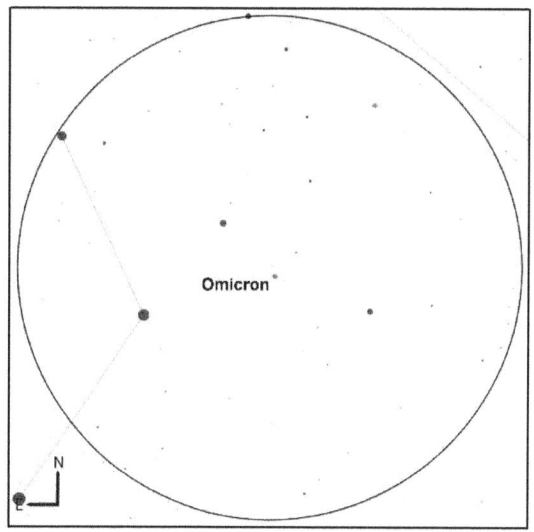

Omicron, binocular view

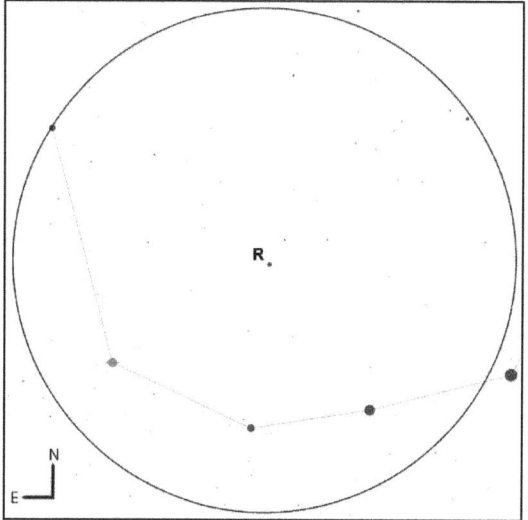

R Coronae Borealis, binocular view

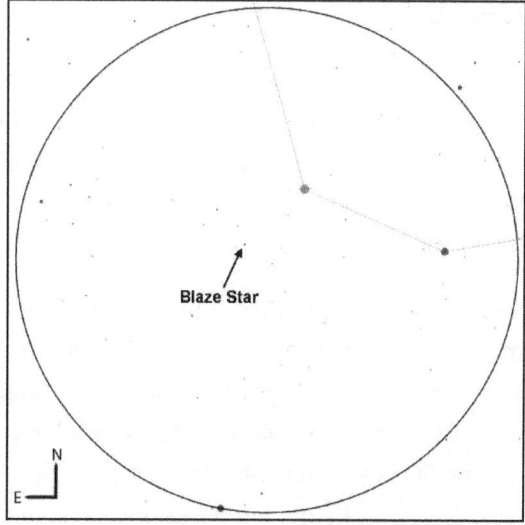

The Blaze Star, binocular view

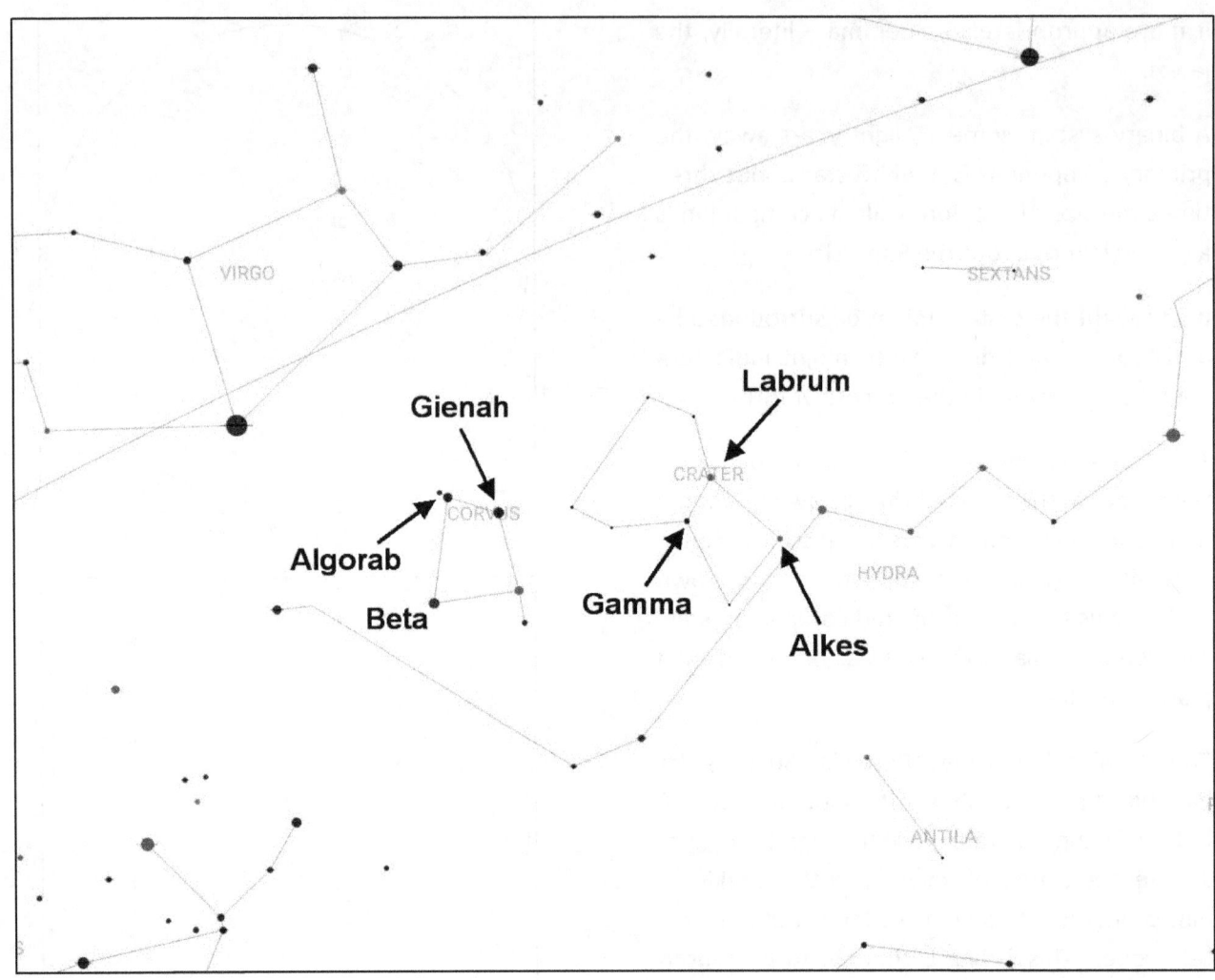

Corvus & Crater

Corvus and Crater are two relatively small constellations predominantly visible in the Spring sky.

Corvus, the brighter and more easterly constellation of the two, represents a crow while Crater depicts a cup. According to myth, the god Apollo sent the bird with the cup to bring him some water.

However, the bird stopped along the way to eat some figs, which delayed his return and made Apollo angry. Furthermore, rather than telling the truth, the bird claimed that a snake had temporarily prevented him from gathering the water. As proof, he held in his talons a snake.

Apollo, in his anger, threw the snake into the sky where it can still be seen today. He then cursed the bird to be eternally thirsty before throwing both it and the cup into the sky. The cup is to the west, always out of reach of the thirsty bird, while Hydra, the snake, slithers to the west.

Of the two constellations Corvus is smaller, being ranked 70th in size out of the 88 recognized constellations. Its four main stars – Beta, Gamma, Delta and Epsilon – form an asterism known as Spica's Spanker. An odd name, a spanker is actually a sail flown from the aftmost mast of a ship or yacht. Spica refers to the brightest star in the constellation of Virgo, the Virgin.

At magnitude 2.6, the brightest star in Corvus is Gamma. Also known as Gienah, the name is appropriately derived from the Arabic for "the right wing of the crow."

Like many others, this giant blue-white star is part of a double star system and has a Sun-sized orange companion. The pair orbit one another once every 158 years and lie about 150 light years away.

If you have binoculars, look out for Zeta Corvi, a white magnitude 5.2 star near the southern end of the constellation. Just to the west you should easily see a slightly fainter, unrelated orange companion.

An easier target is Delta Corvi, better known as Algorab from the Arabic word for "crow." Binoculars will show the wide, magnitude 4.3 unrelated star Eta Corvi, but turn a telescope toward it with a medium power eyepiece (about 50x) and Algorab may reveal a close companion. The primary is a brilliant white and the much fainter secondary appears bluish.

If you want a challenge, try your hand at spotting Messier 68, a small and relatively faint globular cluster that's just over the border in neighboring Hydra. You can use Algorab and Beta as pointers and the cluster is conveniently located within the same binocular field of view as the latter star.

Meanwhile, over in Crater, its brightest star is Delta, which glimmers at magnitude 3.5. Named Labrum, "the lip (of the cup)", it's an orange giant star just under 200 light years away.

At magnitude 4.1, Alpha Crateris is fainter and is known by the name Alkes, which is derived from the Arabic word for cup. Like Delta, it's an orange giant star, but slightly closer to Earth at 174 light years.

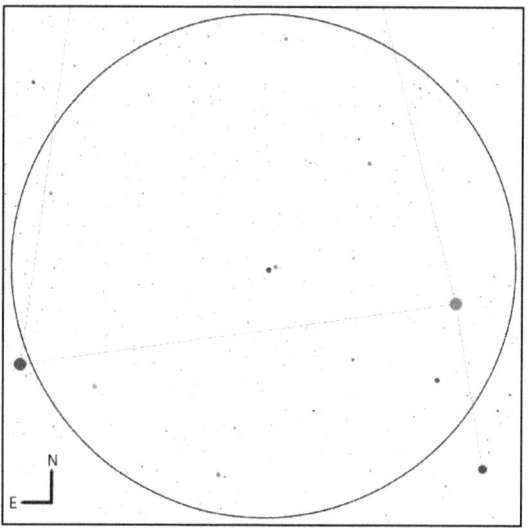

Zeta, binocular view

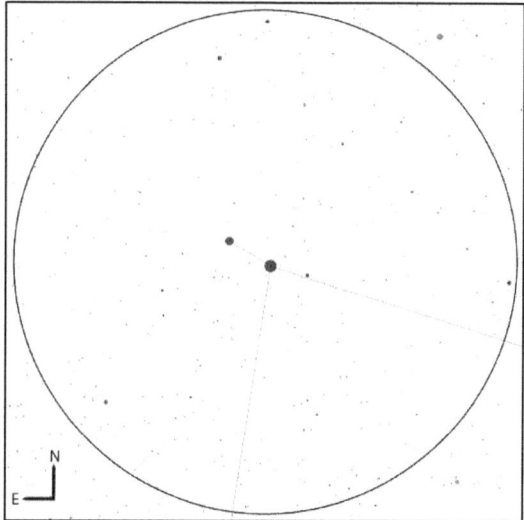

Algorab, binocular view

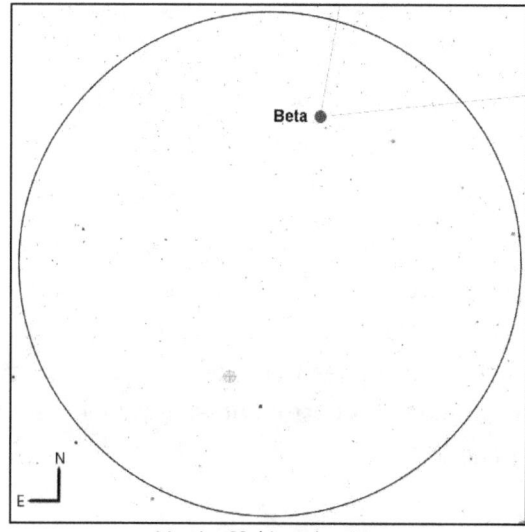

Messier 68, binocular view

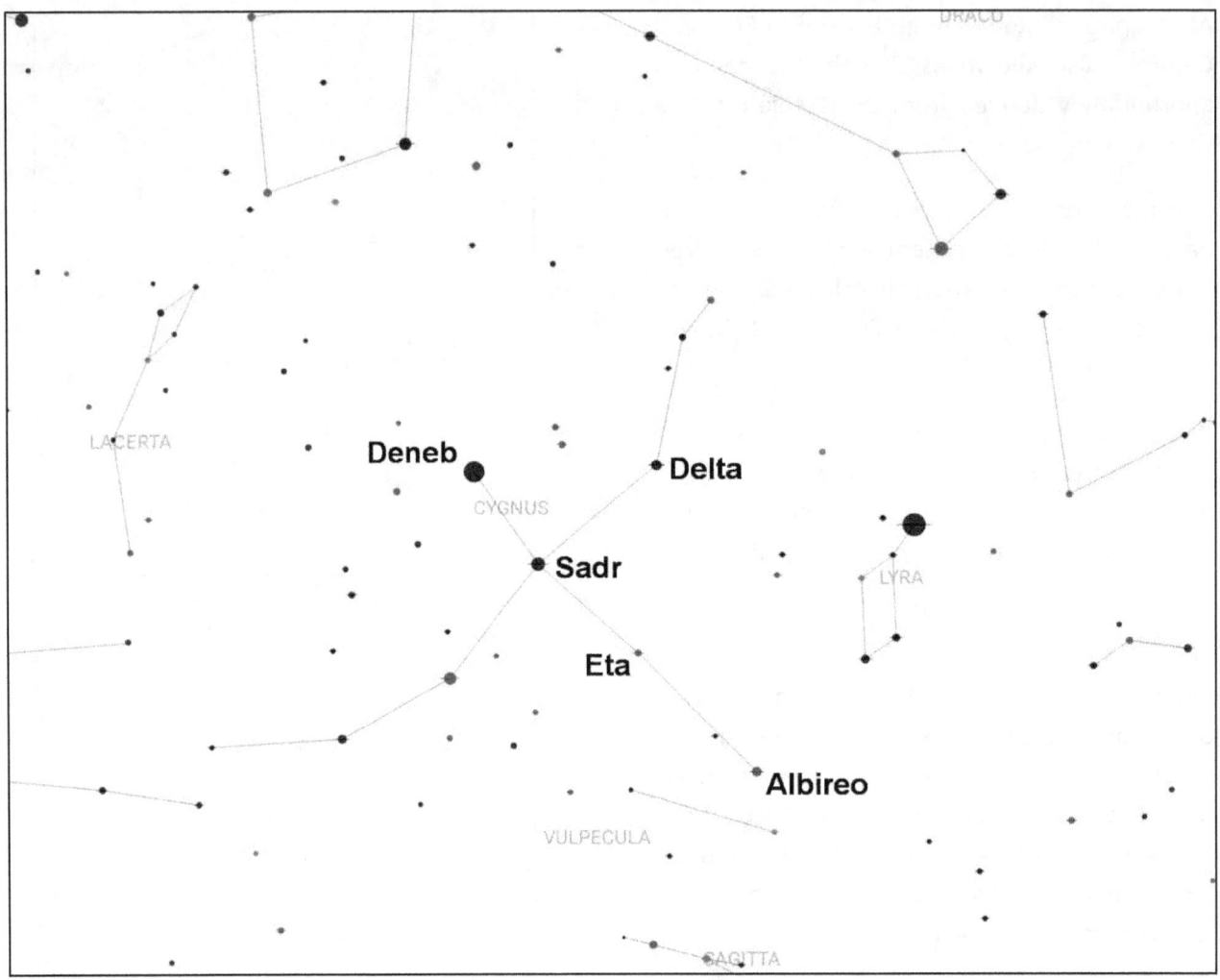

Cygnus

Cygnus, the Swan, is a large, bright constellation easily seen throughout the Summer and Autumn months. From the northern hemisphere the celestial bird flies high overhead and its distinctive cross-shape makes it an unmissable sight on any warm summer night.

There are a number of myths associated with the constellation but one of the most common concerns yet another amorous adventure of Zeus. Transforming himself into a swan, he seduced Leda, Queen of Sparta, who later gave birth to the twins Castor and Pollux. (See Gemini on page 144)

However, the constellation is known for something far more spectacular. Get yourself away from the city lights on an August night and you'll see our own Milky Way galaxy streaming down toward the south like a river. Under the right conditions, it is literally breathtaking.

Cygnus is rich with stars and sweeping the constellation with binoculars can be a rewarding experience. At the northern tip is the constellation's brightest star, Alpha, more commonly known as Deneb.

Part of the Summer Triangle of stars (see page 182), Deneb is the 19th brightest star in the sky and, at about 800 light years away, one of the most luminous known. Estimates vary, but it may shine with a light nearly 200,000 times brighter than the Sun.

Now move halfway toward Delta on the western wing to find Omicron Cygni. If you're using 10x50 binoculars, you can have Deneb just on the edge of your view and you should easily see two golden stars nearby. However, the real treat comes from the southern component, Omicron[1] Cygni.

Look carefully and you'll see a third, slightly fainter pale blue star very close to it. It makes for a nice sight through binoculars but a small telescope at low power only improves the view.

Move back toward the center of the cross to find magnitude 2.2 Sadr. This area of Cygnus is particularly rich with stars and provides a very nice view in binoculars. Can you see tiny Messier 29 close by?

Further south along the neck of the swan, we come to Chi Cygni. Appearing close to Eta, it lies about midway between Sadr in the center and Albireo at the bottom.

Chi is a red giant variable star which typically ranges between roughly magnitude 5 to magnitude 14 over a period of about 408 days. However, on rare occasions it has been known to flare up to magnitude 3, making it easily visible to the naked eye.

Lastly, no words on Cygnus would be complete without mentioning Beta, also known as Albireo. One of the most popular double stars in the sky, every amateur astronomer knows it as a Summer showcase for small telescopes. Low power will reveal a golden primary with a stunning sapphire blue companion. But try your hand with binoculars — held steady under excellent conditions, you might get lucky!

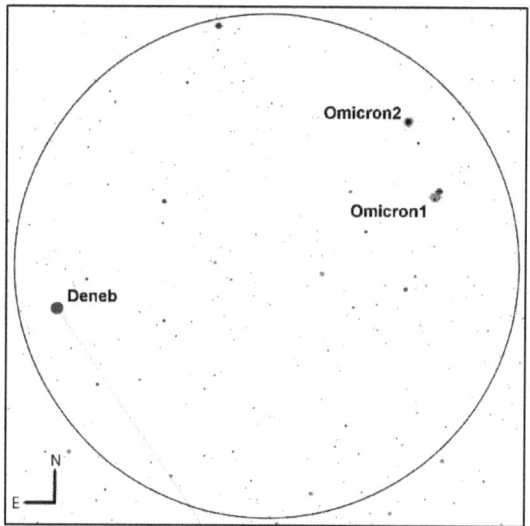

Omicron Cygni, binocular view

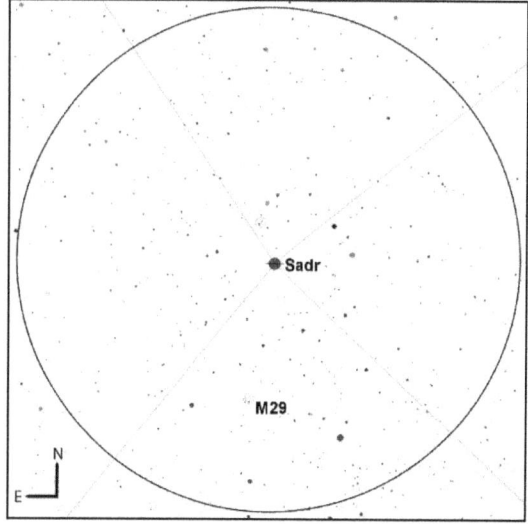

Sadr and M29, binocular view

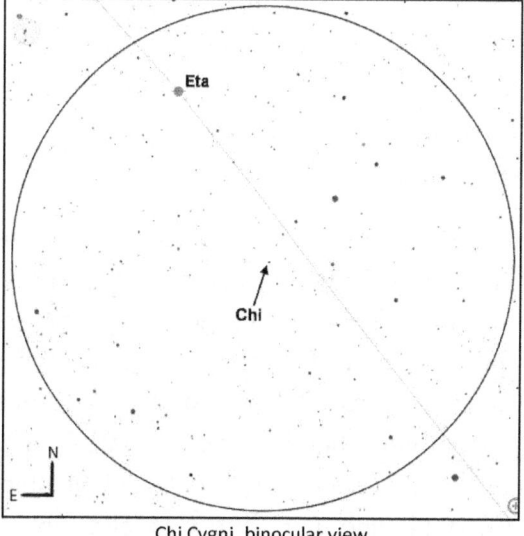

Chi Cygni, binocular view

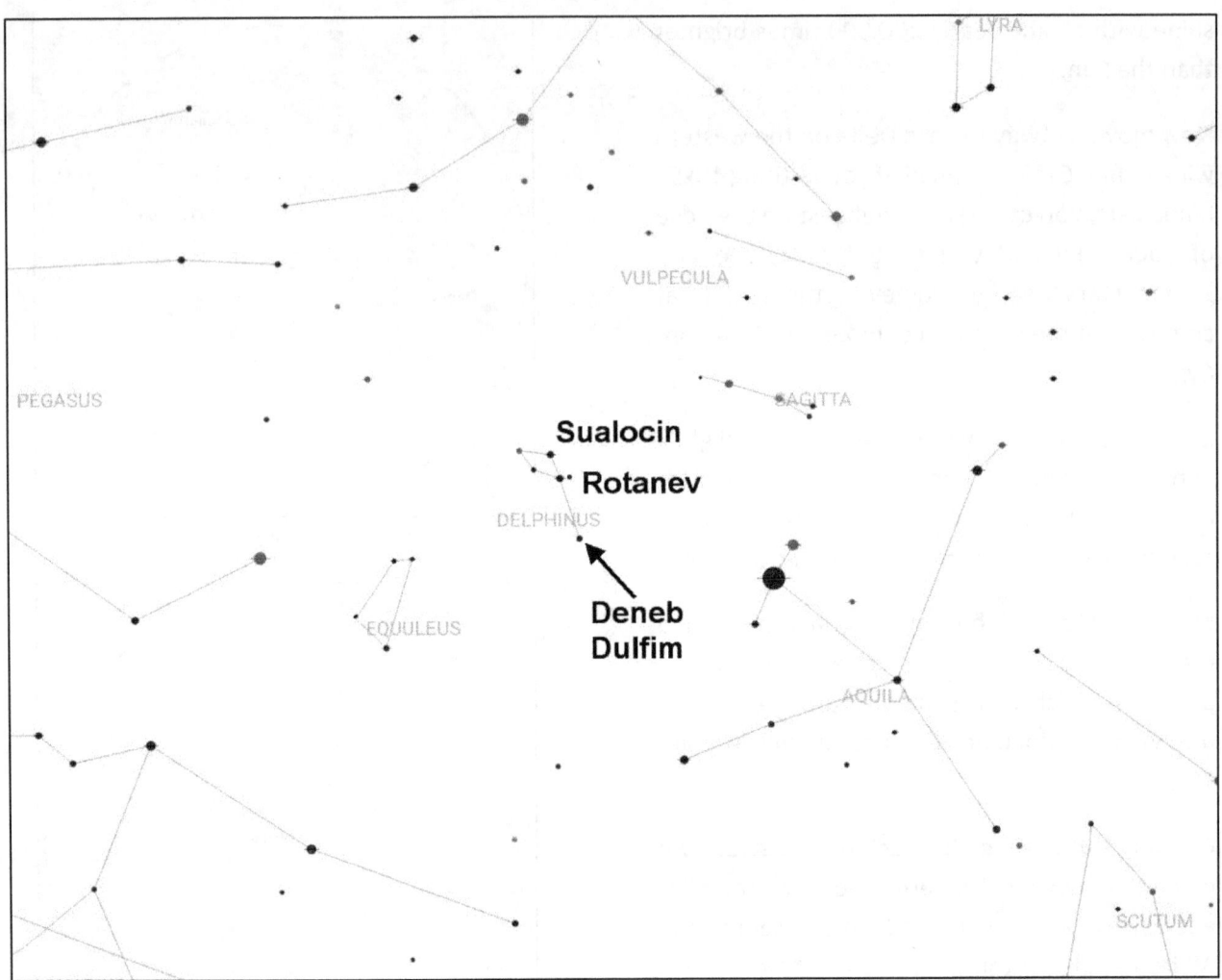

Delphinus

One of the smallest constellations in the sky, it ranks 69th in size and the entire constellation can comfortably fit into the field of view of a pair of 10x50 binoculars. Delphinus represents a dolphin and is one of the few constellations that at least partially looks like the creature or object it's meant to depict.

One legend tells of how Poseidon, god of the sea, fell in love with the sea nymph Amphitrite but unfortunately, not feeling quite the same way, Amphitrite fled to the mountains to escape his advances. Being a god, this didn't deter Poseidon, who then sent out a number of sea creatures to find her. Not only did Delphinus successfully find her, but was also able to

persuade the nervous nymph to return to the god. In gratitude, Poseidon placed the dolphin amongst the stars.

Delphinus is not particularly bright, but thanks to its position, just east of Cygnus the Swan and north-east of Aquila the Eagle it's relatively easy to find and has a distinctive shape. Visible throughout the summer and autumn, I tend to think of it as being the last of the summer constellations and the first of the aquatic autumn constellations.

At magnitude 3.6, its brightest star is Beta, a double star system just slightly more than 100 light years away that consists of a white giant star with a subgiant companion. Unfortunately, the pair are too close to be split with all but the largest of amateur telescopes.

Second brightest is Alpha. It appears as a single magnitude 3.8 star to the unaided eye but is actually a complicated system comprised of seven stars some 74 light years away.

Both these stars have unusual names that deserve a mention. Alpha is also known as Sualocin while Beta is also known as Rotanev. Like Cor Caroli in Canes Venatici these are not names with Arabic or Greek origins but date back to the early 19th century. The names were given to the stars by the Italian astronomer Niccolò Cacciatore who was helping to compile the second edition of the Palermo Star Catalogue.

Taking the Latinized version of his name, Nicolaus Venator, he simply reversed the names before applying them to Alpha and Beta respectively. The names were consequently printed in the catalogue and their origins were a mystery for years.

Alpha and Beta are not the only stars with odd names; Epsilon, sometimes known as Deneb (not to be confused with the brightest star in Cygnus the Swan) was known to the Chinese as Pae Chou, "the rotten melon."

Alpha, Beta, Gamma and Delta collectively form an asterism known as Job's Coffin; despite being an obvious reference to the Biblical character, the exact origins of its name remain a mystery.

If you have binoculars, take a look at Sualocin and you should see a fainter, magnitude 6.0 white star just to the west of it. Meanwhile, to the south of Epsilon, also known as Deneb Dulfim, lies NGC 6934, a small and faint globular cluster of magnitude 8.9 that should prove a worthy challenge on a clear, dark night.

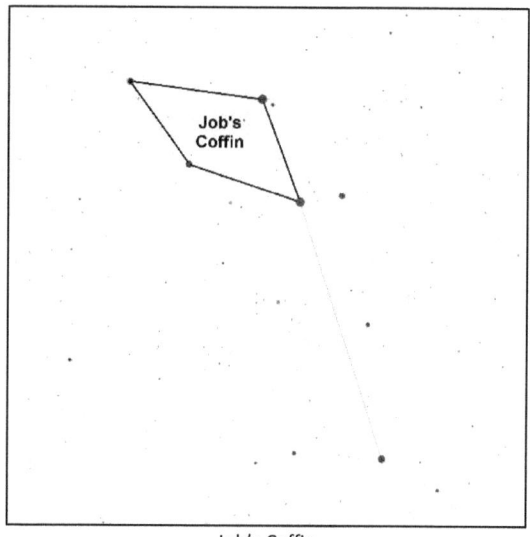

Job's Coffin

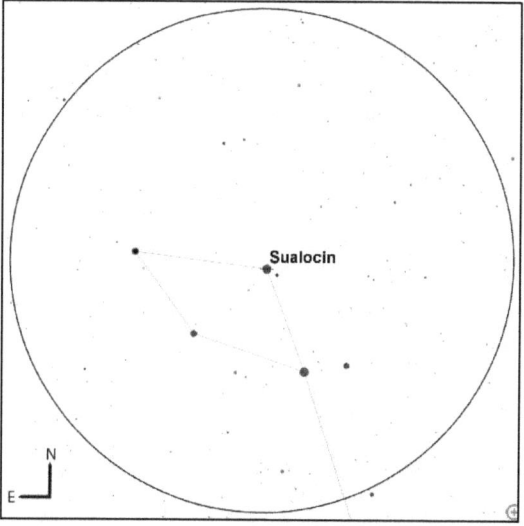

Sualocin, binocular view

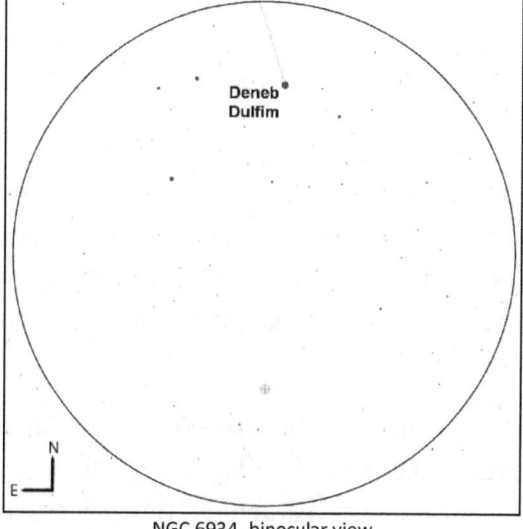

NGC 6934, binocular view

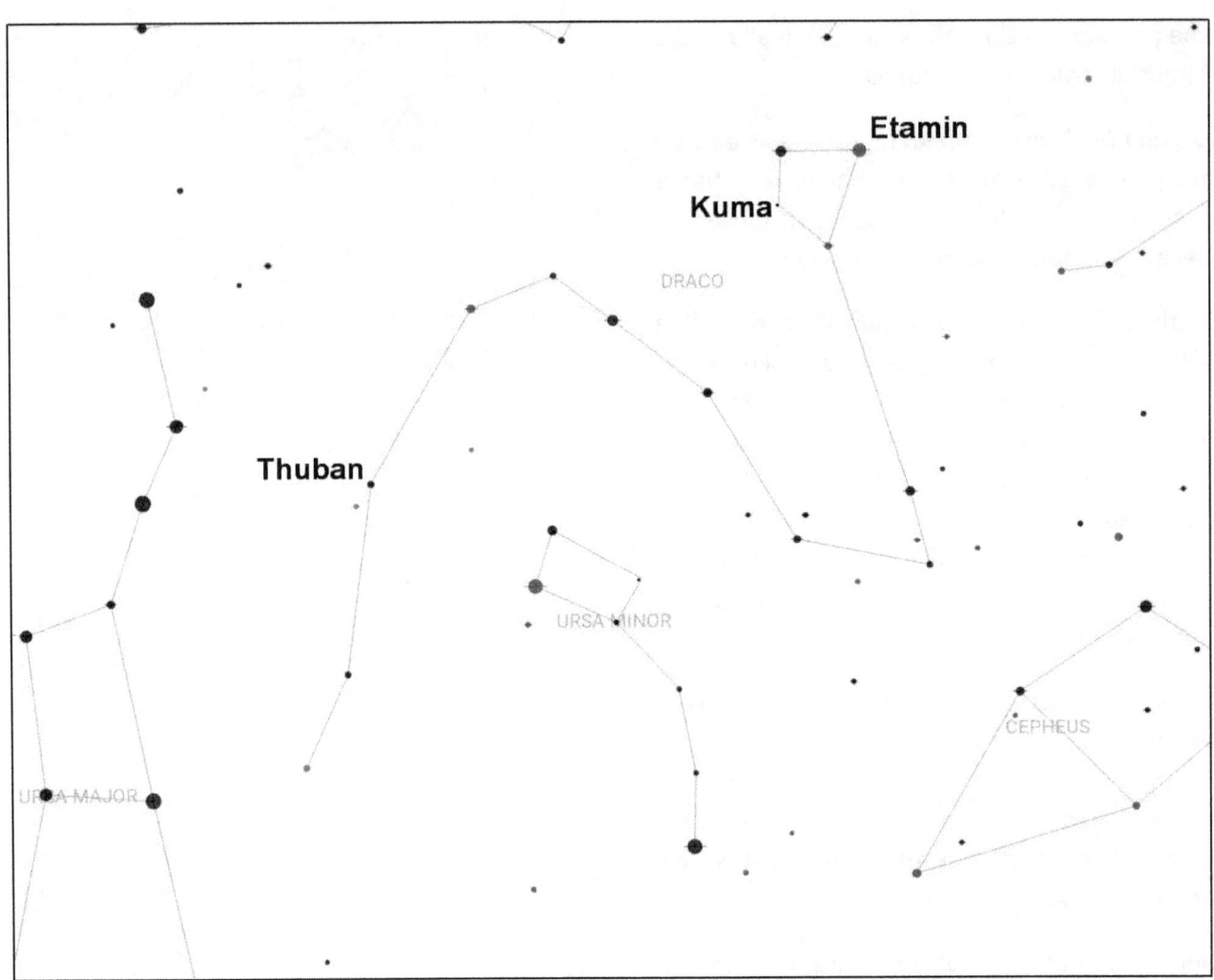

Draco

Draco, the Dragon, is an unusual constellation in that you could start to observe it in the spring, when it's tail appears high above the northern horizon, work your away along its body throughout the summer and then finish with its head in the autumn.

That being said, perhaps the best time to spot the constellation as a whole (as depicted in the image above) is early July, when the entire constellation arches over Ursa Minor at about 10pm.

In Greek mythology, Draco represents Ladon, the dragon tasked with guarding the golden apples that grew in the garden of Hesperides. The garden contained an orchard owned by Hera, wife of Zeus, who had placed the dragon there to stop the Hesperides from taking the apples for themselves. Unfortunately, the dragon was slain by Hercules who was tasked with stealing the apples as one of his twelve labors.

Draco is a very large constellation, 8th in size and covering over 1,000 square degrees of sky. But, with the possible exception of its head, it's not easy to spot from the city, a fact compounded by its long, twisting serpentine body of relatively faint stars.

Its brightest star is Gamma, also known as Etamin, from the Arabic for "the great serpent." An orange giant star nearly fifty times the size of the Sun, it shines at magnitude 2.2 and lies about 150 light years away.

It has a suspected red dwarf companion and the pair are gradually moving closer toward us. In another 1.5 million years, it will only be about 28 light years away and will appear to be the brightest star in the sky.

Meanwhile, the fainter Alpha Draconis, also known as Thuban (from the Arabic for "the snake"), has its own claim to fame. As every astronomer knows, our north pole star is the appropriately named Polaris, the brightest star in the constellation of Ursa Minor.

This is the star that appears directly overhead from the north pole, but this wasn't always the case. Over the course of 26,000 years, the Earth actually wobbles slightly in a circular motion on its axis. Known as the precession of the equinoxes, this causes the north pole star to change over the course of those years.

So, for example, right now the closest star to the celestial north pole is Polaris. But from 3942 BCE to 1793 BCE the pole star was Thuban and all the other stars appeared to spin about it. In approximately another 7,000 years, this will be the case again (but not before other stars take their turn!)

Lastly, close to Etamin, in the head of the dragon, is Nu Draconis, also known as Kuma. This is a favorite double star for both binoculars and small telescopes. A regular pair of 10x50 binoculars, held steady, should be able to split this magnitude 4.9 star into two brilliant white components of equal brightness.

A true double star system about 100 light years away, this famous double has rightfully earnt itself the nickname "the dragon's eyes." Take a look for yourself and you'll soon see why!

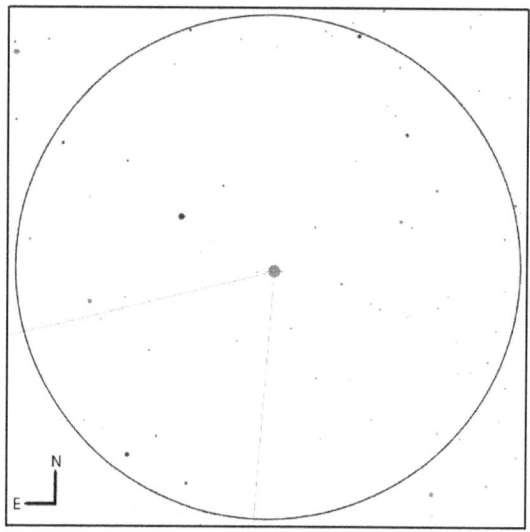
Etamin, binocular view

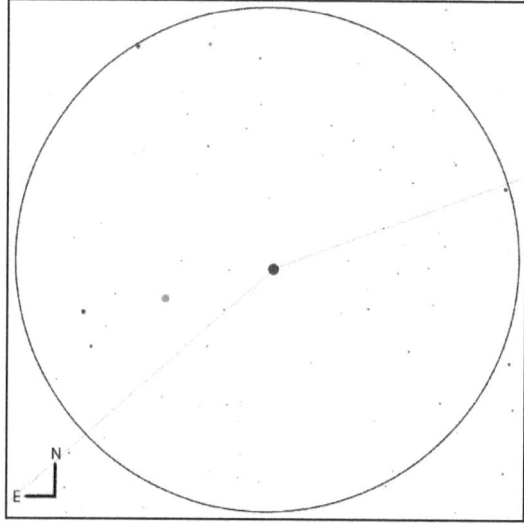
Thuban, binocular view

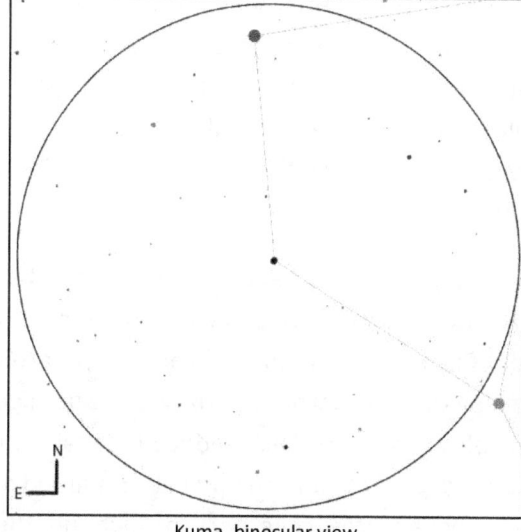
Kuma, binocular view

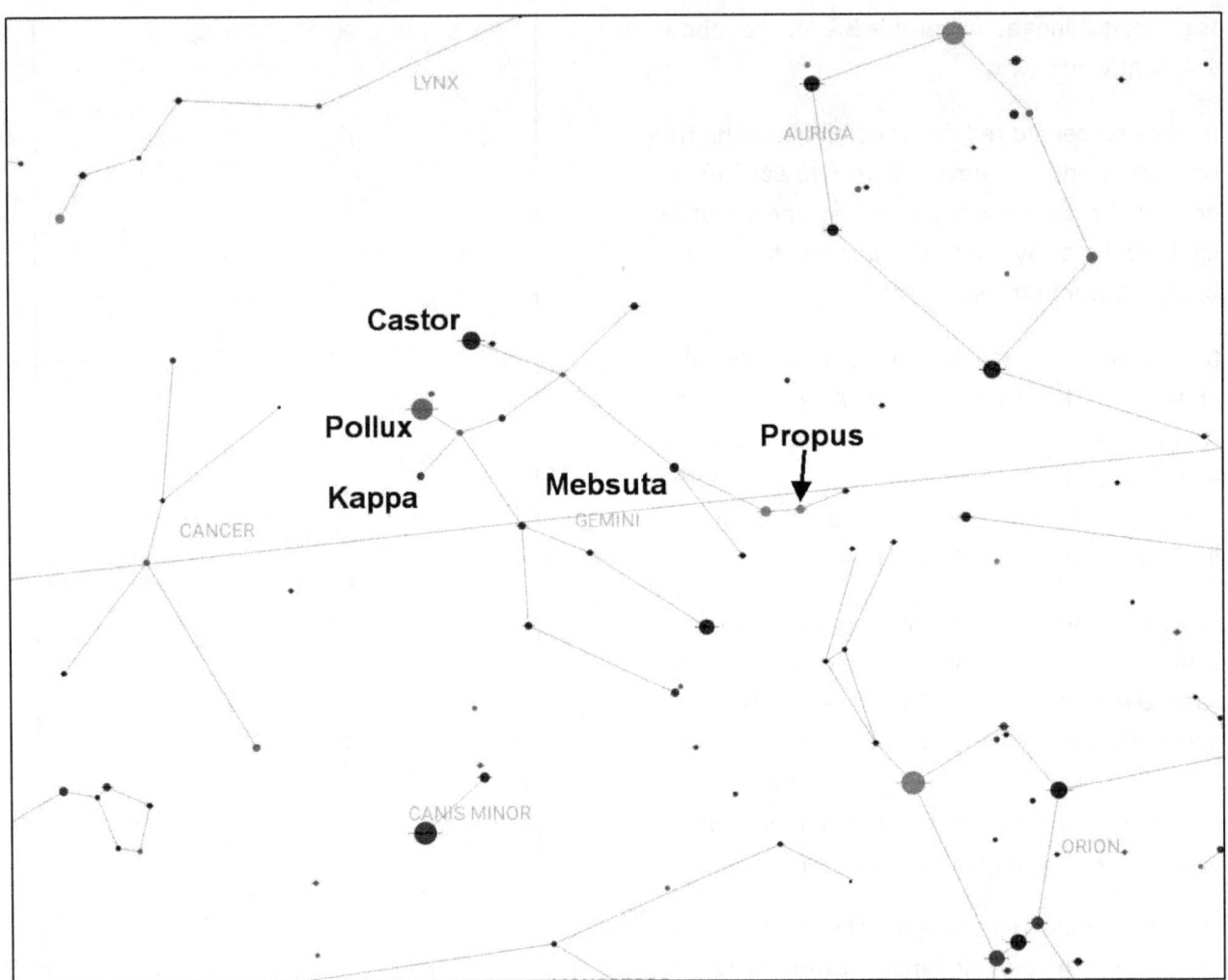

Gemini

Gemini, as almost everyone knows, is one of the twelve signs of the zodiac and is a prominent feature of the winter and early spring skies. Thanks to its two brightest stars, it's been associated with twins across a number of civilizations but the constellation is now most commonly linked to the Greek myth of Castor and Pollux.

Curiously, despite being twins, they have different fathers. Although both were born to Leda, Castor was the mortal son of Tyndareus, the King of Sparta, while Pollux was the demigod son of Zeus who had seduced Leda while disguised as a swan. The twins embarked on many adventures together throughout their lives, including rescuing their sister Helen and joining the crew of the *Argo*.

Despite all their heroic escapades, the end came tragically while the twins were attempting to steal cattle from their cousin. During the raid, Castor was fatally wounded, causing the distraught Pollux to plead with his father Zeus to help them both. Zeus took pity and immortalized the twins by placing them among the stars.

Gemini is a bright, mid-sized constellation, easily found close to Orion the Hunter with the two bright stars of Castor and Pollux marking the heads of the respective twins. At magnitude 1.1, Pollux is the brighter and closer of the two. The 17th brightest star in the sky, it's an orange giant some 34 light years away with a single planet, twice the mass of Jupiter, unofficially named Thestias. The planet takes 590 days to orbit its parent star.

Meanwhile, Castor is a complex multiple star system about 51 light years away. To the naked eye, it's a single magnitude 1.6 white star but, in reality, there are three pairs of stars, making for a system of six in all.

Observers with a small telescope can see the two brightest components for themselves, but it requires a reasonably high magnification of about 100x to split the primary.

Near Castor's foot is Eta, also known as Propus. This variable star is easily found but its changes in brightness are subtle and may require regular observations over a prolonged period to be noticed.

The reason for this is it only ranges in magnitude from 3.1 to 3.9 over a period of 233 days. Take a close look at this orange star and then compare it to Epsilon (Mebsuta) and Kappa. Which star is closest to Propus in magnitude? Mebsuta is magnitude 3.0 while Kappa is magnitude 4.0. Record your observation and then come back in about a month. What has changed?

The constellation isn't devoid of objects for binocular observers either. Close to Propus is Messier 35, a large, reasonably bright open cluster of magnitude 5.3.

Discovered by the French astronomer Phillippe Loys de Chêseaux in 1745, it lies about 2,800 light years away and contains approximately 200 stars. Through 10x50 binoculars under suburban skies I've seen it as a faint, grey, hazy hourglass shaped misty patch with a number of individual stars being visible.

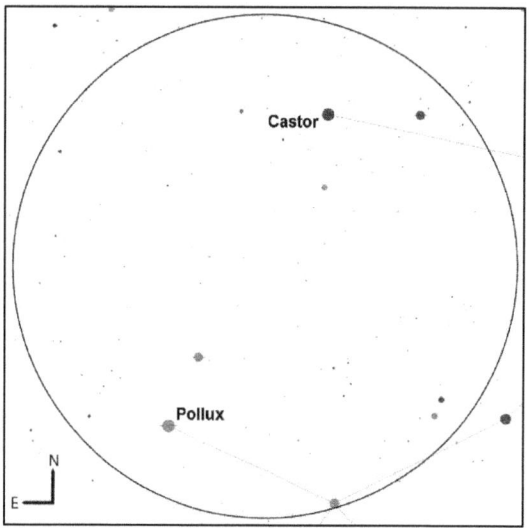
Castor and Pollux, binocular view

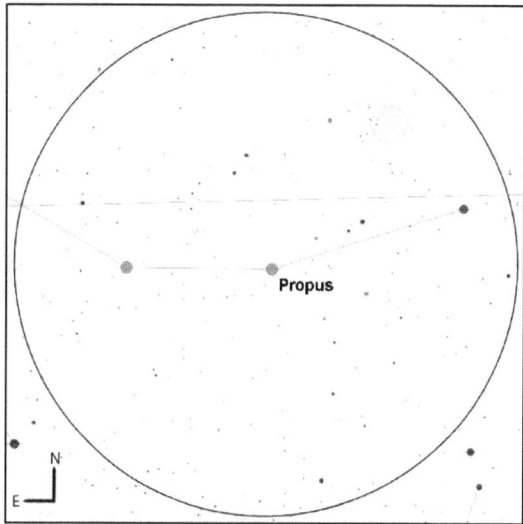
Propus, binocular view

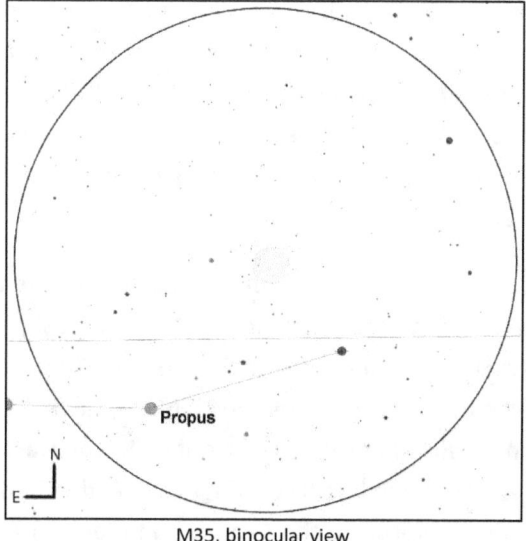
M35, binocular view

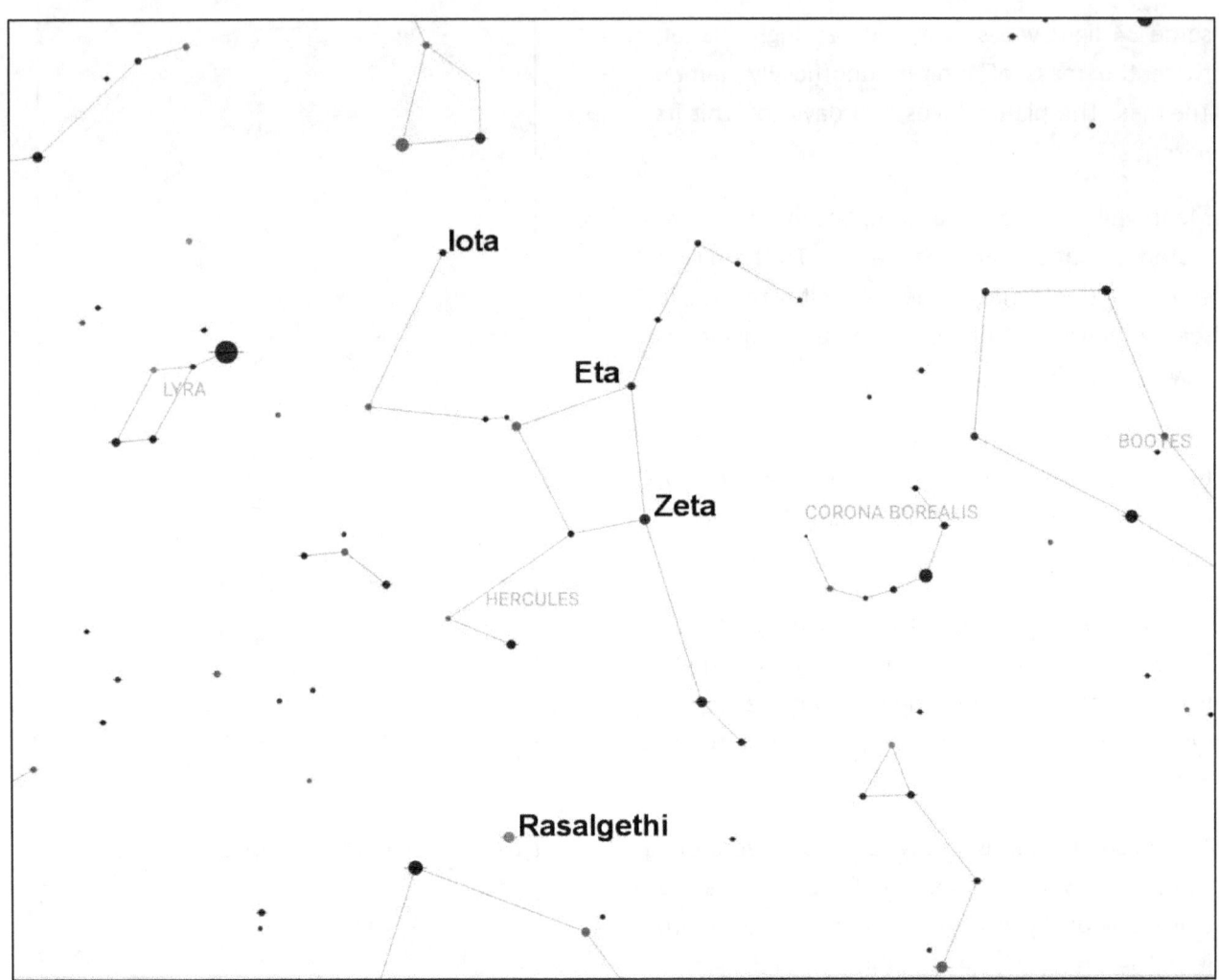

Hercules

As many know, this constellation represents the hero Hercules (originally known as Heracles in Greece), son of the god Zeus and the mortal woman Alcmene. Poor Heracles did *not* have a happy life, mostly as a result of the anger directed at him by Hera, the wife of Zeus, who had decided to punish Zeus for his infidelity by persecuting his illegitimate son.

Heracles served King Eurystheus for ten years and it was during this time that the King assigned Heracles the labors for which he is well known. Some of these, like Heracles himself, have been immortalized in the sky – specifically, the slaying of the Nemean lion (Leo), the nine-headed hydra (Hydra), capturing the Cretan bull (Taurus) and also stealing the golden apples of the Hesperides after first slaying Ladon, the dragon guarding them (immortalized as the constellation Draco.)

Hercules is surprisingly large. Covering approximately 1,225 square degrees, it's ranked 5th out of the 88 recognized constellations making it one of the largest in the sky.

Despite its size, it has no stars brighter than magnitude 2 and even its brightest – Alpha, also known as Rasalgethi – barely scrapes through magnitude 3 by shining at magnitude 2.9. Its name is Arabic and means "the head of the kneeler," which seems a little odd given that the star appears at the bottom of the constellation as seen from the northern hemisphere.

This is because, traditionally, Hercules is often depicted upside down on star charts (although no one seems to know why!)

A red giant star some 350 light years away, it has a companion that may be revealed with amateur telescopes. I've found that under steady skies and medium to high power the orange primary appears to have a fainter, pale blue companion.

A far easier target is the famous globular cluster, Messier 13. Known as the Keystone Cluster, it gets its name from the distinctive asterism of four stars that marks Hercules' torso. At magnitude 5.8, the cluster may be seen by sharp-sighted observers under clear, dark skies but for the rest of us, a pair of binoculars is required.

Discovered by Edmund Halley (of comet fame) in 1714, the cluster appears as a small, hazy circular patch with a star-like core. It can be found between Eta and Zeta when observed with 10x50 binoculars.

Through a small telescope – even at low power – it can be stunning. Depending upon your conditions and equipment, it's easy to see chains of stars within the cluster, snaking away from the cluster's center like the tentacles of some deep sea creature.

Also within binocular reach is its smaller neighbor, Messier 92. At magnitude 6.3 it's a little fainter but can be conveniently found within the same binocular field of view as Iota.

If nothing else, it's worth seeking out for the challenge and, like Messier 13, can be glimpsed with binoculars under suburban skies.

Of the two, Messier 13 is the closest at 22,000 light years while Messier 92 is 5,000 light years more distant.

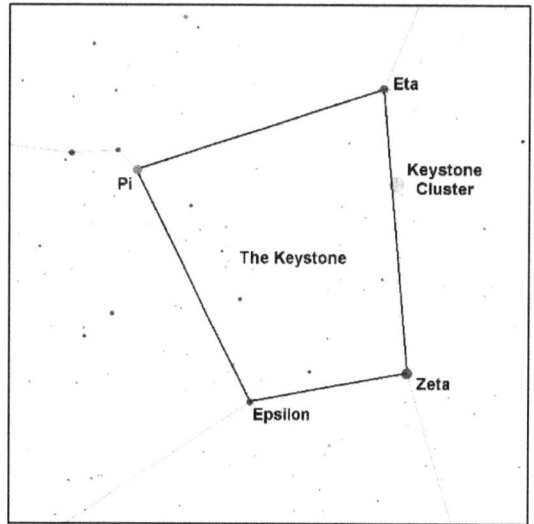

The stars of the Keystone

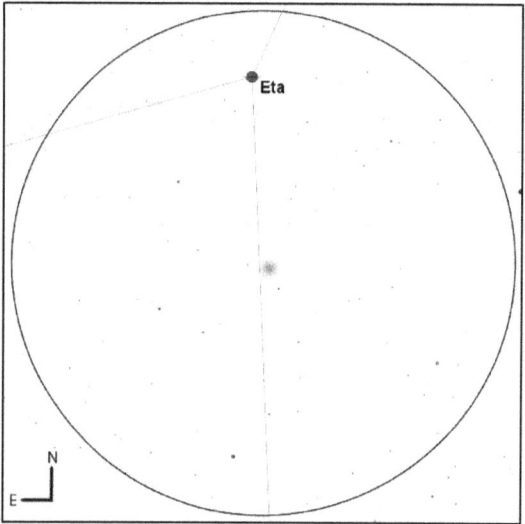

The Keystone Cluster, binocular view

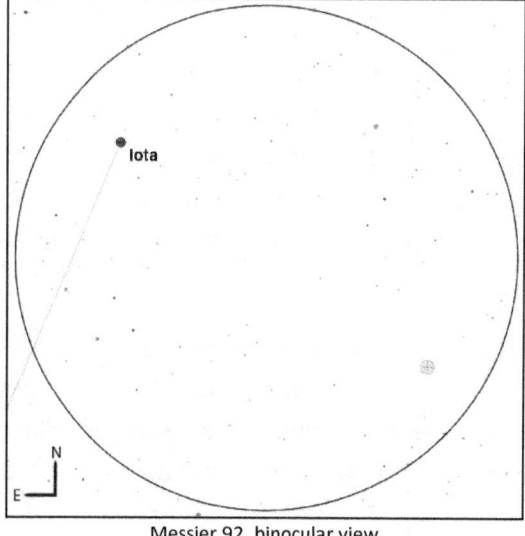

Messier 92, binocular view

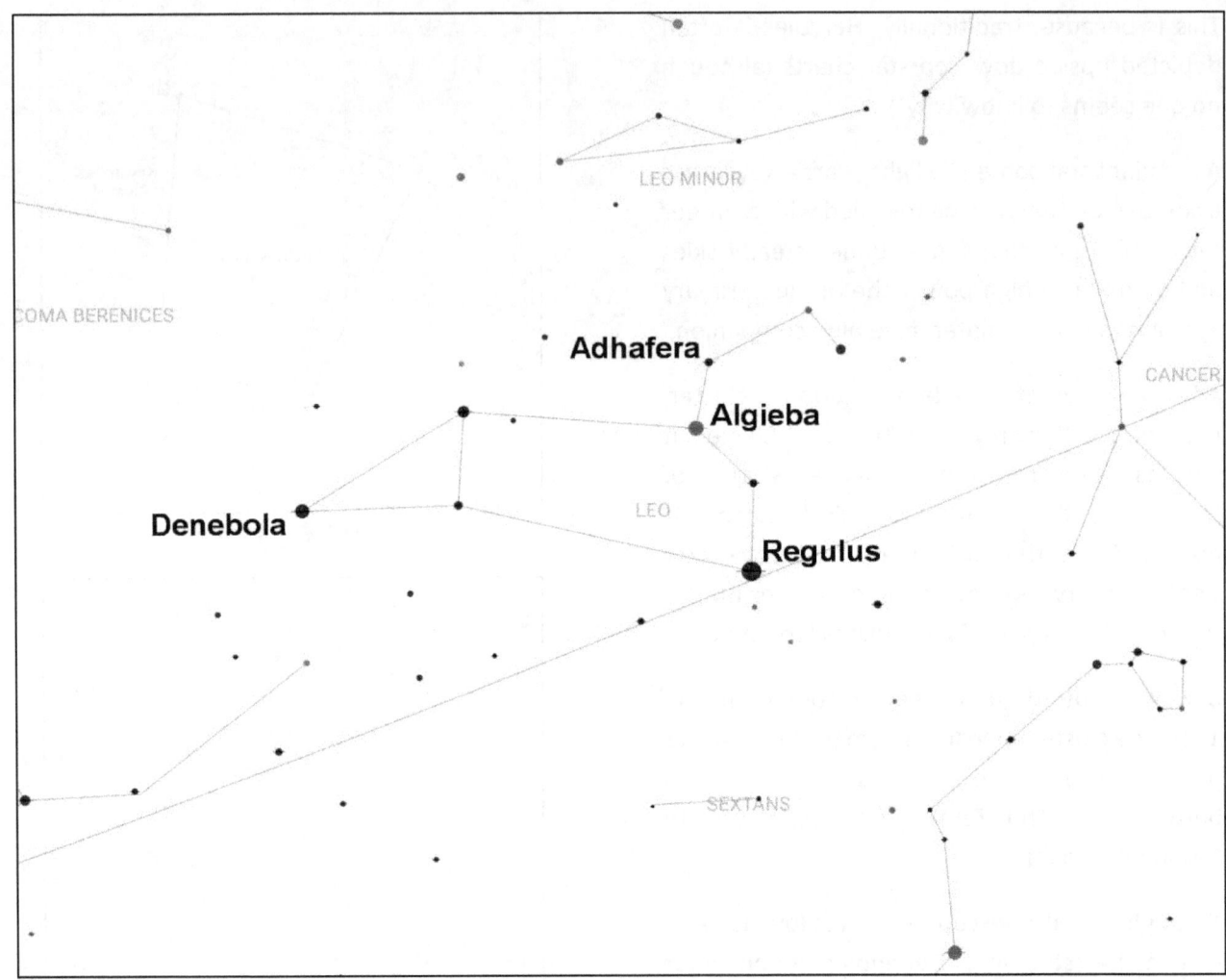

Leo

Leo, a constellation of the zodiac, ranks fifth in overall size and is prominent throughout the spring. (See page 48 for a guide on how to use Ursa Major to locate it.)

It also has the rare distinction of resembling the creature it represents, which is, as almost everyone knows, a lion. In fact, its shape is so distinctive that it's hard to imagine it as anything else.

To the ancient Greeks it was the Nemean lion impervious to all weapons, thereby almost making the beast undefeatable. After killing a large number of warriors, the lion was eventually slain by Heracles (aka, Hercules) who killed it with his bare hands as one of his labors.

The constellation contains a number of bright stars that make it easily recognizable. In particular, look out for the backwards question-mark asterism that forms the head and front of the lion. Also known as the Sickle, it's punctuated by the constellation's brightest star, Alpha (Regulus) at the bottom.

Regulus is the 21st brightest star in the sky and, at only 79 light years away, one of the nearest. A magnitude 1.4 blue-white star about 3½ times the size of the Sun, it's a multiple star system made up of two pairs of stars that take several million years to orbit one another.

Unfortunately, you won't see all four stars with amateur equipment, but binoculars will reveal a wide, faint bluish companion.

Regulus appears only half a degree away from the ecliptic, which is the path the Sun, Moon and planets take to traverse across the sky. What does this mean? Well, depending on the position of the planets, you can get some rather nice sights as the Moon and planets will sometimes group together near the star.

Equally entrancing is an occultation of Regulus by the Moon. This happens when the Moon appears to pass in front of the star, thereby hiding it from view. It's not uncommon, but your location and timing will determine its visibility and it's best to check online for upcoming events.

Another double appears nearby in the Sickle asterism. Gamma, (aka Algieba, from the Arabic for "the forehead") is a famous multiple star best observed with a small telescope. A medium to high magnification will reveal two gold stars of equal brightness.

Binoculars won't allow you to split the pair, but you *can* see an unrelated magnitude 4.8 star, 40 Leonis, just to the south of Algieba itself. If you have good eyesight, you may even be able to spot it without optical aid.

Within the same binocular field of view is Zeta, commonly known as Adhafera. This lemony-white magnitude 3.4 star has a white, unrelated magnitude 5.8 companion.

At the opposite end of the constellation, Beta (aka, Denebola, from the Arabic for "tail of the lion") is a white magnitude 2.2 star. Like Algieba, binoculars will reveal an unrelated companion but with a third, fainter star between them.

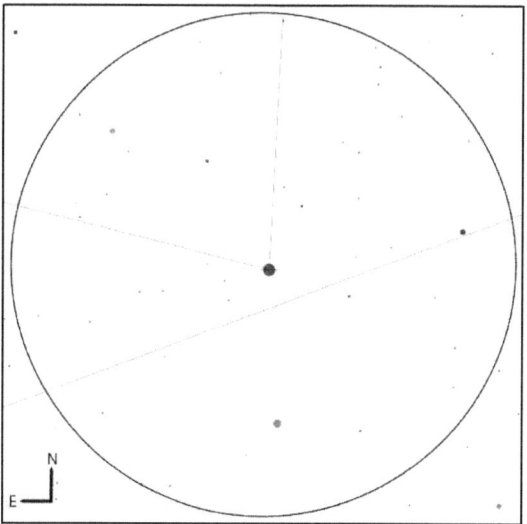
Regulus, binocular view

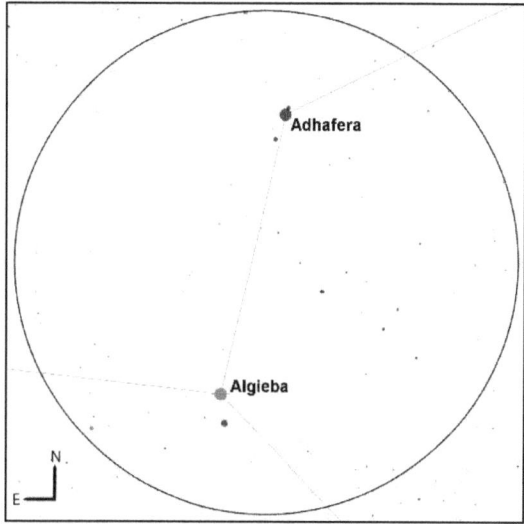
Algieba and Adhafera, binocular view

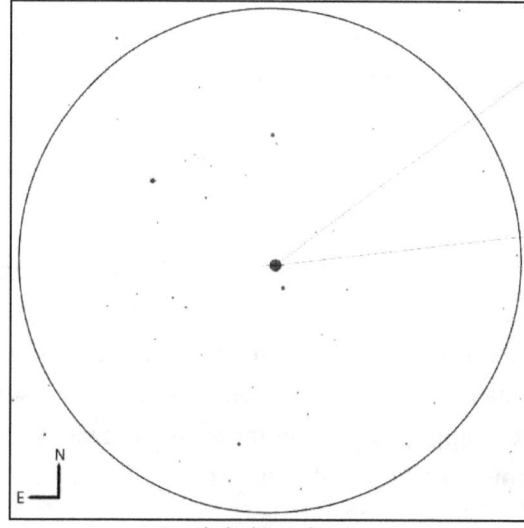
Denebola, binocular view

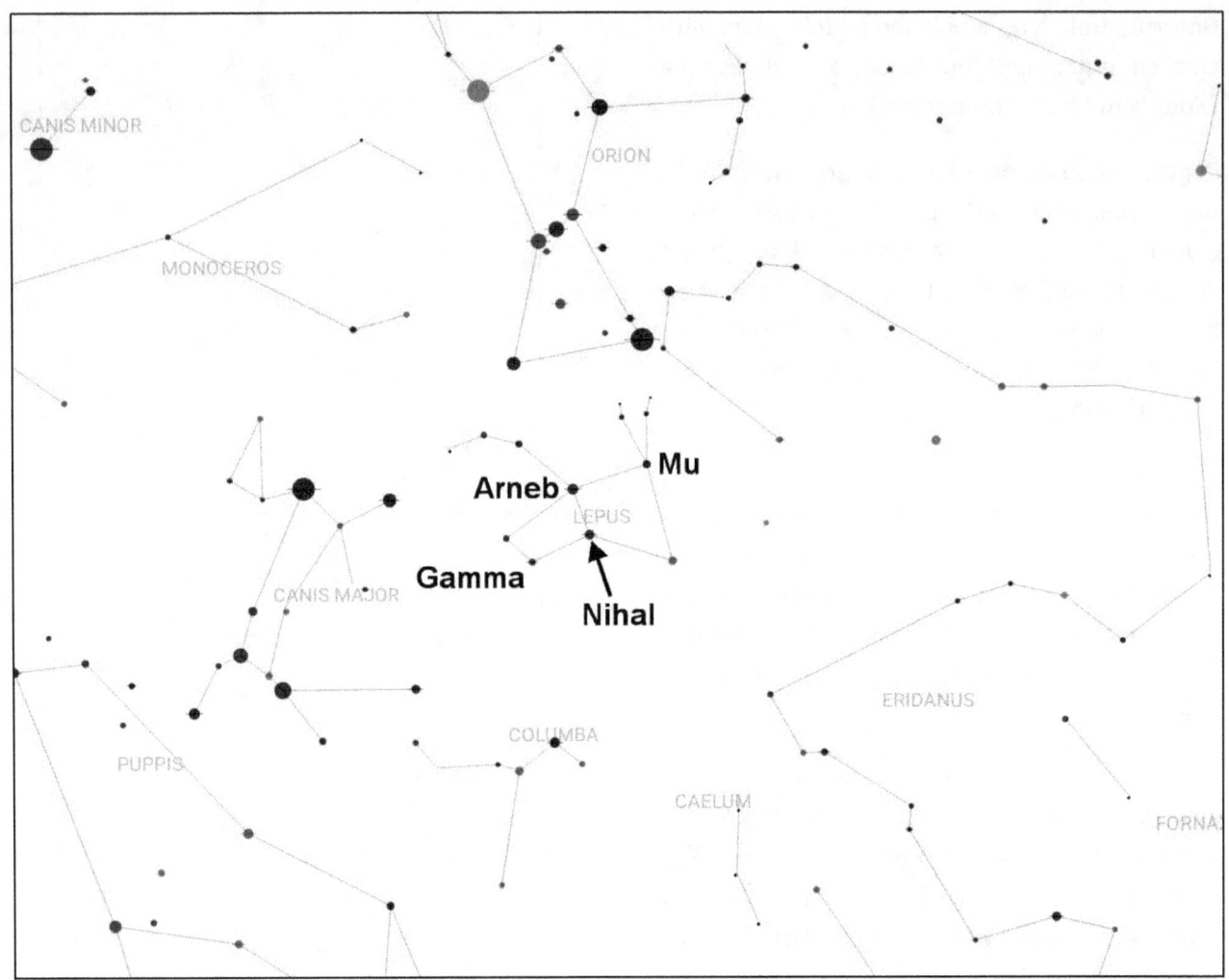

Lepus

Lepus represents a hare, the hunting target of Orion and his dogs Canis Major and Canis Minor.

The constellation is thought to originate from the Greeks of ancient Sicily, which once suffered from an infestation of the furry critters, but there are no clear myths or legends associated with it.

Perhaps one of the reasons Sicily suffered was because the hare was able to out-smart the hunters. I say this because Lepus is clearly hiding in plain sight, directly beneath Orion himself! (Although, in Orion's defence, he appears to be fending off Taurus the Bull at the same time.)

It's a md-sized constellation (51st in ranking) but given its location, it's fairly easy to spot. Four of its stars, Alpha (Arneb), Beta (Nihal), Gamma and Delta form an asterism known to the Arabs as the Chair of the Giant while the Egyptians knew it as the Boat of Osiris (as represented by Orion above it.)

At magnitude 2.6, Alpha, also known as Arneb (derived from the Arabic for "hare") is the constellation's brightest star. Despite its modest magnitude, Arneb is a yellow-white giant star, nearly 130 times larger than the Sun and about 14 times more massive. It lies a whopping 2,200 light years away, which means it also has a very high luminosity – 32,000 Suns!

In other words, if Arneb were only one parsec away (equivalent to 33 light years), it would shine at magnitude -6.5 – far brighter than anything else except the Sun and the Moon!

Lepus is home to several sights worthy of binocular observers. Firstly, there's R Leporis, also known as Hind's Crimson Star – and for good reason, as it has a deep red color that its discoverer, John Russell Hind, compared to blood. You can find it within the same binocular field of view as Mu.

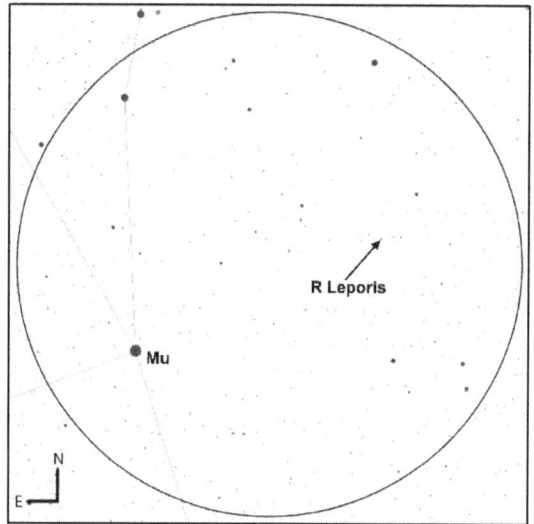

R Leporis, binocular view

That in itself would make the star worth a look, but like many other red giant stars in the waning years of its life, it's also a variable. At its brightest, it glows at a reasonable magnitude 5.5 but will then fade to magnitude 11.7, making it invisible in binoculars. It period has been known to vary, from 418 to 441 days, but is currently estimated at 427 days.

If you're under dark skies, trying looking for M79, the brightest deep sky object in an otherwise fairly barren constellation. Unfortunately, M79 is not only fainter but less than half the size of M13, its summer cousin.

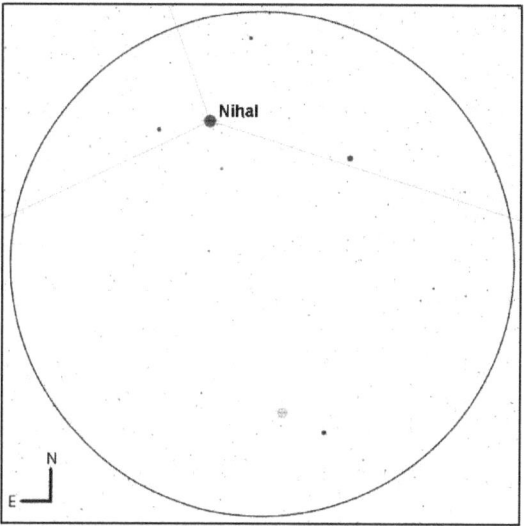

Messier 79, binocular view

However, it can be glimpsed close to Nihal as a small, hazy star, almost comet-like in appearance. At 41,000 light years away, M79 might not belong to our Milky Way galaxy at all. Studies have shown that it might be a member of the Canis Major Dwarf Galaxy, which just so happens to be passing close to our own, much larger galaxy, at this time.

Lastly, there's Gamma, a fairly well-known double that can be split with 10x50 binoculars. The primary appears creamy and much brighter than the coppery companion, which appears to the south-east. If you have a small telescope, two other components may also be visible.

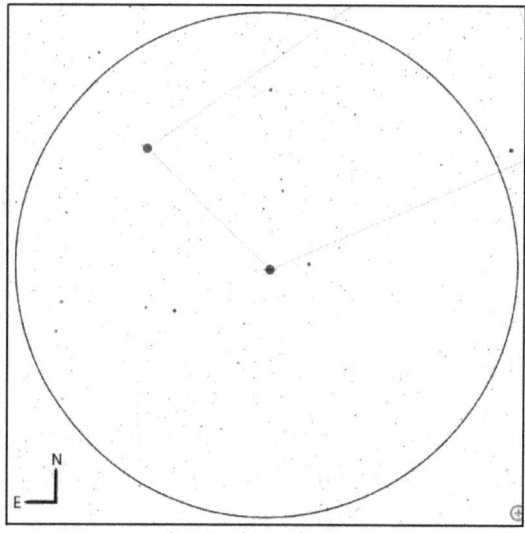

Gamma, binocular view

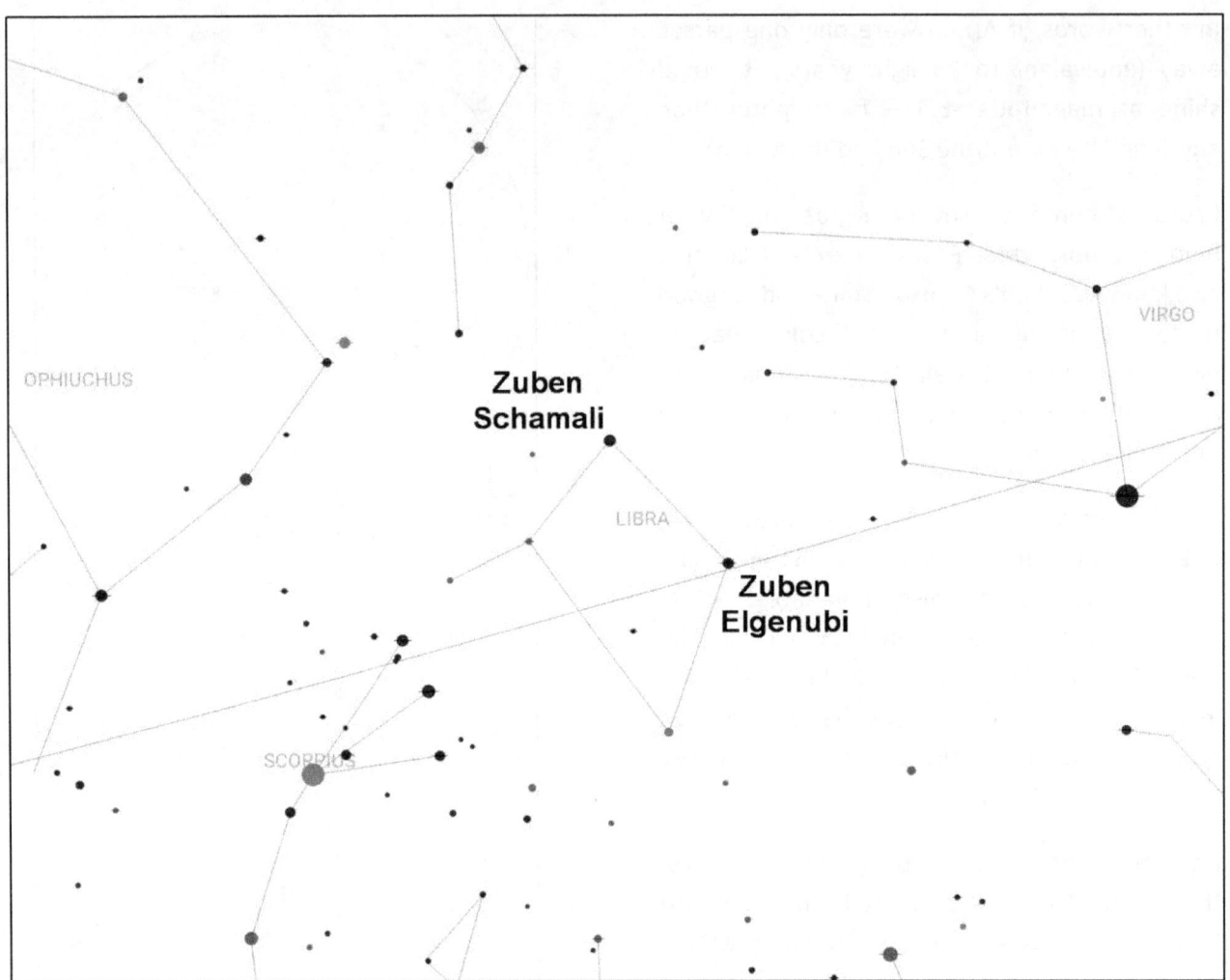

Libra

Perhaps not surprisingly, Libra, the Scales, doesn't have any myths or legends associated with it as the ancient Greeks identified the stars as the claws of nearby Scorpius.

It's thought the constellation became known as the Scales (or the Balance) at a time when the Sun would pass through the constellation on the spring equinox. With the days and nights being of equal length on that date, it must have seemed an appropriate way for the ancients to commemorate the event.

Libra is also a curious constellation in that it probably wasn't one of the original signs of the zodiac but may have been added by the Romans as part of their Julian calendar.

It's a mid-sized constellation, found midway between white Spica in Virgo and orange-red Antares in Scorpius. Unfortunately, it's not particularly bright, with no stars brighter than magnitude 2.

Its brightest star is Alpha, also known as Zuben Elgenubi from the Arabic for "southern claw" – an obvious reference to when the star was still a part of Scorpius.

To almost everyone, it appears as a single star of magnitude 2.7 but a standard pair of 10x50 binoculars will easily split the star in two.

What you'll see is a pair of white stars of almost equal magnitude. This is a true multiple star system, some 77 light years away, with both components forming their own binary star systems. Beyond that, a theoretical fifth component may be a part of the system, making a total of five stars in all, but only the two binocular stars are visible to amateurs.

What color are the stars? Many are white, or blue-white, with the occasional yellow star or even an orange or red giant thrown in for good measure. So why are there no green stars?

The reason is a little complicated (and I simply don't have the room to explain it here!) – suffice it to say, a star's color depends upon its surface temperature. A hot star will literally glow white hot (just the same as metal does) but as it cools, the color fades to yellow, orange and then red. Just as metal (typically) doesn't glow green when it's hot, neither do stars.

Except, perhaps Beta Librae, also known as Zuben Schamali, from the Arabic for "northern claw." Nicknamed "the emerald star," numerous observers have noted a pale green color, visible with just the unaided eye.

In reality, it's a blue-white star, five times larger than the Sun and about 185 light years away. Not everyone sees its green light, but it's worth taking a look for yourself!

Lastly, Libra is home to Delta Librae, a variable star much like Algol in Perseus. At its brightest, it glows at magnitude 4.4, which might make it tricky to spot without optical aid from the light polluted skies of a town or city.

If you can see it, come back to it tomorrow night and look for it again. It'll drop to magnitude 5.8 and then return to its former glory over a period of 2.3 days.

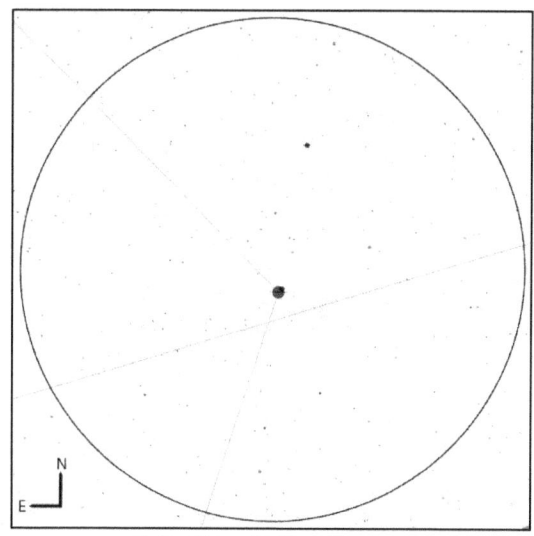

Zuben Elgenubi, binocular view

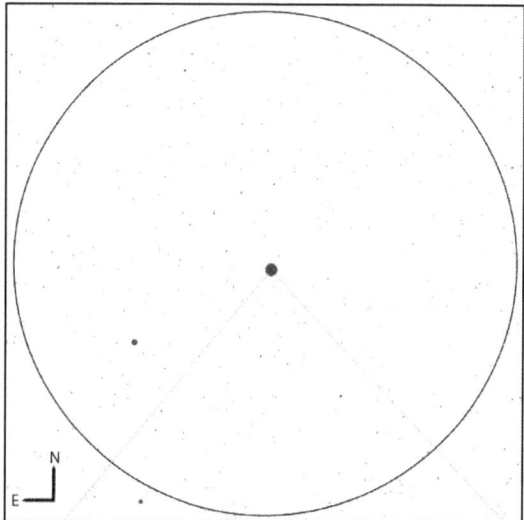

Zuben Schamali, binocular view

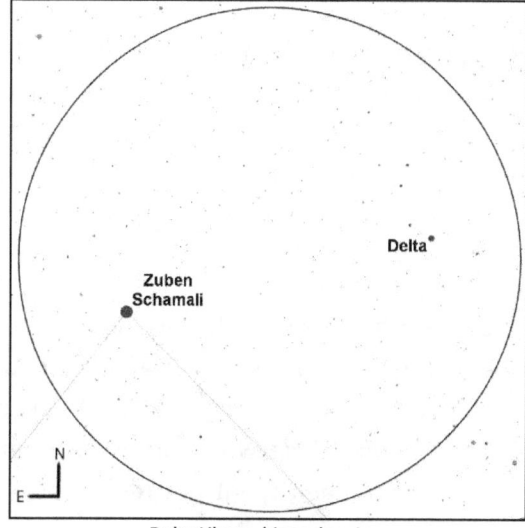

Delta Librae, binocular view

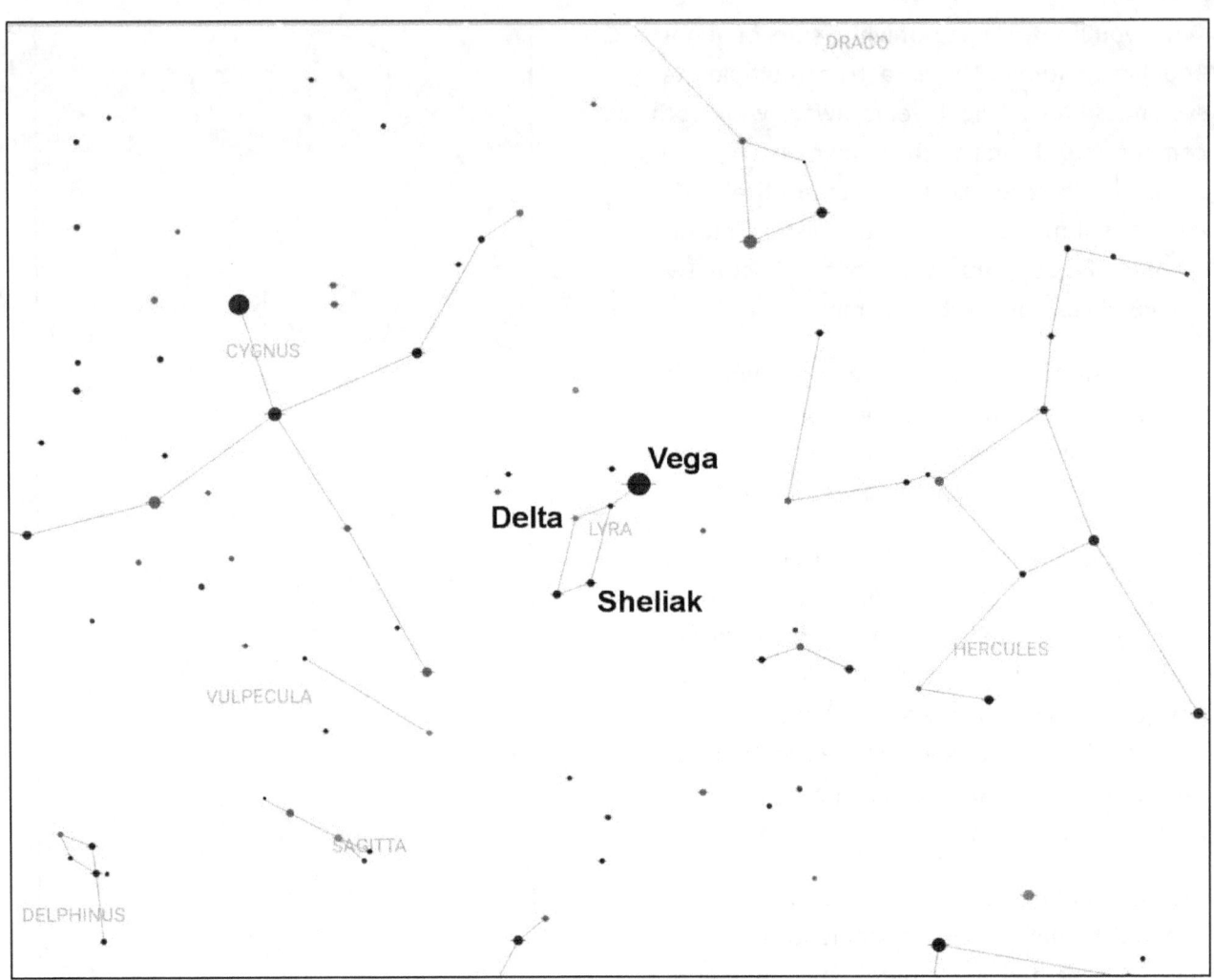

Lyra

Lyra represents the lyre given to Orpheus by his father, the god Apollo. With this musical instrument, Orpheus was able to charm pretty much everyone and everything, including Hades, the god of the Underworld. When Eurydice, beloved wife of Orpheus, was killed by a snake, Orpheus entered the Underworld and used his lyre to charm Hades into releasing her.

Hades only agreed on the condition that, as Orpheus left, he would not look back. However, Orpheus could not resist temptation and did indeed look back, causing Hades to keep Eurydice in the Underworld forever. Heartbroken Orpheus then spent the rest of his life

wandering throughout the land, playing his lyre and forever mourning her loss.

This summer constellation is relatively small, but bright and easily found, midway between Hercules the Hero and Cygnus the Swan. Its brightest star is Alpha, a brilliant white star of magnitude 0.0 also known as Vega, the fifth brightest star in the entire night sky and one of the three stars of the Summer Triangle (see page 182.)

At only 25 light years away, it's one of the closest stars to Earth and, at just under 500 million years old, one of the younger stars known. Like Altair in Aquila the Eagle, another bright star of summer it spins rapidly on its axis, causing the star to be flattened at the poles.

Another fascinating feature of Vega is the disk of dust that surrounds the star. No one knows for sure, but the disk may be a planetary system in the early stages of formation.

Close to Vega is another fascinating star system, Epsilon Lyrae. Popularly known as "the double double" this is a multiple star system some 160 light years away. Its two brightest components are easily seen in binoculars as a pair of identical, bright white stars. Turn a small telescope toward them and crank up the magnification to about 150x and you'll see that each star is itself a double with all four stars appearing white and of almost equal brightness. In all, there may be as many as ten stars in this system!

Meanwhile, just a little further east (but still within the same binocular field of view) is Delta Lyrae, another binocular double. In reality, the two stars merely appear close to one another in the sky. Delta[1], the fainter of the pair, is approximately 1,100 light years away while Delta[2] is about 200 light years closer.

Move to the south-west and look out for Beta Lyrae, known as Sheliak. Like the autumn and winter star Algol (see page 162) Sheliak is an eclipsing binary, which means a fainter star regularly passes in front of the brighter main star. Seen from Earth, the star will fade for a few days before brightening again. Sure enough, Sheliak shines at magnitude 3.25 at its brightest, fades by a full magnitude and then brightens again over a period of 12.9 days.

Lyra is home to several other sights, most notably the Ring Nebula and Messier 56 - but both require a small telescope to be fully appreciated.

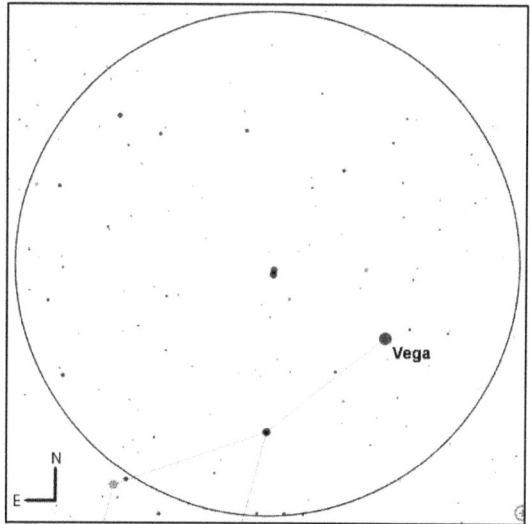
The Double Double, binocular view

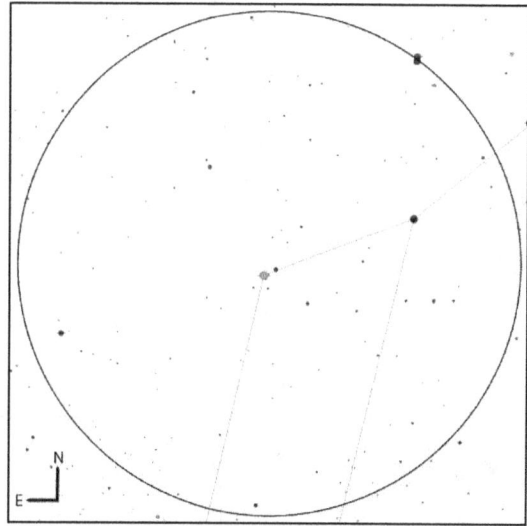
Delta Lyrae, binocular view

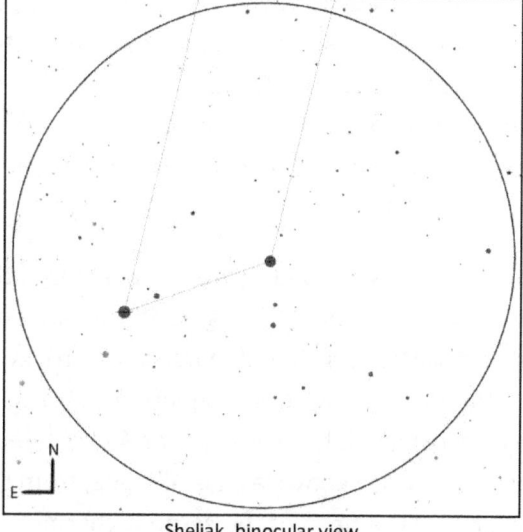
Sheliak, binocular view

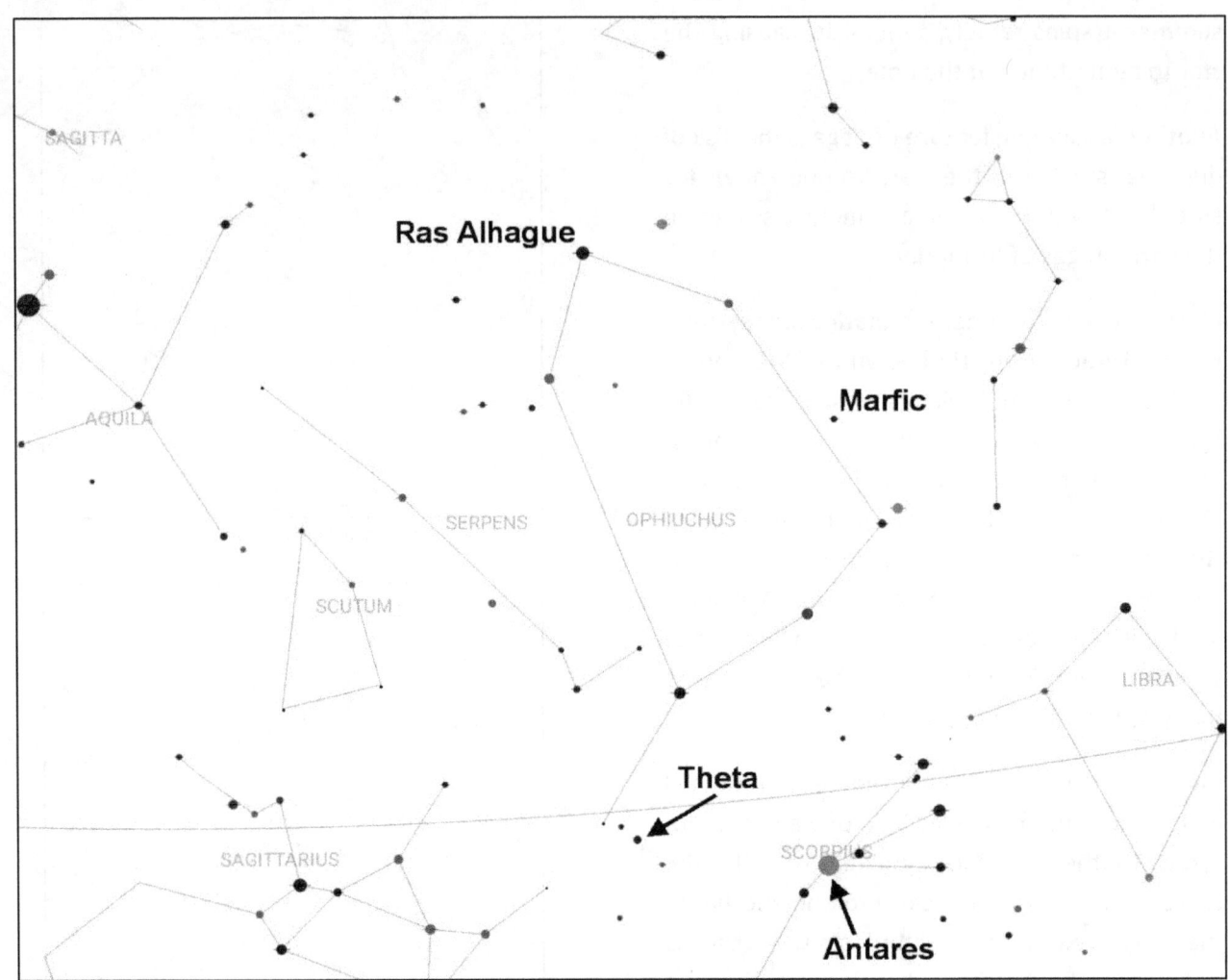

Ophiuchus

Also known as the Serpent Bearer, Ophiuchus has several associations with the ancient Greek gods and heroes. It's been depicted as the god Apollo wrestling with the snake that guarded the Oracle at Delphi and, alternatively, it's also been identified as Asclepius, a famed healer with a close association with serpents (see also Serpens on page 172.)

Covering nearly 950 square degrees, the constellation is the 11th largest in the sky and, with no particularly bright stars, can be tricky to identify, especially from suburban skies. Its brightest star, Alpha, is just under 49 light years away and is also known as Ras Alhague, from the Arabic for "the head of the serpent charmer." A

white star of magnitude 2.1, it's a binary system comprising of a giant white primary and a suspected orange companion. The two orbit one another once every 8½ years but the pair are too close to be split with amateur equipment.

If you have binoculars, take a look at the area close to Theta and especially 36 Ophiuchi. Appearing as a modest magnitude 4.3 star to the unaided eye, binoculars will reveal a wide companion. The primary appears golden and about twice as bright as the white-ish secondary. (Look out for a close, faint magnitude 6.4 companion to bright Theta too!)

Just slightly to the west is Messier 19, one of several reasonably bright globular clusters in the constellation. Discovered in 1764 by Charles Messier, it appears as a slightly flattened, faint comet-like star through binoculars. The effect is certainly more noticeable in a small telescope; during one observation, I noted it almost appeared rectangular.

Look a little to the south and you'll encounter Messier 62. If you're using 10x50 binoculars, you should be able to comfortably fit both clusters in the same field of view with both appearing about the same brightness and size. While you're here, look out for a bright pair of stars to the north-west of M19 and a faint, close pair of unrelated stars just to the north of M62.

Meanwhile, close to Marfic and near the center of the constellation lie Messier 10 and Messier 12, two other bright globulars easily picked out with binoculars. Again, like M19 and M62, the pair appear within the same field of view, slightly to the south-west of a triangle of three stars of almost equal magnitude.

All four clusters are roughly the same brightness but M62 is the most compact while, at 14,000 light years, M10 is the closest to us. All are about 12 billion years old.

Lastly, head south toward Scorpius and you'll find Rho Ophiuchi, a binocular triple star close to Antares. Together, the three stars form an almost equilateral triangle with the primary appearing at the apex and the two much fainter stars forming the base.

Through a small telescope, the primary appears to have an off-white, creamy colour while the two companions, of equal brightness, both appear blue. In all, there are five stars associated with this system, which lies some 360 light years away.

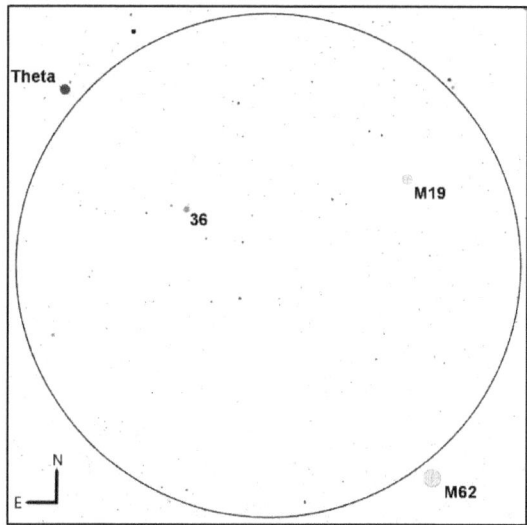

Theta Ophiuchi region, binocular view

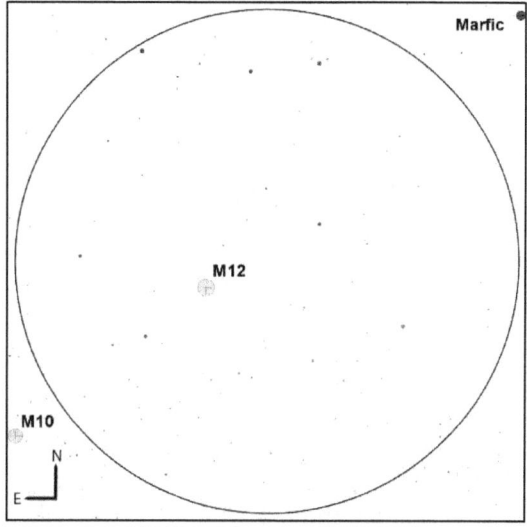

Marfic region, binocular view

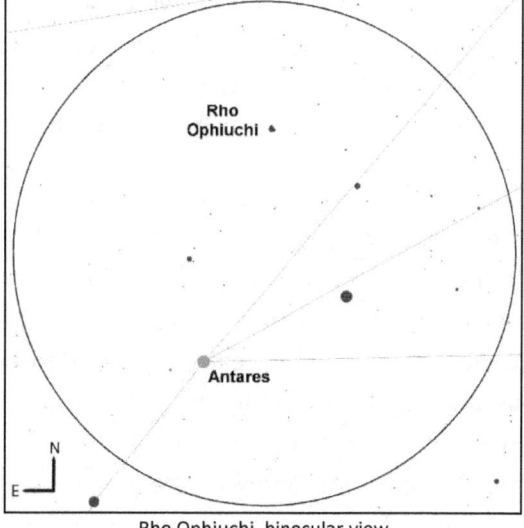

Rho Ophiuchi, binocular view

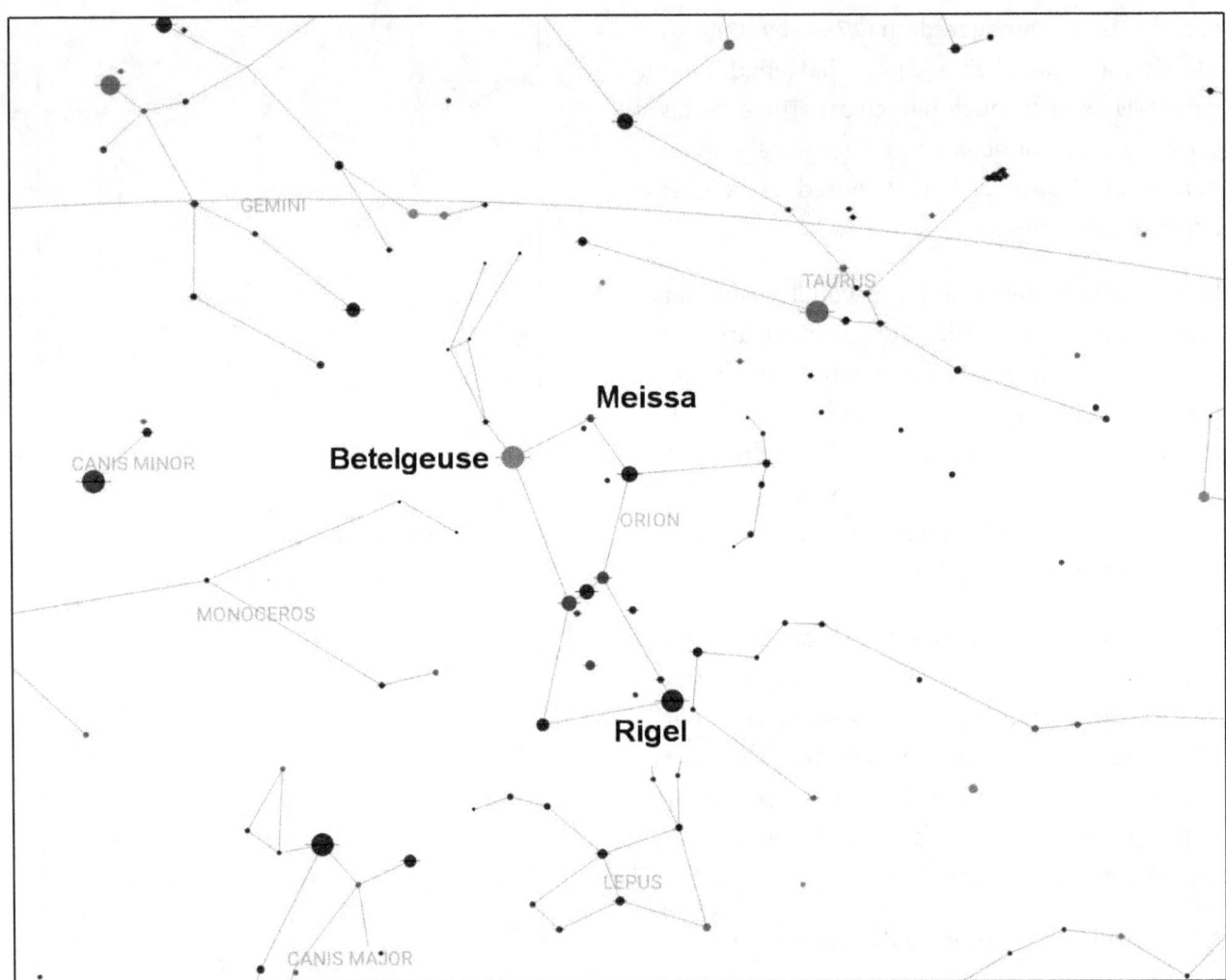

Orion

Orion is arguably the most famous constellation in the sky and is certainly one of the brightest. It's been known since antiquity with many myths and legends associated with it from cultures all over the world. To the Greeks, Orion was a skilled hunter, son of Poseidon, god of the sea.

According to one legend, Orion was hunting with the goddess Artemis and boasted he could kill any creature on Earth. This angered the Earth goddess Gaia, who sent a giant scorpion to kill him. Sure enough, the creature succeeded and Zeus placed the hunter amongst the stars. He also immortalized the scorpion, Scorpius, but placed it opposite Orion in the sky where it would never harm him again. Hence, Orion is a winter constellation while Scorpius rises to prominence in the summer sky.

Orion is a moderately large constellation, 26th in size, with three of its brightest stars forming a distinctive belt across his middle. He's often depicted defending himself against Taurus, who appears to be stampeding toward him.

As Orion is so easily identified, it can be conveniently used to locate other bright stars and constellations (see page 49), most notably Sirius, Canis Minor, Gemini and the afore-mentioned Taurus.

The brightest star is Beta, also known as Rigel, from the Arabic for "the left leg." A magnitude

0.1 star, it's the seventh brightest star in the sky and lies approximately 850 light years away. There are thought to be three or four stars in the system in all, with the primary being a blue-white supergiant nearly 80 times larger than the Sun and about 120,000 times more luminous.

The second brightest star is Alpha, famously known as Betelgeuse, from the Arabic for "the hand of Jauzā (the Arabic name for Orion.)" There's no officially recognized pronunciation of the name - some say "betel-geez" while others say "beetle-juice", like the Tim Burton film of the same name. The ninth brightest star in the sky, it has a distinctly orange glow and lies about 640 light years away.

Betelgeuse is huge. A red supergiant, if it were placed at the center of the solar system, it would extend well beyond the orbit of Mars and is nearing the end of its life. It could, in theory, explode and appear as a supernova at any time. When it does, it will probably be visible in daylight for months.

Slide over to the north-west to find Lambda, also known as Meissa. This magnitude 3.4 star marks the northernmost corner of a triangle that represents the head of the hunter. Through binoculars you'll also see a tiny trail of three stars that form a slightly curved line to the south.

The Great Orion Nebula is the only nebula easily visible to the unaided eye from suburban skies and can be found below the starry belt of Orion. Known and observed for thousands of years, it represents the sword of Orion and appears as a bright, misty patch through binoculars. This is the birthplace of stars and even 10x50's will reveal at least two or three tiny points of light – members of the famed Trapezium cluster – young stars in the process of being born.

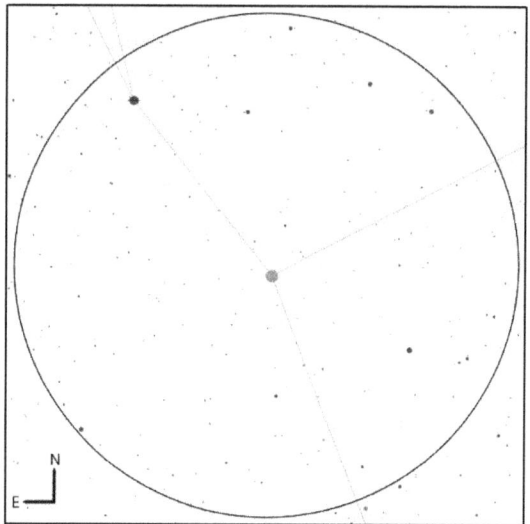
Betelgeuse, binocular view

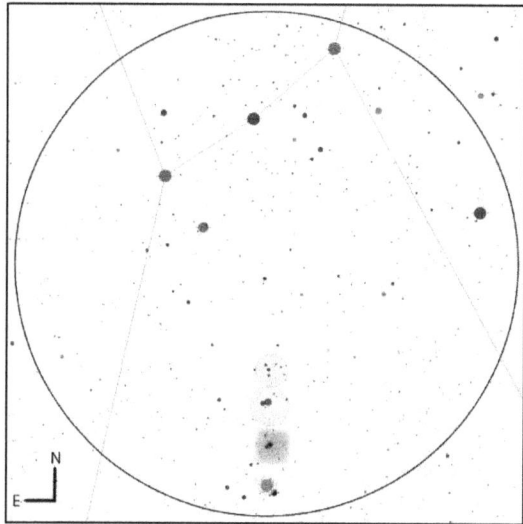
The Great Orion Nebula and belt, binocular view

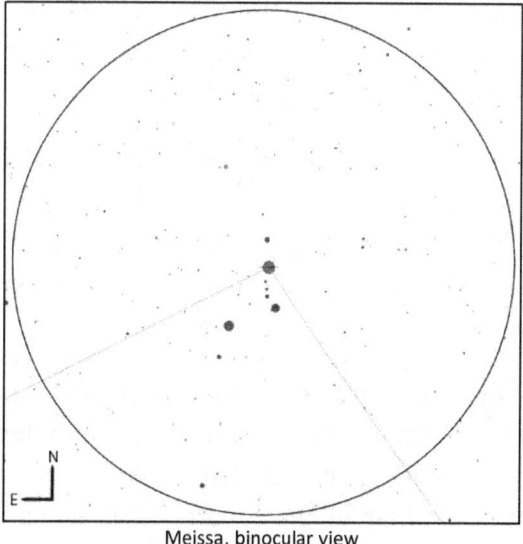
Meissa, binocular view

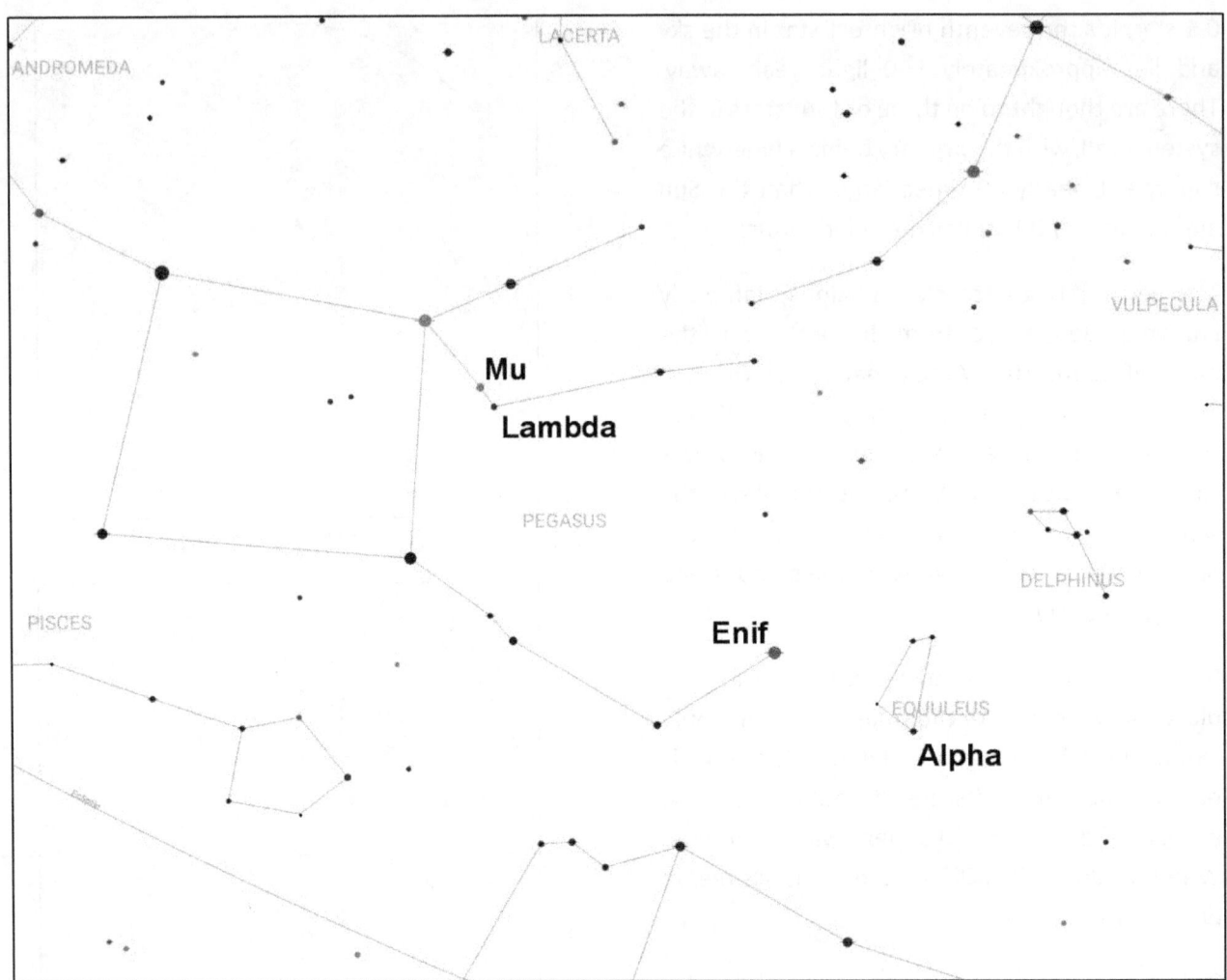

Pegasus & Equuleus

Pegasus, as mythology and film fans know, is the flying horse Perseus rode as he swooped down to rescue the chained princess Andromeda from Cetus the sea monster (see page 104 - or the original *Clash of the* Titans - for the full story.)

It's a large constellation, ranked 7th in size and covering over 1,100 square degrees in a relatively barren area of sky. Its most distinctive feature is the Great Square of Pegasus, a noteworthy asterism for several reasons.

Firstly, Pegasus is one of the few constellations to be directly joined to another, Andromeda. Consequently, the star that marks the north-eastern corner of the square, once known as

Delta Pegasi (or Sirrah) now belongs to Andromeda and is more commonly called Alpheratz.

Secondly, by counting the number of stars you see within the square you can get a good idea of the sky conditions from your location.

About five stars can indicate good conditions while ten or more can indicate very clear skies.

Fortunately, you don't need great skies to spot the constellation's brightest star, Epsilon. Also known as Enif (derived from the Arabic for "nose") it's found some way to the west of the great square (which marks the body of the horse.)

Enif is an orange supergiant, some 185 times the size of the Sun with about twelve times its mass and over 12,000 times its luminosity. It's close to 700 light years away and may be nearing the end of its life.

Close to Enif and within the same binocular field of view is Messier 15, a globular cluster discovered by Jean-Dominique Maraldi in 1746. Shining at magnitude 6.3, it has a reasonably bright core and emits X-ray radiation, hinting at a possible black hole within it. M15 is about 33,000 light years away.

Just outside the Great Square is another notable object, 51 Pegasi. In 1995, this nearby magnitude 5.5 sun-like star was one of the first found to be orbited by a planet. Thought to be a gas giant half the mass of Jupiter, it orbits its parent star once every four days and has since officially been named Dimidium, from the Latin for "half" – a reference to its mass.

(Incidentally, no planet outside our own solar system can be seen with amateur equipment so binoculars or a telescope will only show the star itself. You can, however, see the star 51 Pegasi within the same field of view as Lambda and Mu Pegasi.)

Leaving Pegasus behind, we come to Equuleus the Foal to the west. Ranked 87th in size, this is the smallest constellation visible from the northern hemisphere and the second smallest overall. It has no stars brighter than magnitude 3 and even its brightest, Alpha, is only slightly brighter than magnitude 4.

But if you like a challenge, try tracking down Gamma, a wide double star for binoculars or a small telescope at low power. The primary appears white and about 2-3 times brighter than the bluish companion.

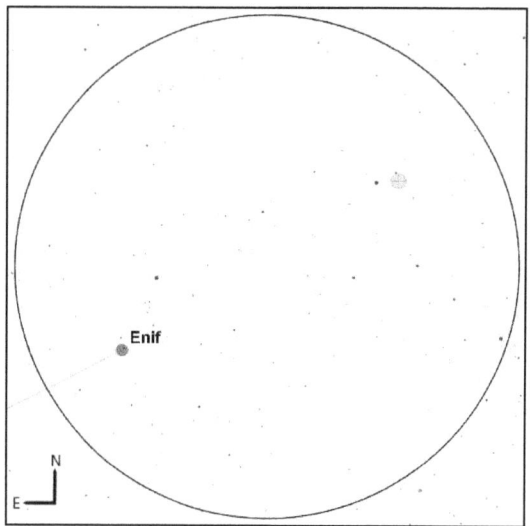
Enif and Messier 15, binocular view

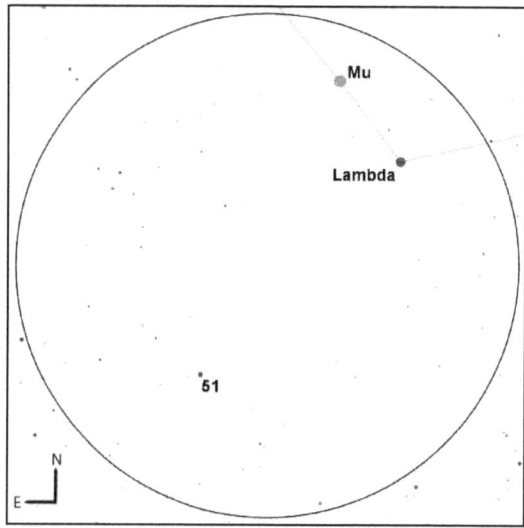
51 Pegasi, binocular view

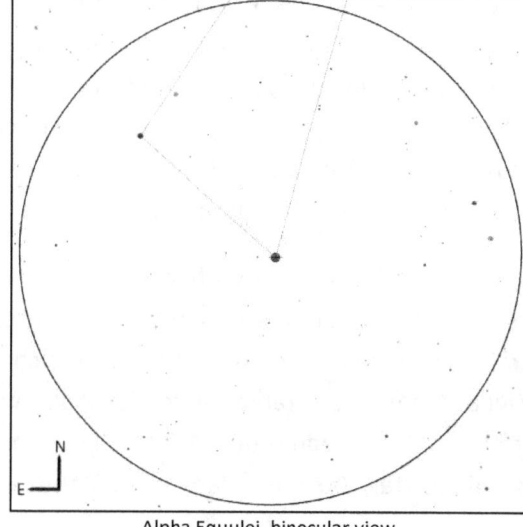
Alpha Equulei, binocular view

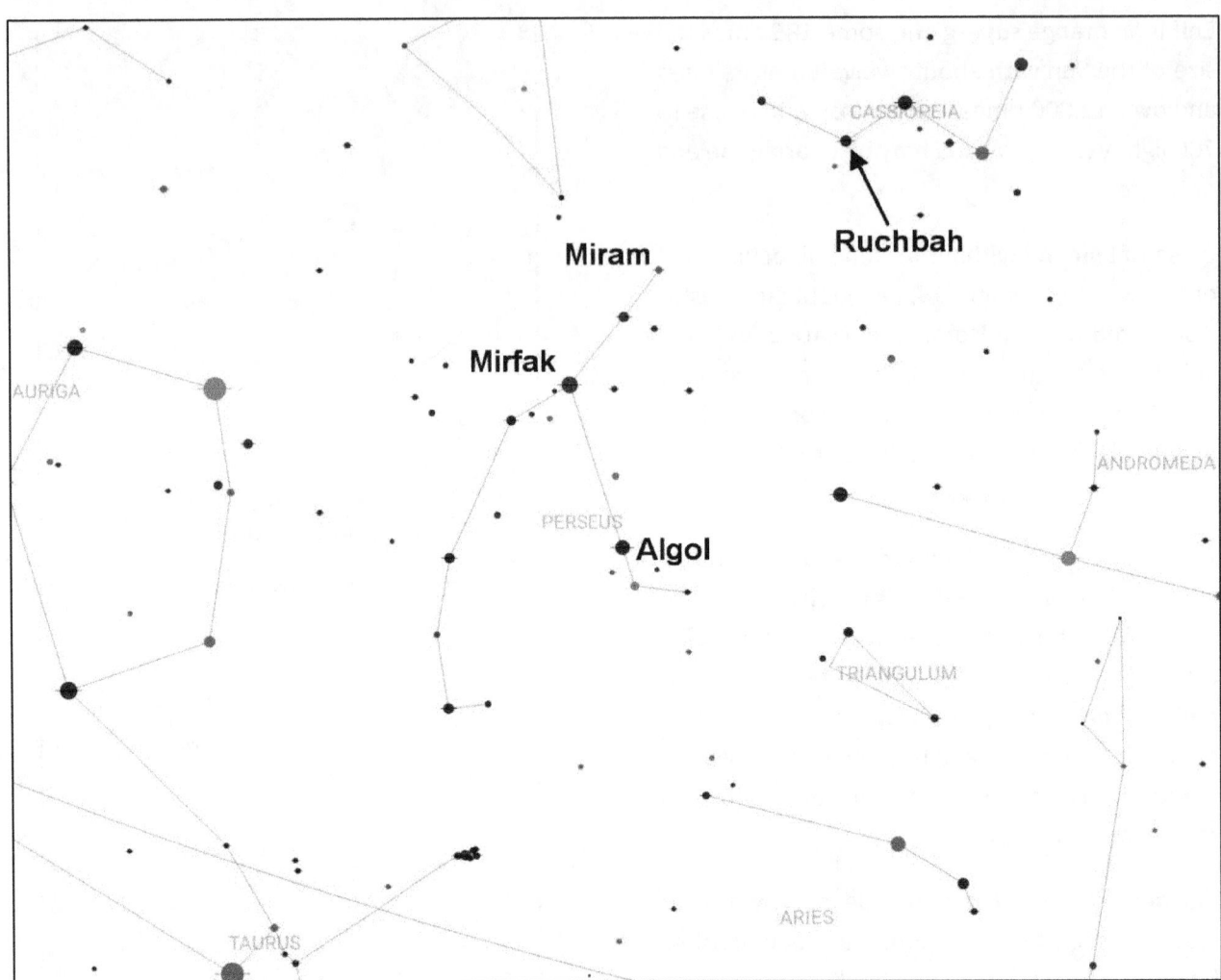

Perseus

Perseus depicts the Greek hero famed for rescuing the princess Andromeda from Cetus, the Sea Monster (see page 104 for the full story.) It's a mid-sized constellation, ranked 26th in overall size and located close to Andromeda and Cassiopeia in the night sky. It has a curved shape, almost like a wish-bone, and can be easily seen all the way from early autumn to early spring.

The constellation has a number of bright stars and objects for the casual observer. Its brightest star is Alpha, also known as Mirfak, which is derived from the Arabic word for "elbow." Mirfak is a magnitude 1.8 yellow-white supergiant star, 60 times larger than the Sun with 8 times its mass and about 5,000 times its luminosity. It lies about 600 light years away.

Mirfak itself isn't so remarkable, but the surrounding sky is worth sweeping with binoculars. The Milky Way flows straight through the constellation, leaving a stream of stars in its wake.

In particular, Mirfak swims amongst a large scattering of blue-white stars, appropriately named the Alpha Persei Cluster. It can be seen with just the unaided eye from a dark location but suburban sky observers can still enjoy the view with binoculars. A true cluster, it covers an area of sky six times the size of the full Moon.

To the south, we find Beta Persei, arguably the most famous variable star in the entire night sky. It's name is Algol, from the original Arabic, Ras Al Ghul, which means "the head of the demon." In older star charts, Perseus is often depicted as holding the severed head of Medusa, the serpent-haired Gorgon whose gaze could turn all living creatures to stone. Algol marks the eye of Medusa and is still appropriately known as "the demon star."

Algol typically shines at a respectable magnitude 2.1 but will reliably (and predictably) fade to magnitude 3.4 every 2 days, 20 hours and 49 minutes. It remains subdued for about ten hours and then begins to brighten again. A triple star system, this variability is caused by a fainter companion passing in front of the brighter primary, thereby causing the star to apparently dim. Algol was the first of its kind to be discovered.

Close to Algol (and within the same binocular field of view) lies Messier 34, a bright open cluster reminiscent of the Beehive. It can be easily seen as a faint, misty patch through binoculars but is best observed through a small telescope; it appears as an elongated X and a number of double stars are also revealed within it. The cluster spans some 7 light years in space and lies at about 1,500 light years away.

If you like Messier 34, you'll love the Double Cluster. This isn't a misnomer; found midway between Eta Persei (Miram) and Ruchbah in Cassiopeia, these two clusters can be seen very close together with just your eyes under dark skies and are great in binoculars.

Through 10x50's I've seen curved lines to the north and south, giving the cluster the form of a bow tie or butterfly. A small telescope at low power will provide simply stunning views.

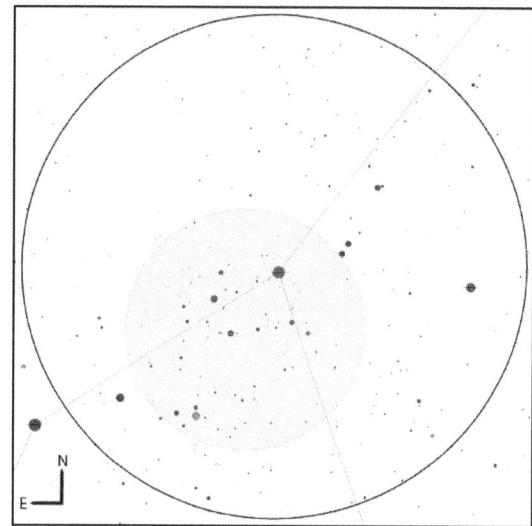

Mirfak and the Alpha Persei Cluster, binocular view

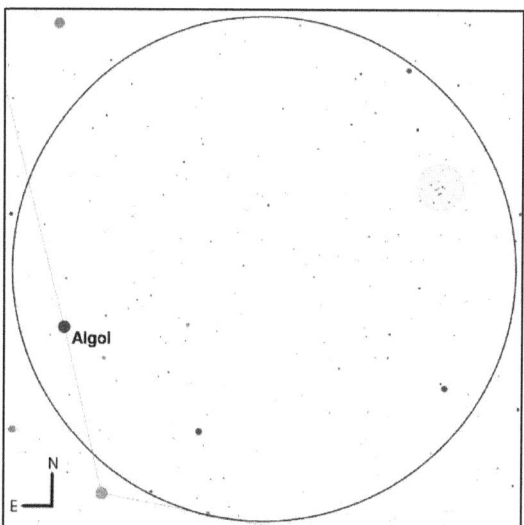

Algol and Messier 34, binocular view

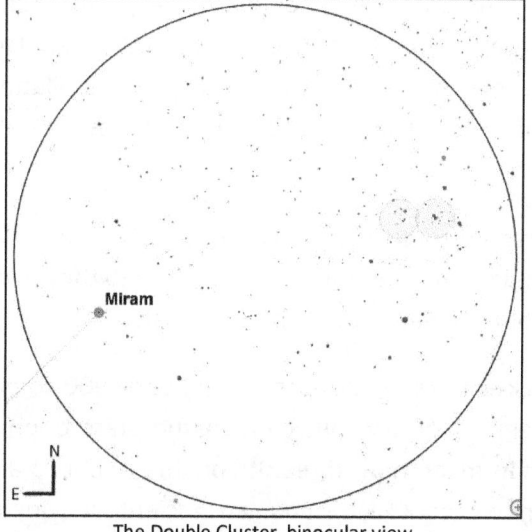

The Double Cluster, binocular view

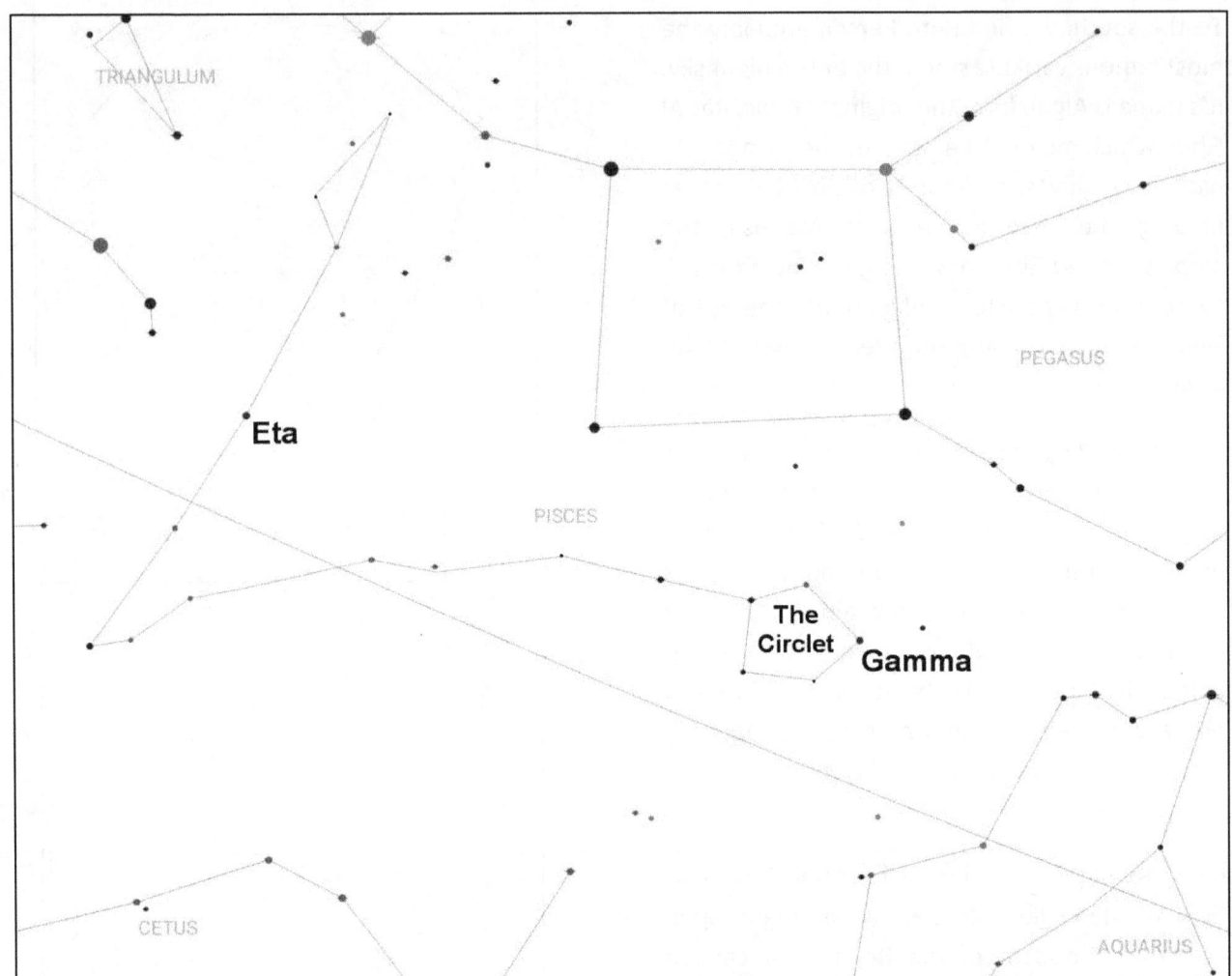

Pisces

Pisces is one of the large, but faint, watery constellations of autumn and depicts two fishes tied together. In Greek mythology, these fishes were Aphrodite and Eros who, like Pan (see Capricornus, another autumn constellation) threw themselves into a river to escape the monster Typhon.

Transforming themselves into fishes, the two tied themselves together with rope so they could not be separated and become lost.

Pisces is 14th in size, covering nearly 900 square degrees of sky, but contains no stars brighter than magnitude three. Its brightest, Eta, is also known as Kullat Nũnu, an ancient Babylonian name meaning "the cord of the fish." A yellow giant, it's about 26 times more massive than the Sun and is approximately 316 times more luminous. It lies just under 300 light years away.

There are no bright deep sky objects in Pisces but it does have at least one famed feature. Below the Great Square of Pegasus is a circle of stars, giving this asterism the appropriate name "the Circlet."

It's too large to fit within the field of view of 10x50 binoculars and you'll need to be under dark skies to discern the Circlet with just your eyes. At magnitude 3.7, the brightest star in the circle, Gamma, is also the second brightest in Pisces as a whole.

Messier 19 in the constellation Ophiuchus. Image by the author using Slooh. (www.slooh.com)

The Double Cluster in the constellation Perseus. Image by the author using Slooh. (www.slooh.com)

The Lagoon Nebula in the constellation Sagittarius. Image by the author using Slooh. (www.slooh.com)

The Trifid Nebula in the constellation Sagittarius. Image by the author using Slooh. (www.slooh.com)

The Pleiades in the constellation Taurus. Image by the author using Slooh. (www.slooh.com)

The Dumbbell Nebula in the constellation Vulpecula. Image by the author using Slooh. (www.slooh.com)

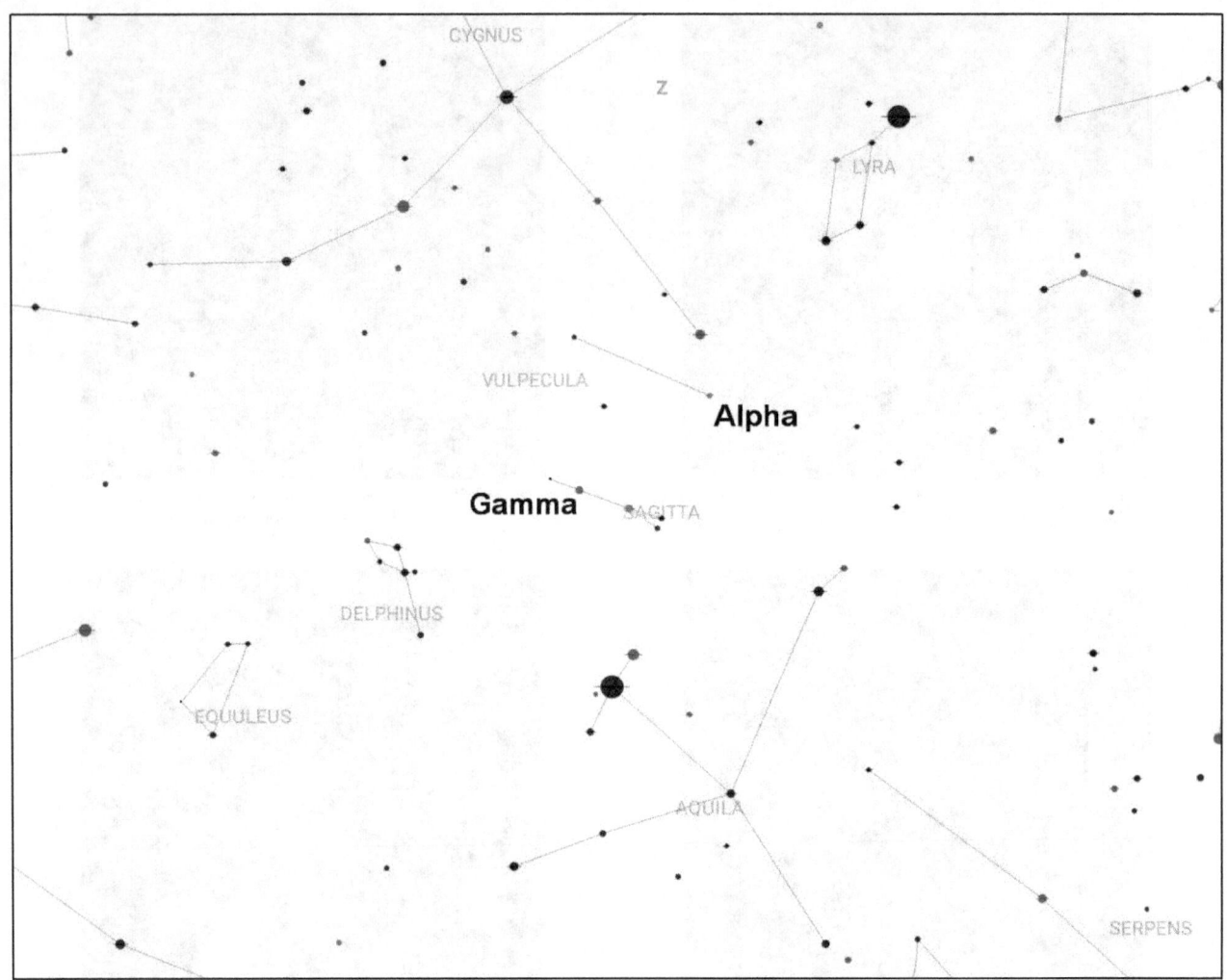

Sagitta and Vulpecula

Sagitta, the Arrow, is an ancient Greek constellation and is thought to represent the arrow that Hercules shot to kill the eagle Aquila. Appropriately, both those characters also appear nearby in the sky; Hercules appears to the west and Aquila appears to the south.

Ranked 86th in size, it's the third smallest constellation in the entire night sky with only Delphinus, the dolphin and the bright southern constellation Crux, the Cross, being smaller. Perhaps not surprisingly, it can quite comfortably fit into the field of view of a regular pair of 10x50 binoculars.

Unfortunately, it's not a particularly bright constellation but can still be easily found thanks to its close proximity to both Cygnus and Aquila. Looking midway between Albireo and Altair should help you to locate it.

Its brightest star is Gamma, a magnitude 3.7 red giant star that lies about 250 light years away.

Sagitta might not be a particularly noteworthy constellation by itself, but it can help us to find a number of deep sky objects in the area.

By placing Gamma near the lower left of your binocular field of view and with Delta just to the lower right, you can catch two deep sky objects in the same binocular field of view.

First, look slightly further north for a small, very faint, circular misty patch. This is the Dumbbell Nebula, and like the Helix Nebula (see page 106) it's a shell of gas thrown off by a dying star. It's about 1,400 light years away and is estimated to be roughly 1½ light years in diameter.

If your skies are dark and clear, you might also catch a glimpse of Messier 71, a small magnitude 7.3 globular cluster nestled between Gamma and Delta.

By placing the two end stars of the arrow (Beta and Alpha, also known as Sham) on the left edge of your field of view, you should easily be able to spot the Coathanger Cluster. You can't miss it; it's large, it looks exactly like its name and it looks great in binoculars.

Visible to the unaided eye under dark skies, this is not a true cluster at all, but merely a chance alignment of stars at varying distances across space. It was first noted by the Persian astronomer Al Sufi in 954 A.D. and has proven to be popular ever since.

Now with the Coathanger at the bottom of our field of view, we can catch a nice double almost due north. This is Alpha Vulpeculae, sometimes known as Anser, the brightest star in the constellation of Vulpecula, the Fox.

A red giant star 300 light years away, it appears as a single magnitude 4.4 star to the unaided eye. But binoculars will show an unrelated star of magnitude 5.8, just to the north-west. To me, Alpha appears a creamy color and about 1½ times brighter than its white companion.

Vulpecula is a relatively modern constellation, invented by the astronomer Johannes Hevellius in the 17th century. Originally known as the Fox and the Goose, it seems only the fox has survived!

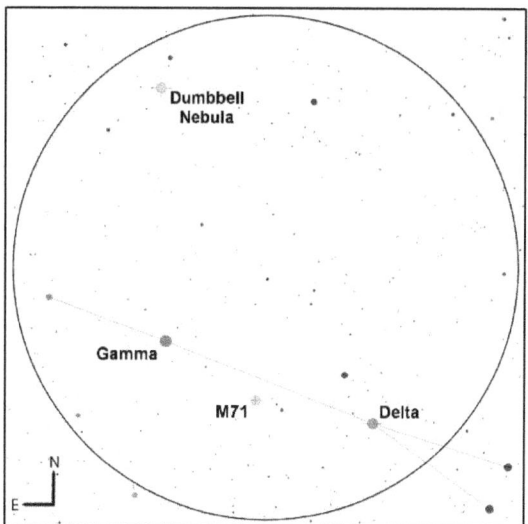
The Dumbbell Nebula and M71, binocular view

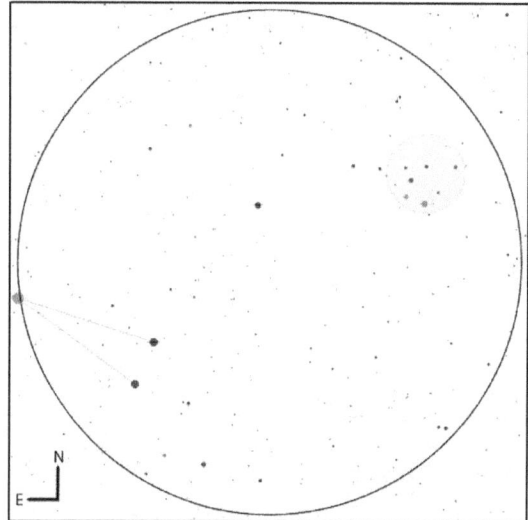
The Coathanger Cluster, binocular view

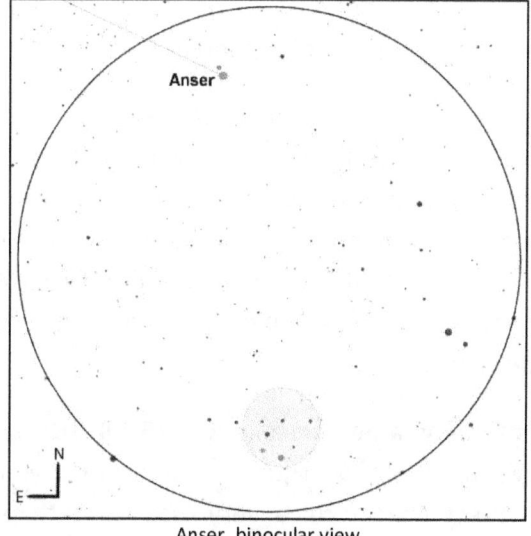
Anser, binocular view

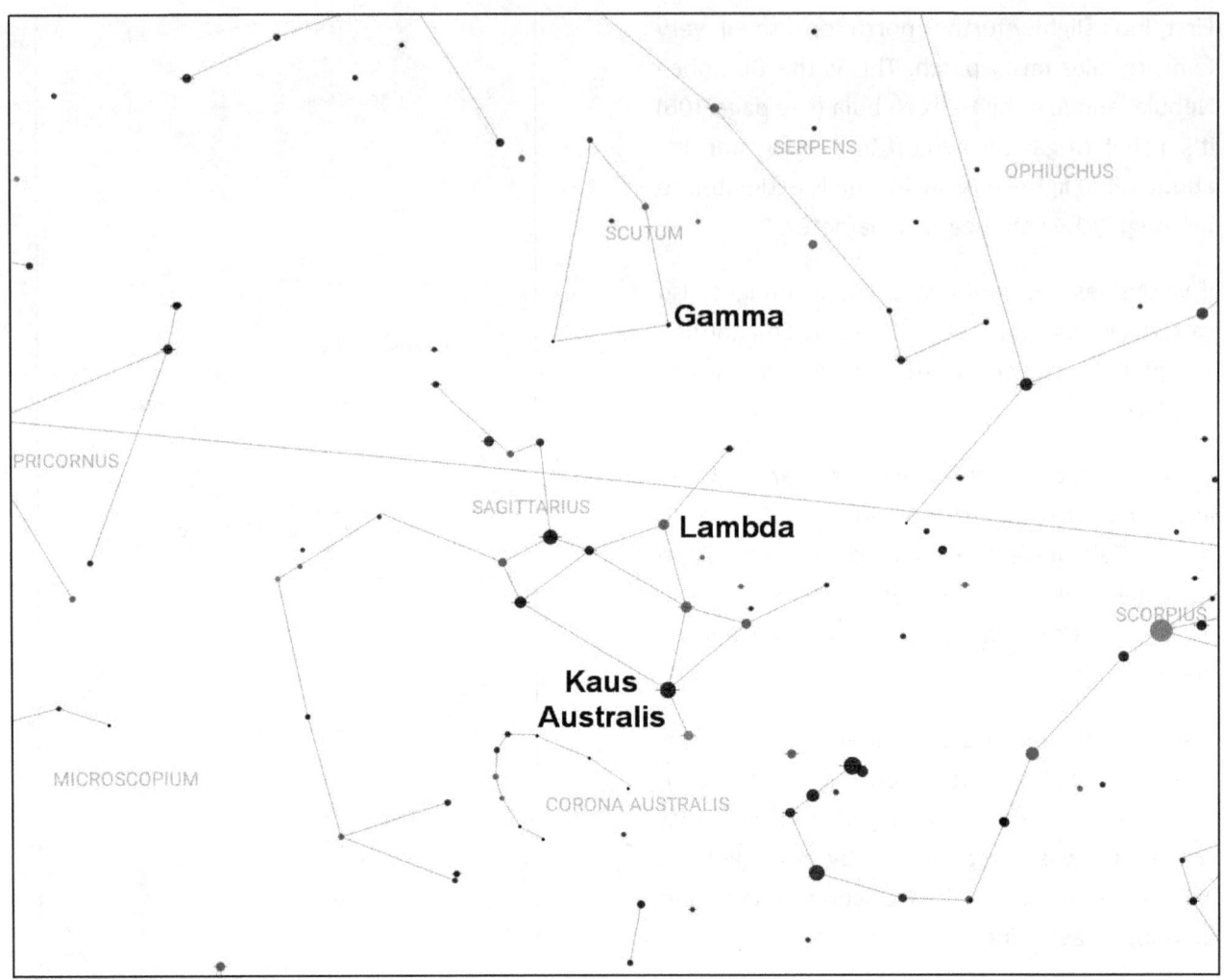

Sagittarius

Sagittarius, like Scorpius just to the west, is one of the larger and brighter constellations of the zodiac but is not well-placed for observation from the northern hemisphere.

The constellation is often associated with Chiron, a half-man, half-horse centaur who was the tutor of Jason. Jason, of course, was the hero who embarked on a quest to find the Golden Fleece with his Argonauts. It's said that Chiron invented the constellation to help guide the Argonauts on their quest.

It's a fairly large constellation, 15th in size and covering over 850 square degrees of sky. Its brightest stars form a very distinctive teapot shaped asterism that's easily identified – if the constellation is high enough above the horizon.

Unfortunately, that can be a problem. Sagittarius barely rises over the horizon for much of the northern hemisphere. But for those lucky observers in the southern hemisphere it's a spectacular sight throughout their winter months with the constellation passing directly overhead for many observers.

Its poor visibility is made all the more unfortunate because of one important fact: the Milky Way passes right through the constellation. Its brightest portion appears right above the spout of the teapot and is sometimes compared to steam escaping the brewing pot!

You'll need to get far away from any city lights to appreciate the view, but once you do, it's a truly breathtaking sight. When you look toward this region of the sky, you're actually looking directly into the heart of our galaxy and, not surprisingly, Sagittarius is rich with deep sky objects as a result. Some are easily seen with binoculars

Its brightest star is Epsilon, known as Kaus Australis, a name derived from a combination of Arabic and Latin and meaning "southern bow." It's a magnitude 1.9 blue-white star that's over three times more massive than the Sun and about 145 light years away.

To the north-west, just above the spout of the tea-pot, we find several nebulae that are relatively easily seen through binoculars from dark sky locations. With Lambda Sagittarii on the edge of the field of view, look out for Messier 8, the Lagoon Nebula on the opposite side.

The nebula can be faintly seen and appears to run in an east-west direction with a tight line of stars within it.

Messier 20, the Trifid Nebula, also appears within the same field of view but is smaller and fainter. In contrast to the Lagoon, it appears to run in a north-south direction and has a tiny group of stars to the north.

Look carefully at the view; can you see Messier 21, an open cluster to the north-east of the Trifid? And what about Messier 28, the globular cluster close to Lambda Sagitarii?

Lastly, try your hand at Messier 17, the Swan Nebula. It's a little out in the wilderness but can be found in the same binocular field of view as Gamma Scutum, at the bottom of the constellation. Under good conditions, it can appear as an elongated number two and is very nicely seen with a small telescope.

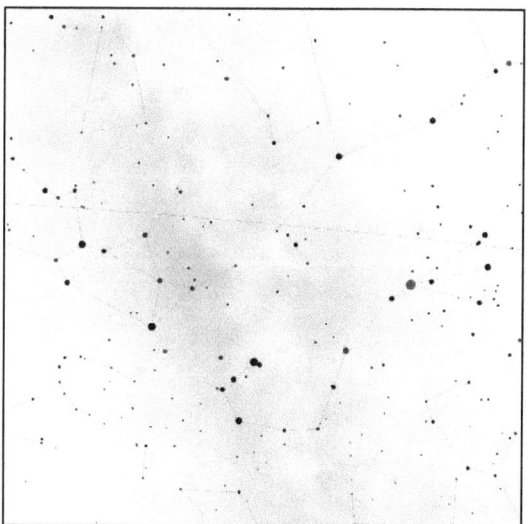

The Milky Way through Sagittarius

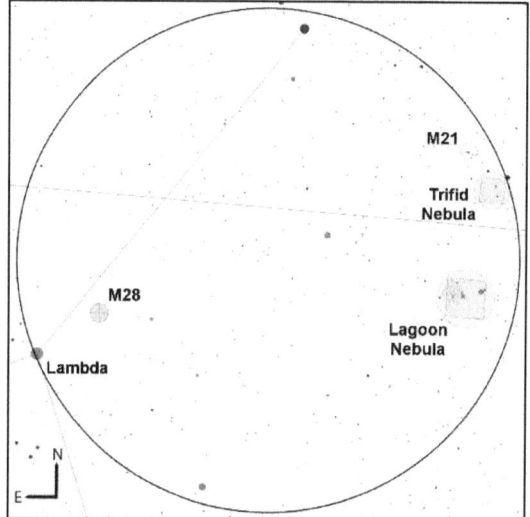

The objects close to Lambda, binocular view

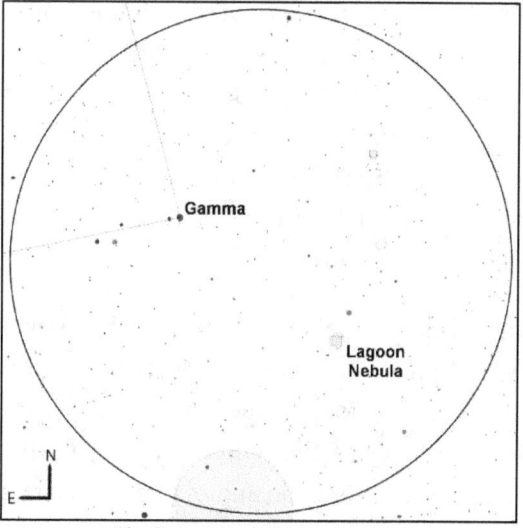

The Swan Nebula, binocular view

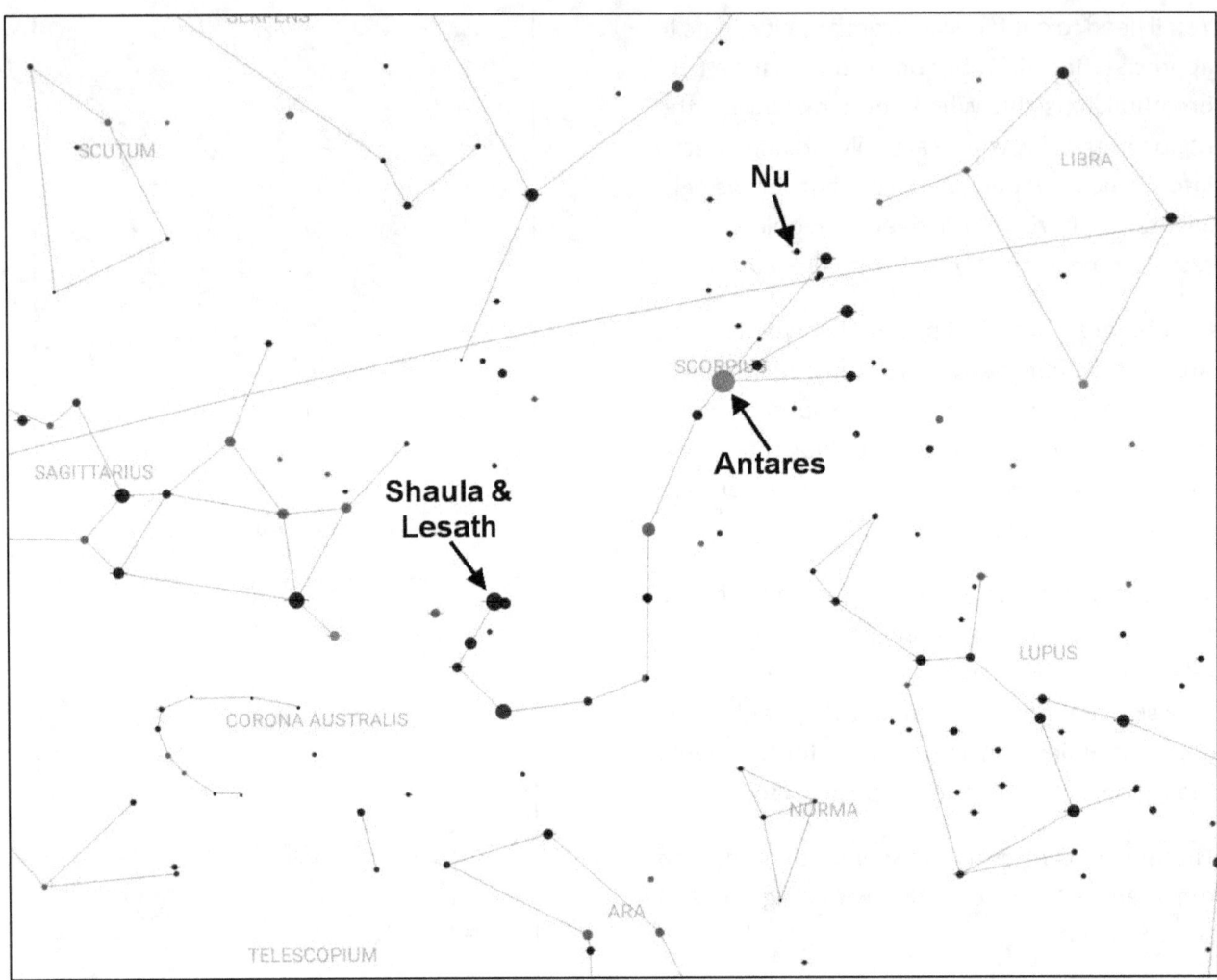

Scorpius

One of the brighter constellations of the zodiac, Scorpius is best seen from the southern hemisphere and never rises completely from the United Kingdom and much of northern Europe. In Greek mythology, Scorpius depicts the scorpion that killed Orion and was consequently placed opposite the hunter in the sky. (There are a number of variations of the story but see page 158 for one version.)

A mid-sized constellation, it ranks 33rd in size and covers nearly 500 square degrees of sky. It has a distinctive S shape, slender and slightly slanted with the claws of the scorpion at the northern end of the constellation and reaching toward the west.

Its brightest star is Alpha, also known as Antares and the 15th brightest star in sky. Unlike most stars, the name is ancient Greek in origin and literally means "rival of Ares" – Ares being the Greek name for Mars - because the star has a strong orange-red hue, similar to our neighboring planet. Its place in the constellation marks the heart of the scorpion.

To say that Antares is a large star would be an incredible understatement. A red supergiant, it's nearly 900 times the size of the Sun and if you were to place it at the center of our solar system it would extend beyond the orbit of Mars. It's also about 15 times more massive and shines with the light of about 10,000 suns.

Antares isn't a solitary star either. It has a blue-white companion that's quite luminous in its own right but its light is overwhelmed by Antares, making it difficult to spot unless you have a larger 'scope. The system lies about 550 light years away.

Scorpius is a summer constellation that has the Milky Way running through it toward the east, between the tail of the scorpion and Sagittarius the Archer. Consequently there are a number of deep sky objects within reach of the binocular observer.

While you're staring at Antares, look out for Messier 4, one of the closest globulars in the sky. You'll need dark skies to properly observe the cluster with binoculars, but you might still catch a glimpse.

Easier targets lie toward the west; Nu Scopii has two very faint companions to the south-east but Omega[1] and Omega[2] form a nice double pair of white stars just to the south-west.

Head east toward the tail of the scorpion for two other fine sights. Messiers 6 (the Butterfly Cluster) and 7 (Ptolemy's Cluster) both appear within the same binocular field of view and are easily found close to Shaula and Lesath, the two stars that form the sting of the scorpion.

Of the two, Messier 7 is larger, brighter and has a close core of possibly hundreds of stars. A line of three stars to the west mimics the head of Scorpius itself.

Despite being half the size, Messier 6 is not without its charms either. Through a pair of 10x50 binoculars it appears compact and forms a triangular shape that points to the east with a bright star on the eastern point. Both M6 and M7 are stunning when observed through a small telescope.

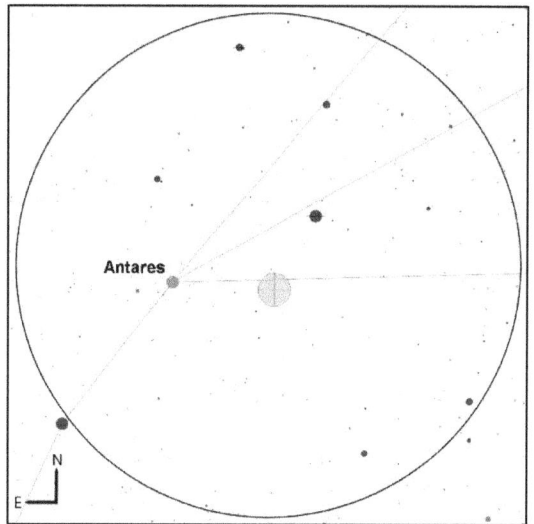

Messier 4, binocular view

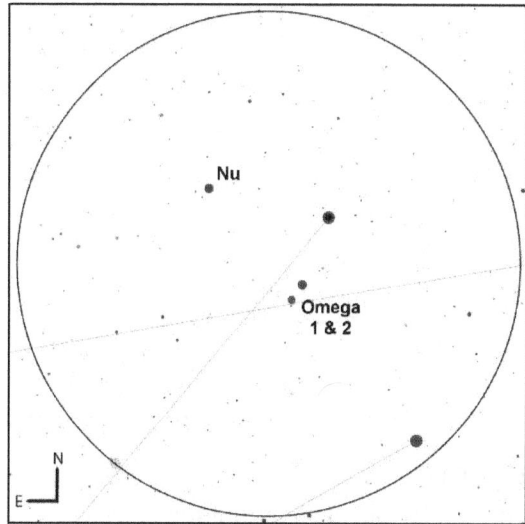

Nu, Omega[1] and Omega[2], binocular view

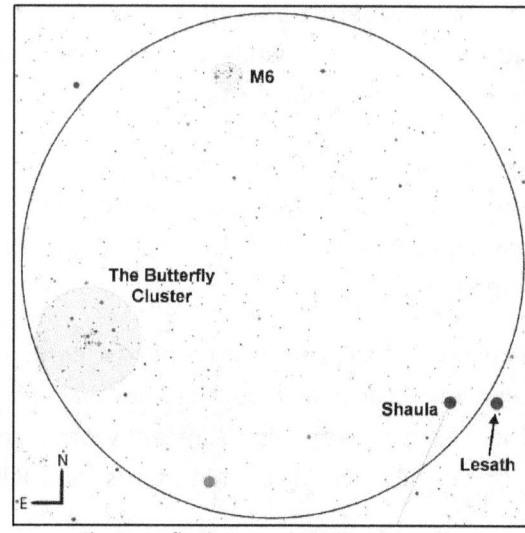

The Butterfly Cluster and M6, binocular view

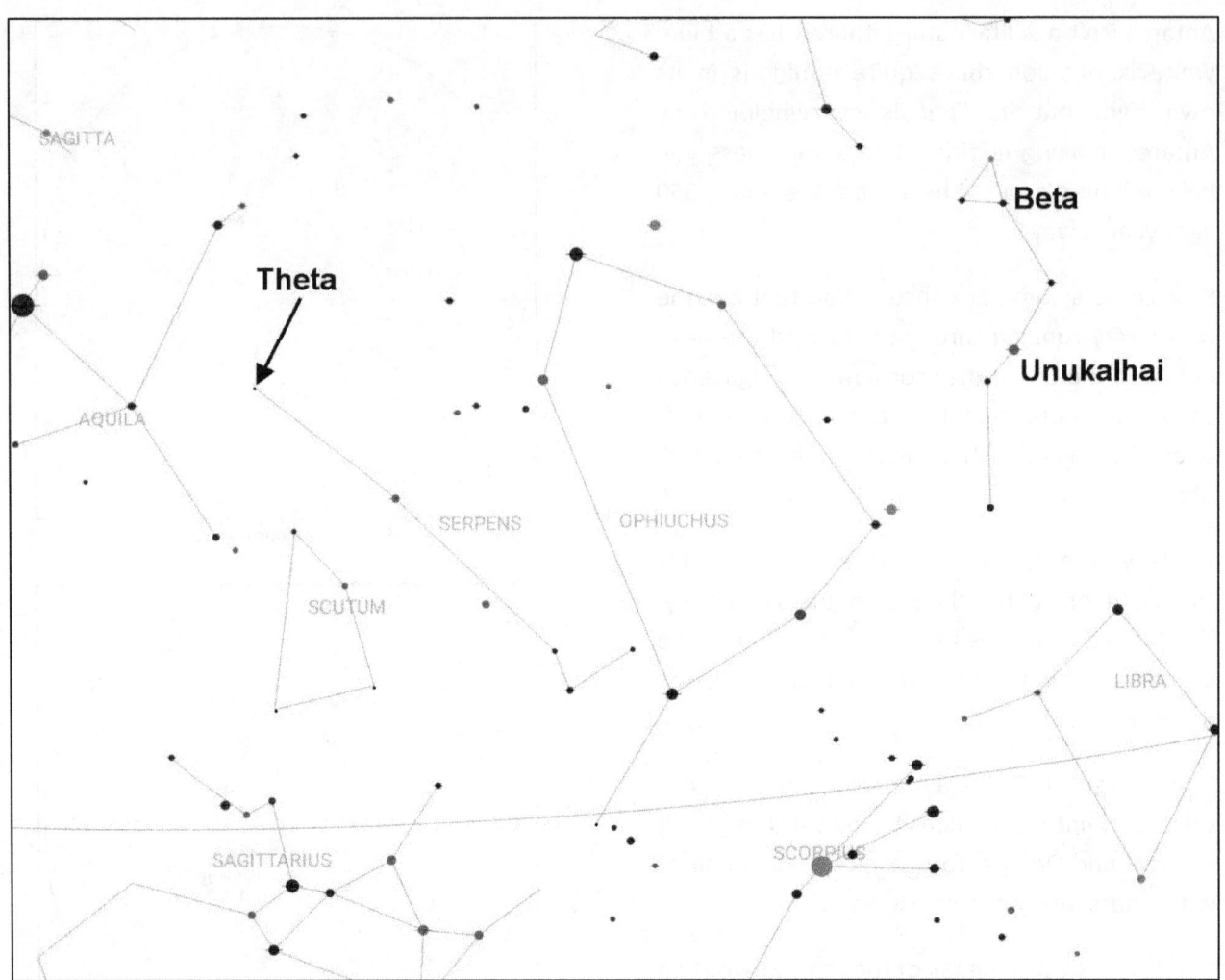

Serpens

Serpens, the Serpent, is unique in that it's the only constellation that's actually divided in half. To the west is Serpens Caput, representing the head of the serpent, while to east is Serpens Cauda which represents its tail.

Between the two halves is Ophiuchus, the appropriately named Serpent Bearer who is often depicted as carrying the serpent as it curls around him.

Ophiuchus is said to represent Asclepius, a healer who learnt to resurrect the dead after witnessing a snake revive a recently deceased serpent using herbs. Given that snakes habitually shed their skins, it's not surprising that snakes were associated with rebirth in ancient Greece.

For many years Serpens did not officially exist and its stars were counted among those of the Ophiuchus himself. It wasn't until the 1920's that the constellation was finally recognized by the International Astronomical Union as it officially defined the constellations.

If you were to combine both halves of the constellation, it would cover more than 600 square degrees of sky, ranking it 23rd in size. It has several reasonably bright stars, with the brightest, at magnitude 2.6, being Alpha.

This star is a part of Serpens Caput, representing the head of the serpent, and sometimes goes by the name Unukalhai, from the Arabic for "the serpent's neck."

An orange giant star, some 75 light years away, it's nearly five times more massive than the Sun and is roughly 38 times more luminous.

Once you find Unukalhai, you can use it to find Messier 5, the constellation's brightest globular cluster. At magnitude 5.7, it lies at the edge of naked eye visibility but sharp-eyed observers may be able to glimpse it under dark skies.

As with all globulars, it's better observed with a telescope but it may still be glimpsed with binoculars. With Alpha on the edge of your field of view, look for 10 Serpentis, a magnitude 5.2 star, on the opposite edge. Place that star in the middle of your field of view and M5 can be glimpsed to the west with 5 Serpentis just slightly to the south-east.

Meanwhile, further north of Unukalhai is Beta Serpentis, one of the stars that forms an asterism representing the serpent's head. A magnitude 3.7 white star, binoculars will reveal a close, fainter, magnitude 6.7 star just to the north of it.

Heading east to Serpens Cauda, we come to Theta Serpentis, a magnitude 4.1 star found just to the west of Aquila, the Eagle. Like Beta, binoculars will also show a close but faint magnitude 6.7 companion just to the north-east.

Lastly, just to the south of Theta and within the same field of view, you'll also see a wider and brighter pair of white stars of slightly unequal magnitude. Can you see a faint star between them? Whether you choose to scan Caput or Cauda with binoculars, either side promises a fascinating and full field of stars.

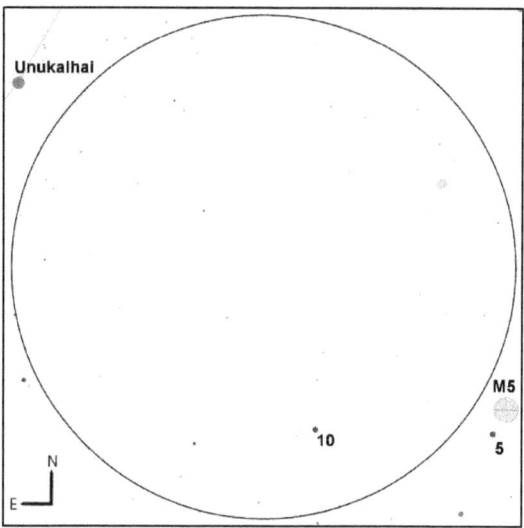

Unukalhai, 10 Serpentis and M5, binocular view

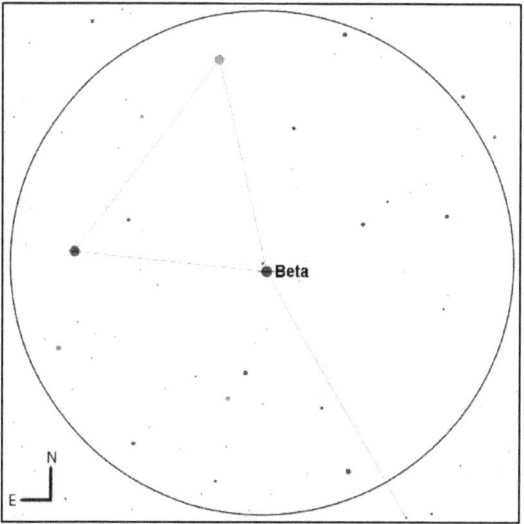

Beta Serpentis, binocular view

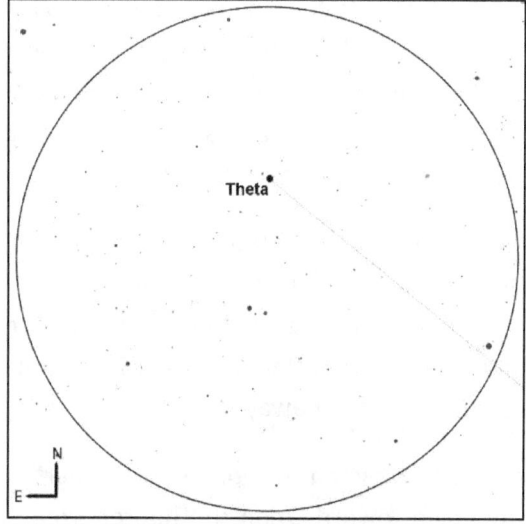

Theta Serpentis, binocular view

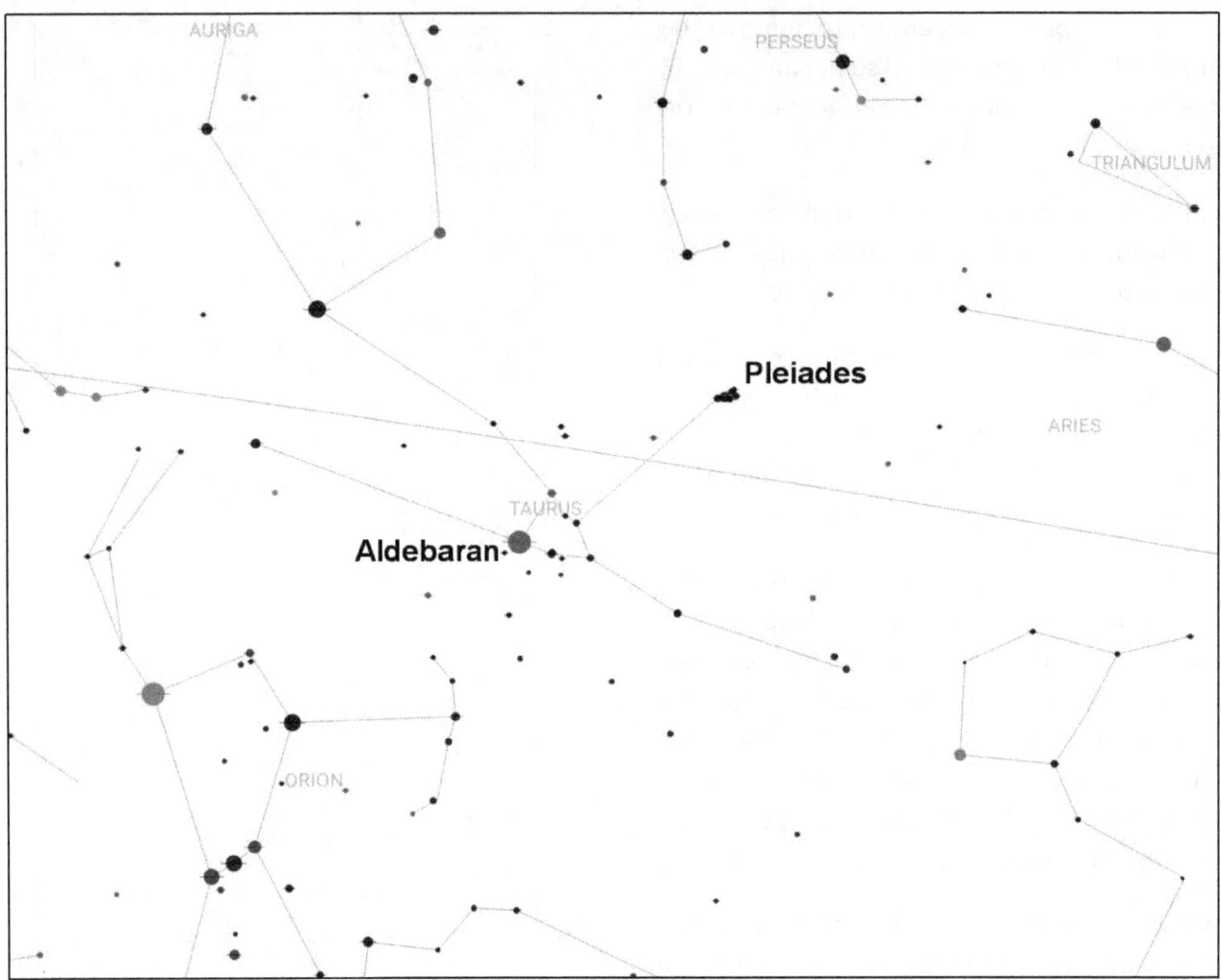

Taurus

Taurus, one of twelve signs of the zodiac, is one of the largest constellations and certainly one of the brightest. It's been known and associated with a bull for thousands of years and across numerous ancient civilizations, including the Mesopotamians, Egyptians and, of course, the Greeks who linked it to several legends.

In one, Zeus became infatuated with Europa, a beautiful princess from Phoenicia. Turning himself into a bull, he hid himself amongst her father's cattle and after she had climbed upon his back, carried her away.

It's often depicted on older star charts as stampeding toward Orion the hunter, who can be seen raising his club and shield to defend himself against it.

Taurus covers nearly 800 square degrees of sky and is ranked 17th in size. Its brightest star is Alpha, more commonly known as Aldebaran, from the Arabic for "the follower." This is a reference to the Pleiades (see below) as the star rises after (and therefore "follows") that bright star cluster.

This famous orange giant star marks the red eye of the bull and, at magnitude 0.9, is the 14th brightest star in the sky. It's a relatively close star, only 65 light years away, and like many giant stars, could easily swallow the Sun. In fact, despite being about 45 times the radius of the Sun it is only about 50% more luminous.

You'll notice that Aldebaran appears at the tip of a V shaped cluster of stars, known as the Hyades. This cluster represents the head of the bull and can be easily observed with the unaided eye, but the best way to enjoy them is with a pair of binoculars. It presents a very attractive sight through a pair of 10x50's and nicely fills the view.

Scanning the area will reveal a multitude of blue-white stars, including several doubles and most notably Theta Tauri. As a challenge, try spotting this double with just your eyes.

Although Aldebaran appears at the edge of the Hyades, it's not actually a member of the group as the cluster is about 90 light years further away.

Taurus is home to another well-known open cluster, the Pleiades or Seven Sisters. Easily seen with the naked eye, these stars have been noted and observed by civilizations for thousands of years.

They appear as a tiny group of blue-white stars, like a miniature Delphinus - but how many can you see with only your eyes? Despite being also known as the Seven Sisters, many people will only see six stars, leading astronomers to wonder if one of them has faded over the intervening millennia. (See Ursa Major for a possible answer!)

Like the Hyades, this cluster is best observed with binoculars and a good pair of 10x50's can provide a stunning view – better, in fact, than a telescope, because the entire cluster will easily fit within the field of view. The Pleiades is one of the not-to-be-missed night sky sights and there are very few astronomers who'll go through the night without giving it at least a passing glance!

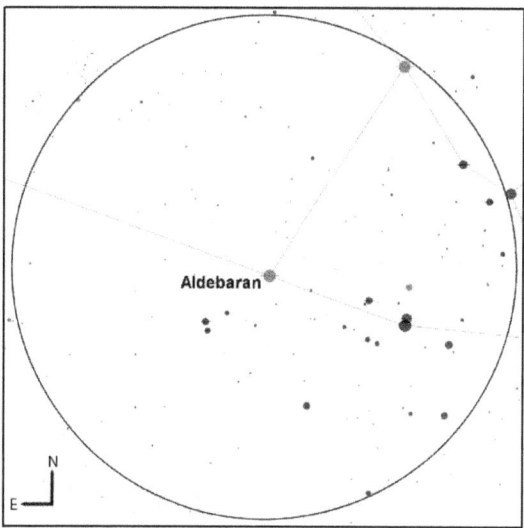

Aldebaran, binocular view

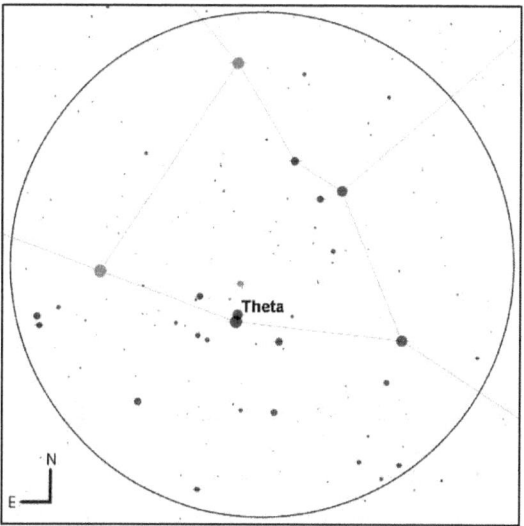

The Hyades, binocular view

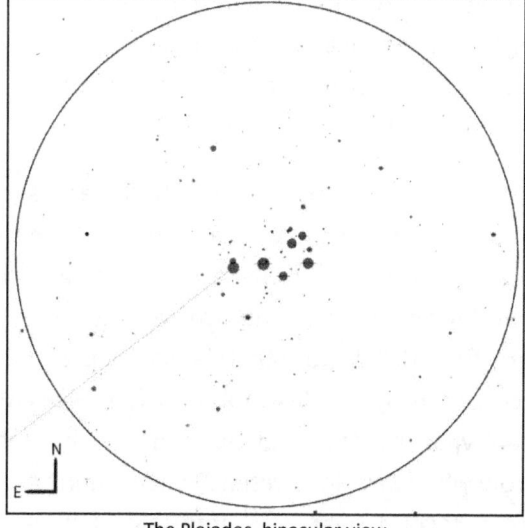

The Pleiades, binocular view

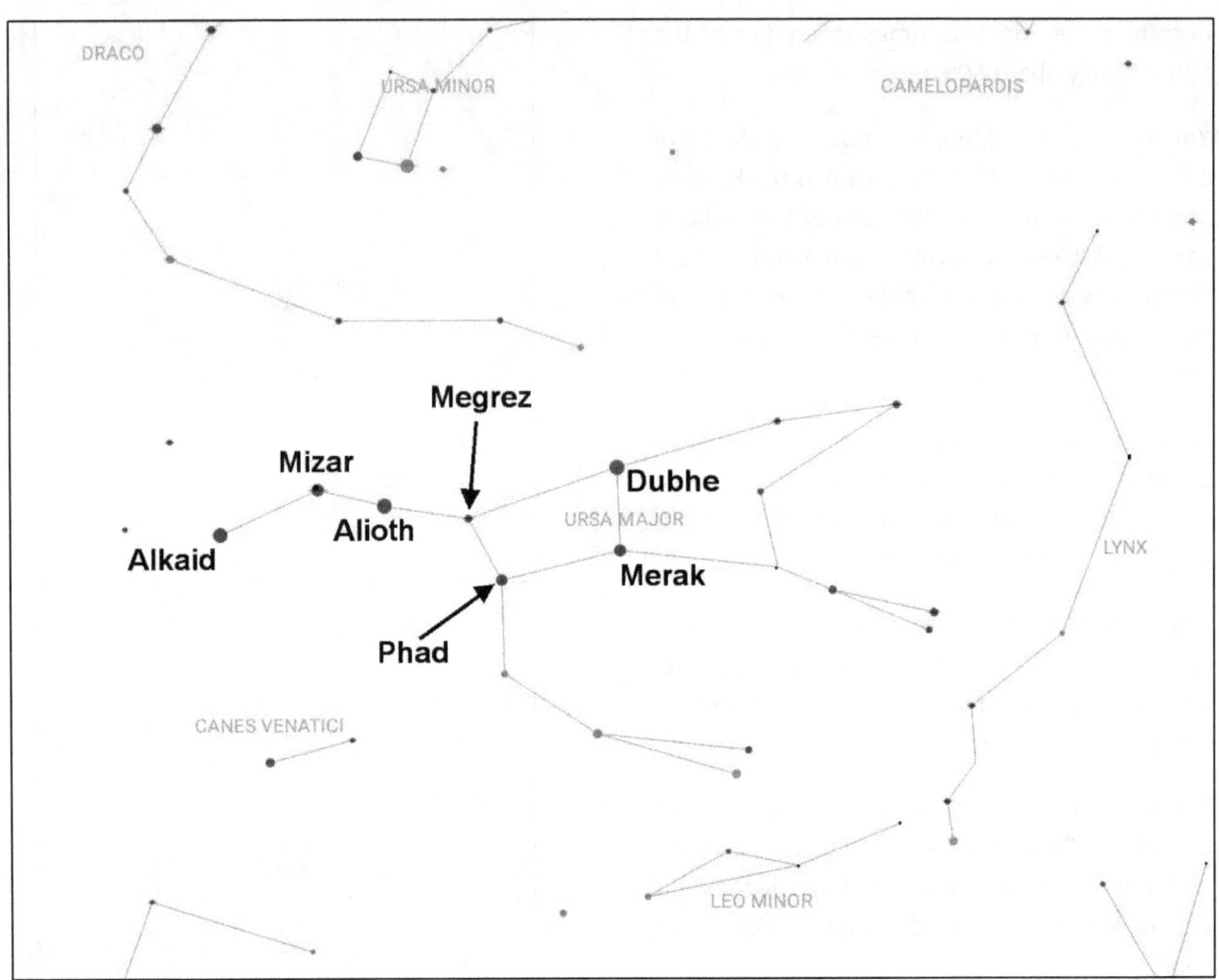

Ursa Major

Ursa Major, the Great Bear, is one of the best known constellations in the entire night sky and is also one of the largest. Third only to Hydra and Virgo, it covers nearly 1,300 square degrees and its stars have been linked to myths and legends across a number of cultures around the world.

To the Greeks, the constellation represents Callisto, a beautiful woman and another of Zeus's potential love interests. Learning of his intentions to seduce the woman, Zeus's wife Hera turned Callisto into a bear. Callisto's son, Arcas, unaware of his mother's fate, sees the bear while hunting and is about to shoot his arrow when Zeus intervenes. Transforming Arcas into a second bear, he immortalized them by placing them among the stars in the sky. Arcas becomes Ursa Minor, the Little Bear while his mother is now Ursa Major.

Its seven brightest stars are circumpolar, meaning they never dip below the horizon for many observers in the northern hemisphere and appear to circle about the pole instead.

These stars form a famous asterism known as the Plough in the United Kingdom and the Big Dipper in North America. This asterism can be used to find other bright stars and constellations (see page 48) and even many non-astronomers know the westernmost stars (Merak and Dubhe) point to the pole star, Polaris.

Its brightest star is Epsilon, also known as Alioth, curiously derived from the Arabic for "fat tail of a sheep!" This magnitude 1.8 star is about three times more massive than the Sun, lies at about 82 light years away and may have an undiscovered low mass companion.

However, almost without question, the most interesting sight out of the seven brightest stars is Zeta, also known as Mizar, from the Arabic for "waistband." This magnitude 2.3 star has a close magnitude 4.0 companion, known as Alcor, that should be a relatively easy target for observers without optical aid – even under suburban skies.

A number of ancient cultures have legends and stories associated with them. To the Greeks, Alcor represented the lost Pleiad, Electra (see page 174), which helps to explain why, despite being known as the Seven Sisters, many observers can only see six stars in the Pleiades star cluster.

The pair make for a nice sight in binoculars with Mizar appearing about twice as bright as Alcor. A small telescope at low power will reveal that Mizar itself can be split into two components. In reality, Mizar and Alcor are unrelated with Alcor being about 82 light years away and Mizar being about five light years further.

Now find the end of the great bear's tail, marked by the magnitude 1.9 star, Eta. Better known as Alkaid, this star serves as a useful marker to help you find Messier 51, the Whirlpool Galaxy.

Technically, this belongs in the constellation of Canes Venatici but it's easiest found by starting with Alkaid. Look out for 24 Canum Venaticorum; it forms a triangle with Alkaid and M51, which should appear as a small, faint misty circle within the same binocular field of view.

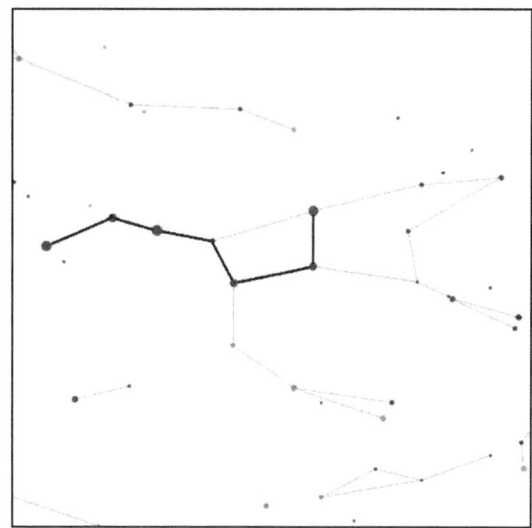
The Big Dipper/Plough asterism

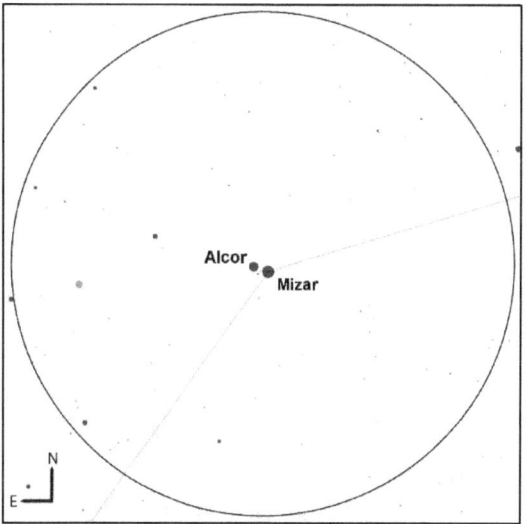
Mizar and Alcor, binocular view

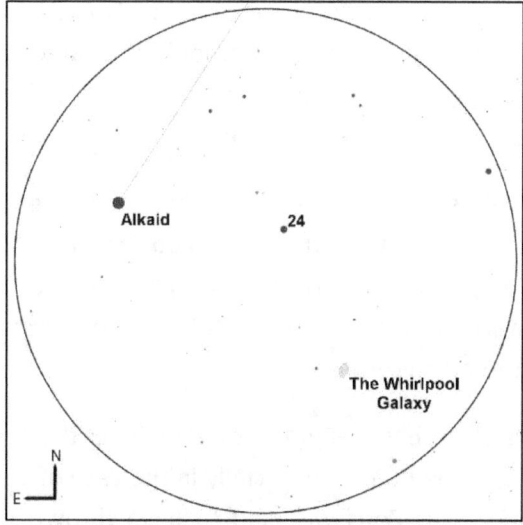
The Whirlpool Galaxy, binocular view

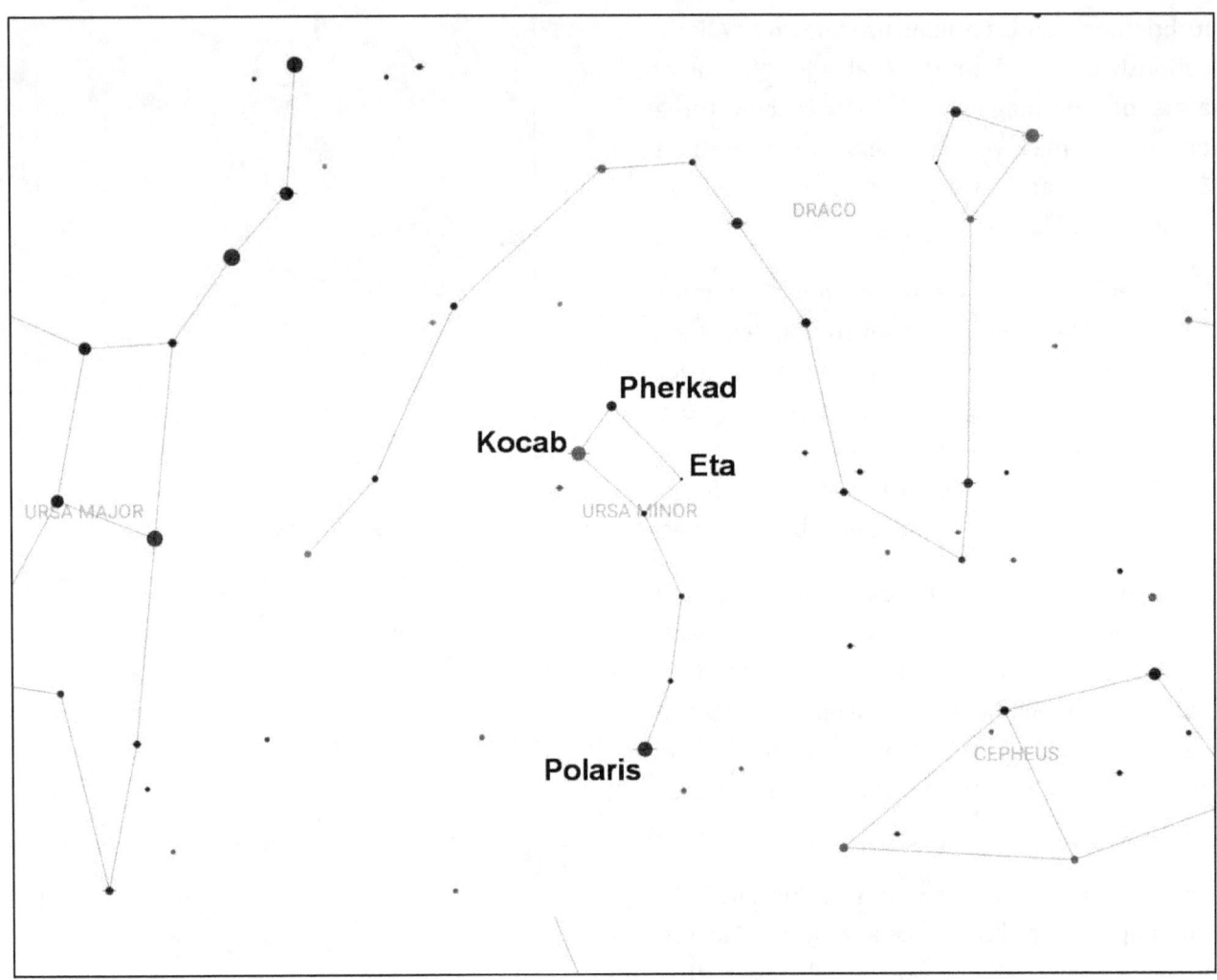

Ursa Minor

Less well known than Ursa Major is its smaller counterpart, Ursa Minor, the Little Bear. Also known as the Little Dipper in North America, the constellation is most commonly linked to the legend of Callisto and Arcas (see page 176) but there's an alternate version.

In this story, the two constellations represent the bears that hid the infant god Zeus from his father, Cronus. When Zeus grew older, he rewarded the pair by immortalizing them as stars in the night sky.

This story has the virtue of explaining the long tails of the bears – especially in the case of Ursa Minor – as Zeus was said to have thrown the

bears towards the heavens by their tails when he immortalized them.

There's little to see here but, like Pegasus, the constellation can be used to gauge your sky conditions. Of the seven stars that form the constellation, only two (Polaris and Kocab) are brighter than magnitude 3 and four are fainter than magnitude four.

Can you see Eta? At magnitude 4.9, this is the faintest of the group while the other stars range from 4.2 to 4.4.

At the other end of the scale is Alpha, the brightest star in the constellation. As many people know, it's more popularly known as

Polaris because it appears almost directly over the Earth's north pole. It never noticeably moves and always marks north (at least in our lifetime and for the foreseeable future) and has been used as a navigation aid for millennia. You can easily find it by drawing a line through Merak and Dubhe in Ursa Major (see page 176.)

Glowing at a modest magnitude 2.0, Polaris is a double star, split with a small telescope, and slightly variable. It lies about 350 light years away.

Take a look with binoculars; you might not see its true companion, but you should still see another, fainter white star close to it. This star should easily be visible, even from suburban skies.

Both Polaris and this companion form part of a faint circle of stars, which will prove harder to spot from suburbia. Polaris is the brightest star in the circle thereby giving the asterism its popular name of "the engagement ring" with Polaris as its diamond.

After enjoying the view, move down to Beta and Gamma at the bottom of the constellation. Respectively, these are the second and third brightest stars of Ursa Minor and shine at magnitudes 2.1 (Beta) and 3.1 (Gamma.)

With Beta more commonly known as Kocab and Gamma known as Pherkad, these two stars are also sometimes called "the Guardians of the Pole" and can always be seen to circle Polaris from the vast majority of the northern hemisphere. (Only observers at latitudes below 20° north will see the stars disappear below the horizon.)

Lastly, binoculars will reveal a magnitude 5 companion close to Pherkad while sci-fans might consider Kochab (Beta Ursa Minoris) to be the home of *The Hitchhiker's Guide to the Galaxy.*

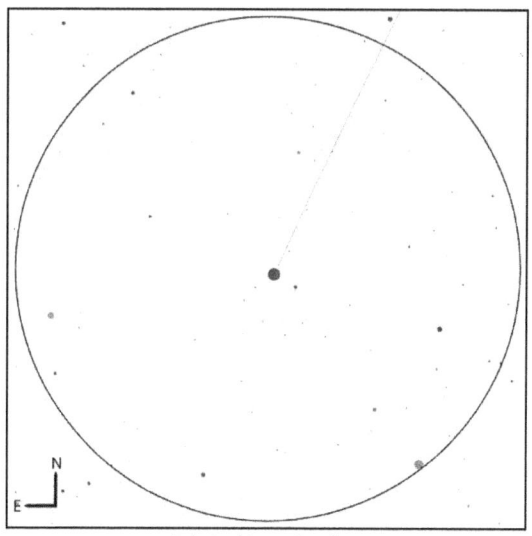
Polaris, binocular view

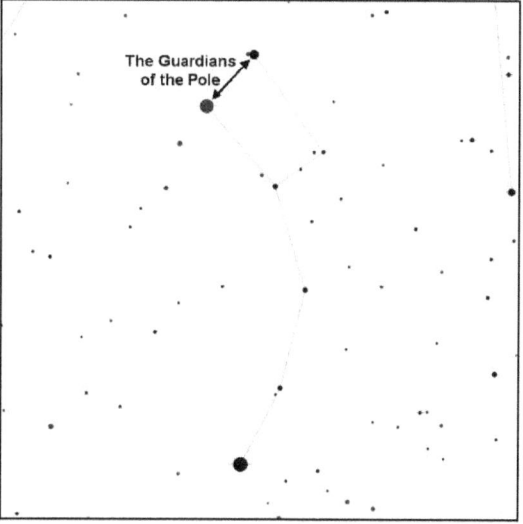
The Guardians of the Pole

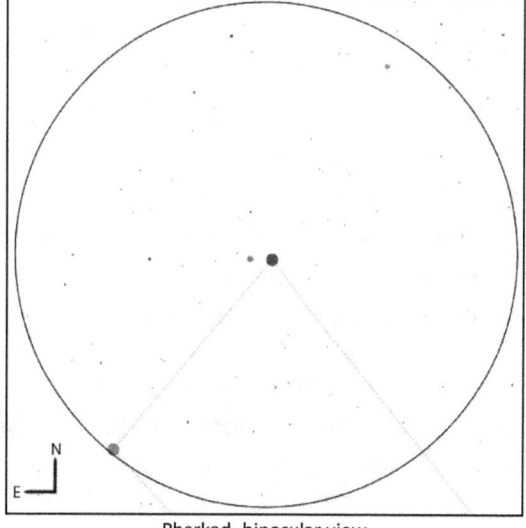
Pherkad, binocular view

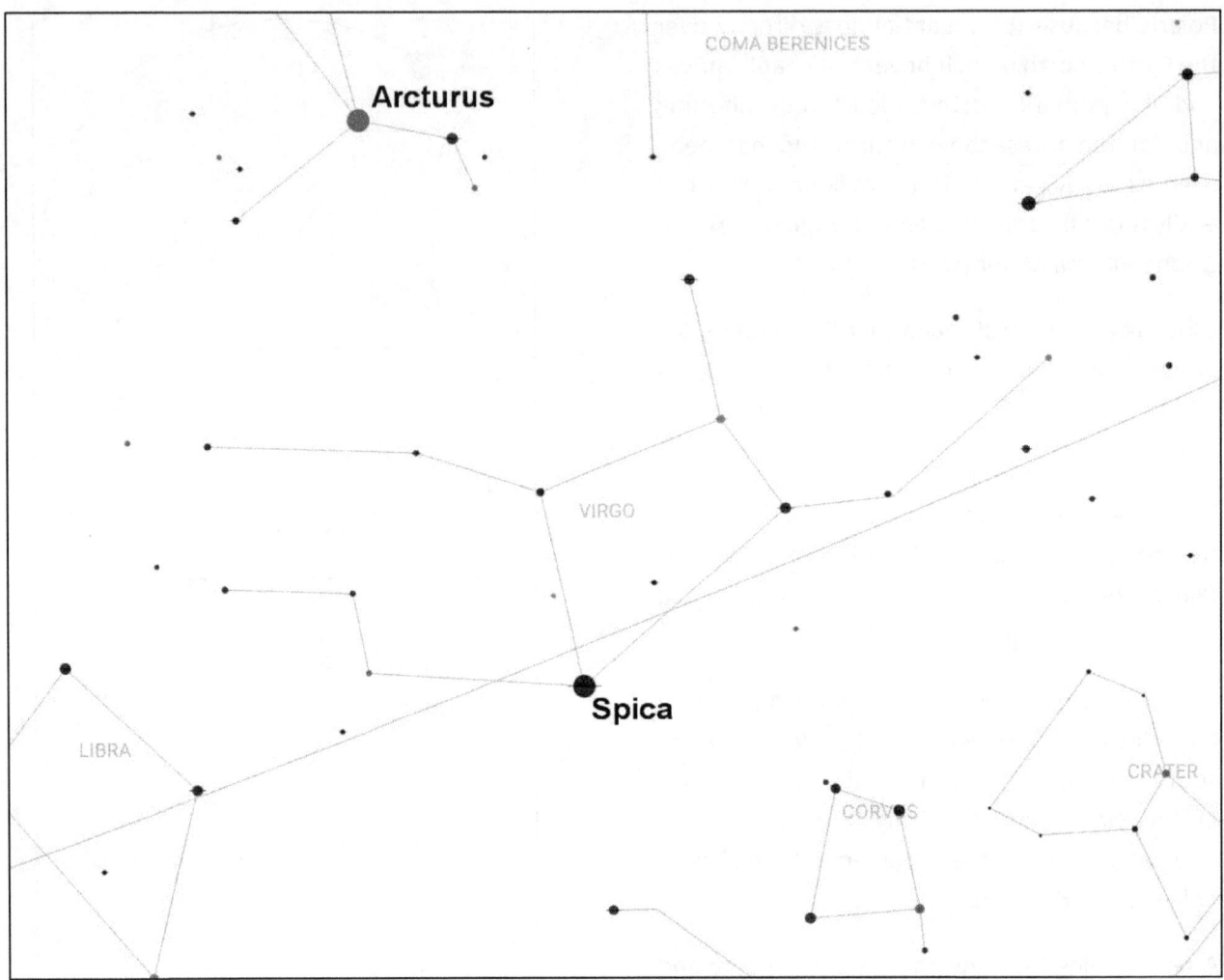

Virgo

Virgo, one of the twelve signs of the zodiac, covers nearly 1,300 square degrees of sky and, after Hydra, is the second largest constellation. (Virgo is only nine degrees smaller.)

Despite being associated with agriculture across different cultures, it has no strong mythological connections. The ancient Greeks linked the constellation to Demeter, the goddess of wheat and harvests, and on older star charts the maiden is depicted as holding an ear of grain.

This depiction is epitomized in the naming of its brightest star, Spica, whose name is derived from the Latin for "ear of grain." The 15th brightest star in the sky, it shines at magnitude 1.0 and is easily found by using the three stars that form the tail of Ursa Major (see the next page for more details and a graphic.)

Spica is a double star system where the components are so close they're actually egg-shaped (and therefore indivisible with amateur telescopes.) This pair of blue giants orbit one another every four days and lie at a distance of about 250 light years.

Virgo is also home to a number of exoplanet star systems and an entire swarm of galaxies. Unfortunately, the galaxies may be hard for binocular beginners to spot and the exoplanets are beyond every amateur's reach!

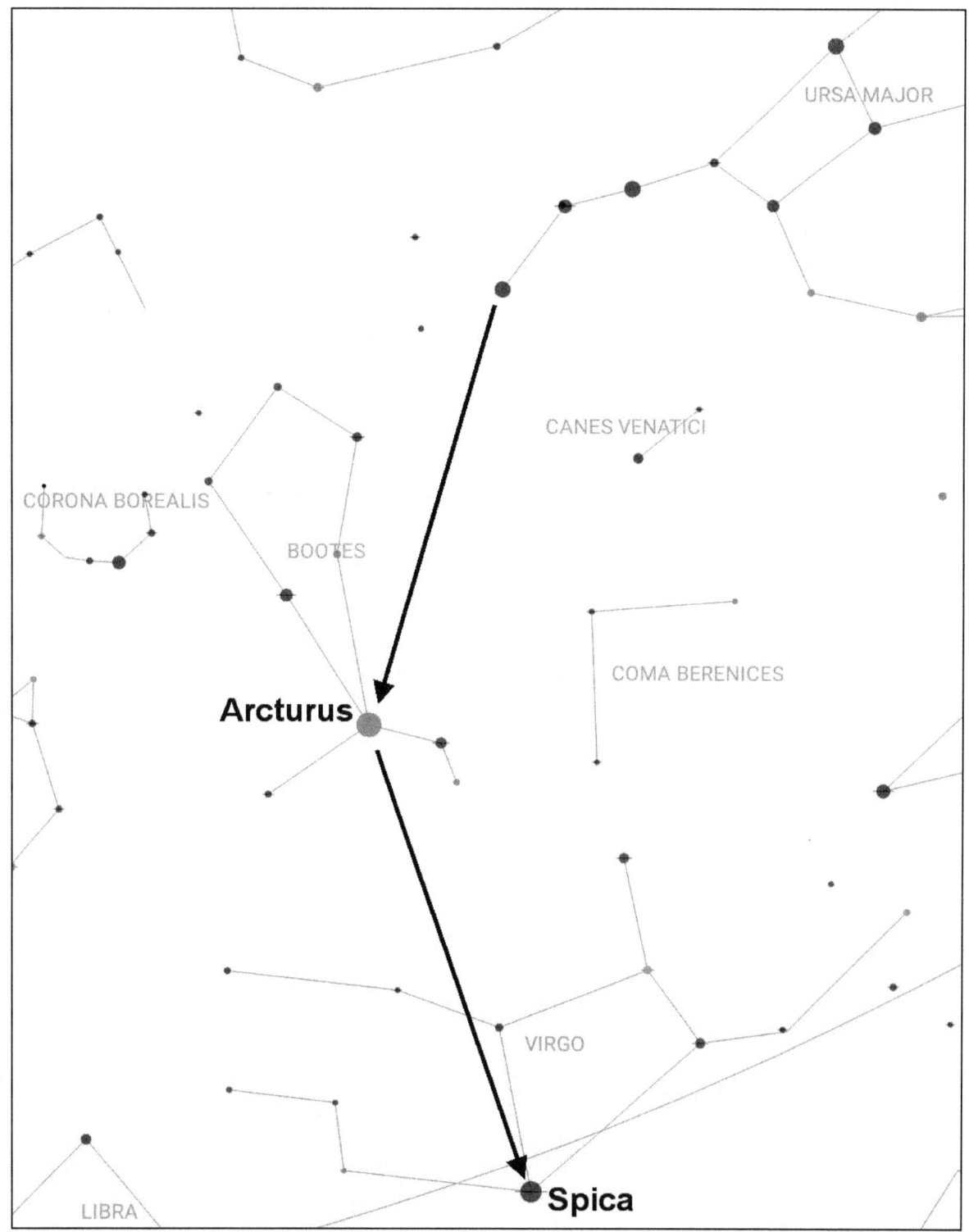

Here's a well-known and handy way to find Spica using the stars of the spring sky. We can follow the tail of Ursa Major, the Great Bear and curve down to ruddy Arcturus, the brightest star in Boötes.

Next, continue the line toward the south until you come to white Spica, the brightest star in constellation of Virgo, the Virgin. Hence the popular phrase "arc down to Arcturus and speed on to Spica!"

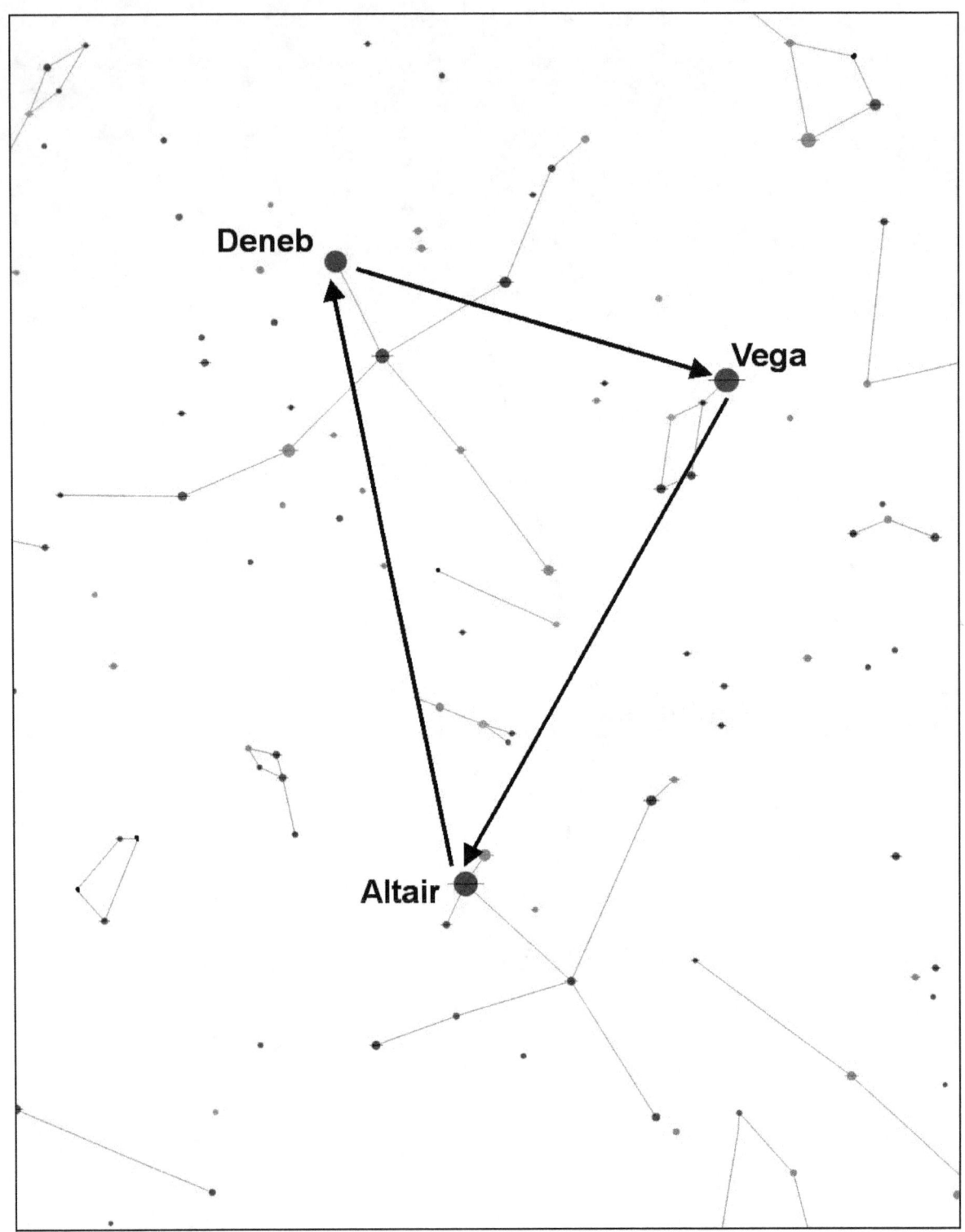

The Summer Triangle is a well known sight for astronomers across the world and features three of the sky's brightest stars. Brilliant Vega in Lyra, the brightest star in the northern celestial hemisphere, shines at magnitude 0.0 and forms the most easterly point of the triangle. Deneb in Cygnus marks it's north-eastern point while Altair in Aquila lies to the south.

Easy Objects for Small Telescopes

Pi Andromedae

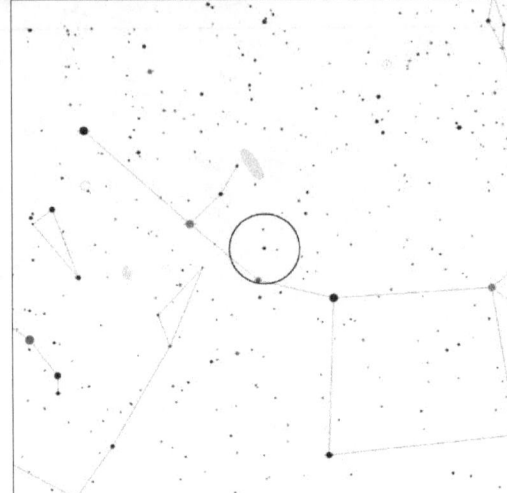

Map courtesy *Mobile Observatory*

Designation(s):	Pi Andromedae
Constellation:	Andromeda
R.A.:	00h 36m 53s
Declination:	+33° 43' 10"
Object Type:	Multiple Star
Location:	★ ★
Rating:	★ ★
Best Seen:	Autumn and Winter

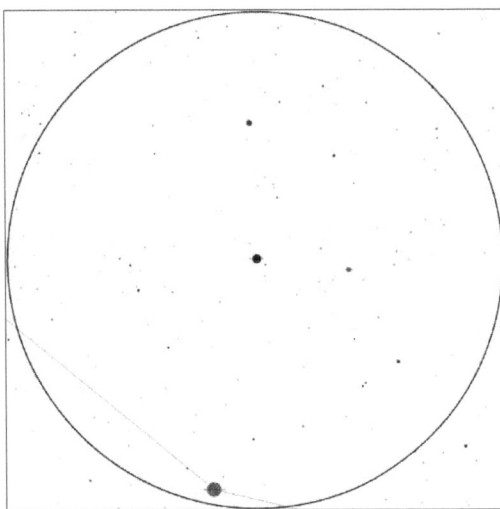

Finderscope view courtesy *Mobile Observatory*

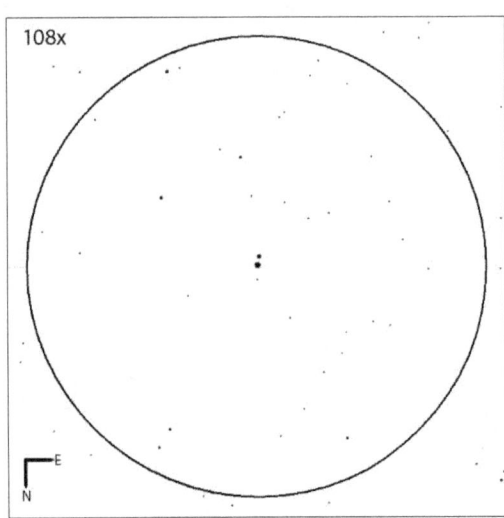

Eyepiece view courtesy *Sky Tools 3*

Pi Andromedae is not one of the brightest stars in the constellation but is still quite easily found from suburban or dark sky locations.

Andromeda is always depicted as being a curved line of stars extending east from the square of Pegasus. However, if you look carefully, you'll see a second curved line of fainter stars of which Pi is the first. (The second, Mu Andromedae, is fairly close to the Andromeda Galaxy.)

Binoculars won't split the star but a small telescope at low power should easily do the trick. At 26x I saw a bright white primary with a very faint bluish secondary. From the city I noted that the secondary appeared to have a coppery color, but I suspect that's more likely due to the city air.

There is a third, much fainter and wider companion but I've been unable to see it from either the suburbs or from the city.

The components we see with our telescopes are not members of a true multiple star system but are simply a chance alignment of stars at differing distances from us. However, the primary is actually a very close pair of nearly identical stars that each shine with the light of a thousand Suns. The pair orbit one another once every 144 days and are approximately 600 light years away.

The Andromeda Galaxy

Designation(s):	Messier 31
Constellation:	Andromeda
R.A.:	00h 42m 44s
Declination:	+41° 16' 09"
Object Type:	Spiral Galaxy
Location:	★ ★
Rating:	★ ★
Best Seen:	Autumn and Winter

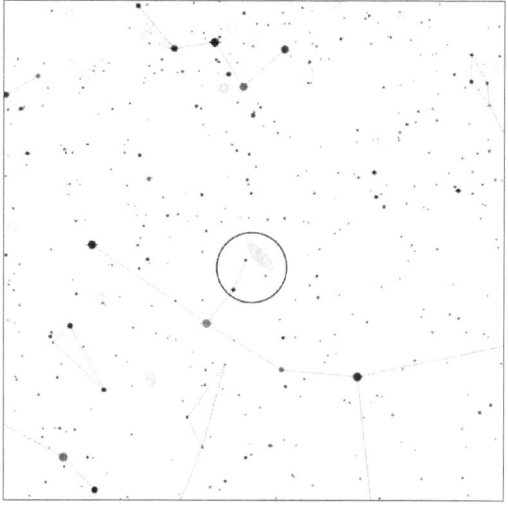

Map courtesy *Mobile Observatory*

Known since antiquity, the Andromeda Galaxy is one of the most fascinating objects you can see – but if you own a small telescope, it can also be one of the most disappointing.

The reason is simple. Although it's visible to the unaided eye under dark skies, its light is spread over a large area and, consequently, it usually doesn't appear that bright (or detailed) through a small telescope.

So what's the fascination? Well, for one thing, you're looking at a whole other galaxy.

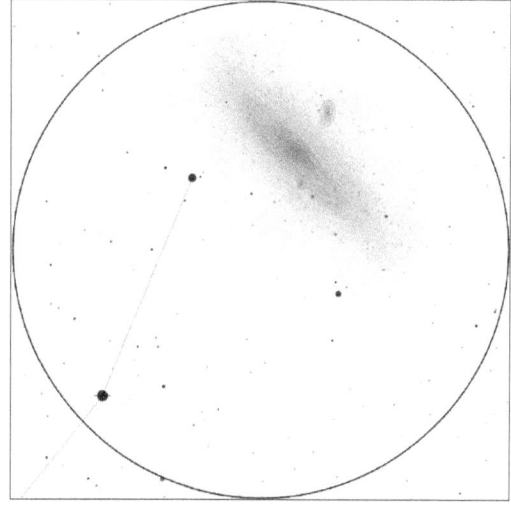

Finderscope view courtesy *Mobile Observatory*

Our own Milky Way galaxy has about 200 billion stars and the Andromeda Galaxy is even bigger. And at a distance of 2 ½ million light years, it's the most distant object you can easily see with just your eyes. It's also worth remembering that, as you stare at it through the eyepiece, the light that's now hitting your eyes first started out millions of years ago.

What does this mean? You are, in fact, looking at the galaxy as it once was. You're literally looking into the past.

Under suburban skies at 35x, it appears as a moderately bright, misty oval but in the city it was hard to see at all. In fact, I had to use averted vision to even see its shape. Otherwise, only the core is visible. Look for it, enjoy it, but don't get your hopes up too much!

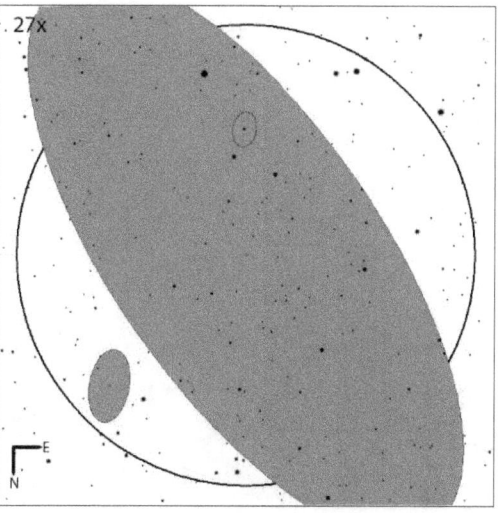

Eyepiece view courtesy *Sky Tools 3*

NGC 752

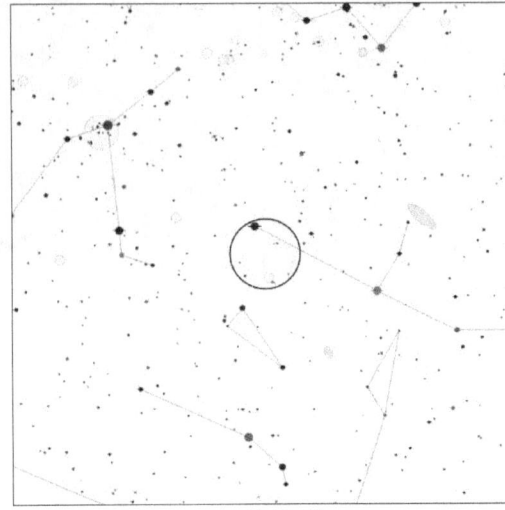

Map courtesy *Mobile Observatory*

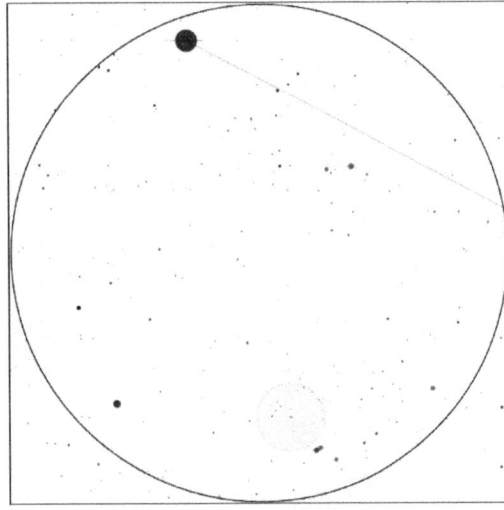

Finderscope view courtesy *Mobile Observatory*

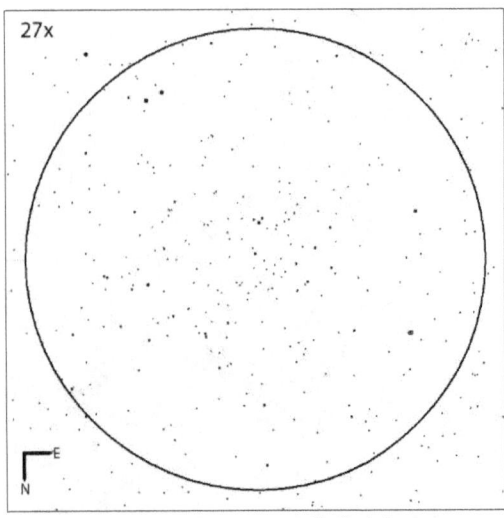

Eyepiece view courtesy *Sky Tools 3*

Designation(s):	NGC 752
Constellation:	Andromeda
R.A.:	01h 57m 55s
Declination:	+37° 51' 57"
Object Type:	Open Cluster
Location:	★ ★
Rating:	★ ★
Best Seen:	Autumn and Winter

NGC 752 is one of the larger open clusters and its light is consequently scattered across a relatively large area of the sky.

That being said, it should be visible in binoculars as a faint misty patch and using averted vision may help to resolve some individual stars.

This is one of those clusters that looks good under both suburban and city-based skies. You should be able to locate it with Almach on one edge of the finderscope field of view and the cluster on the opposite edge.

At 35x, it appears as a large, sparse cluster and denser toward the core. The stars are predominantly blue-white but you'll also notice some scattered orange stars throughout.

The double star 56 Andromedae lies on the edge and makes a convenient marker for the cluster. When viewed at this magnification, the entire cluster fits into the field of view with an orange star lying just outside the main group.

On occasion, when observing the cluster as it sets in the west, it takes on the appearance of Boötes with this star taking the place of Arcturus.

Designation(s):	Gamma Andromedae
Constellation:	Andromeda
R.A.:	02h 03m 54s
Declination:	+42° 19' 47"
Object Type:	Multiple Star
Location:	★ ★ ★
Rating:	★ ★ ★
Best Seen:	Autumn and Winter

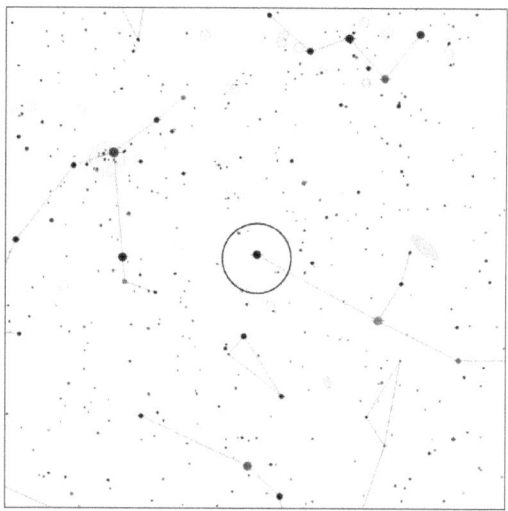

Map courtesy *Mobile Observatory*

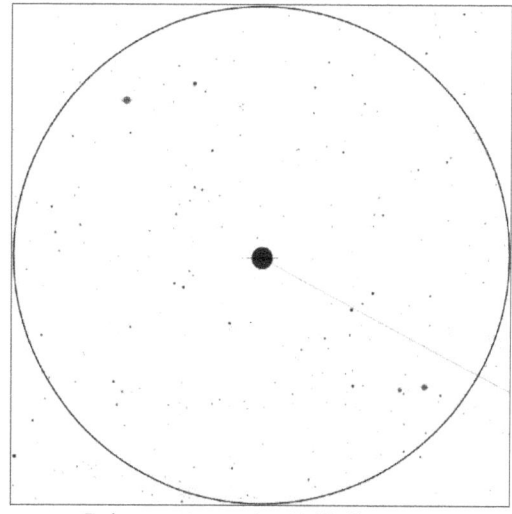

Finderscope view courtesy *Mobile Observatory*

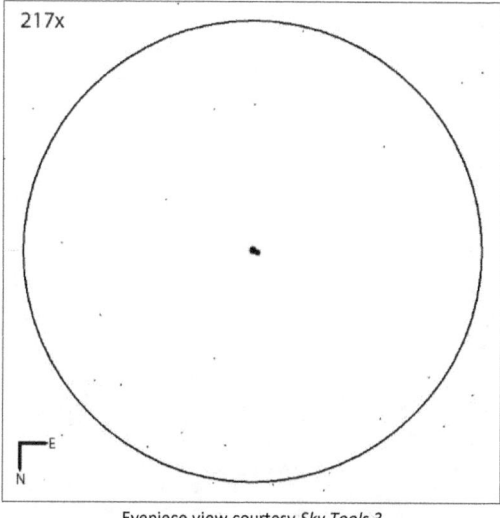

Eyepiece view courtesy *Sky Tools 3*

Almach, the third brightest star in Andromeda, is a true multiple star system located some 350 light years away. Although only two stars can easily be seen with a small telescope, there are, in fact, four stars present with the fainter, blue component being its own triple star system.

Definitely a favorite for the autumn, this double star can be split with low power and is reminiscent of Albireo in Cygnus.

But whereas Albireo is easily resolved at low power, you'll probably need at least 50x to properly enjoy the view. I've barely split it at 26x from the city, but this is one double where greater magnification definitely comes in handy. I've found that a magnification of about 65x or 70x works quite nicely.

The primary is a pale yellow-white gold and about four times brighter than the pale blue secondary. On a few occasions, I've noted that the secondary will flash purple and violet, but these colors may only be apparent at higher magnification (around 90x or 100x.)

One last thing – I've noticed that Almach is one of those pairs where the colors were more apparent at higher magnifications. What do you see?

Almach was first seen as a double in 1778 by the German physicist Johann Tobias Mayer.

Achird

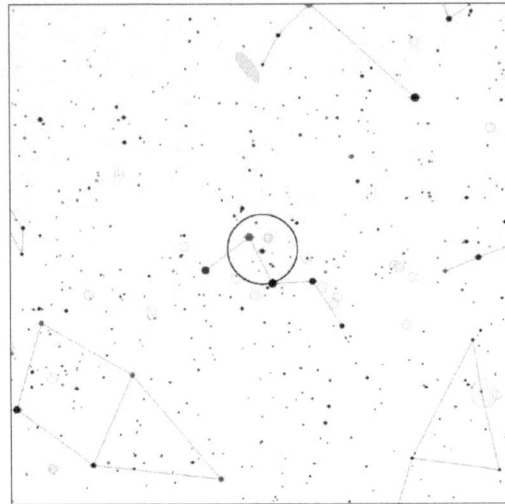

Map courtesy *Mobile Observatory*

Designation(s):	Eta Cassiopeiae
Constellation:	Cassiopeia
R.A.:	00h 49m 06s
Declination:	+57° 48' 55"
Object Type:	Multiple Star
Location:	★ ★ ★
Rating:	★ ★ ★
Best Seen:	Autumn and Winter

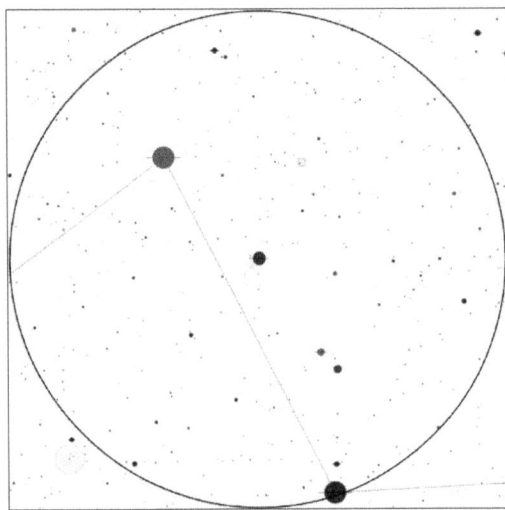

Finderscope view courtesy *Mobile Observatory*

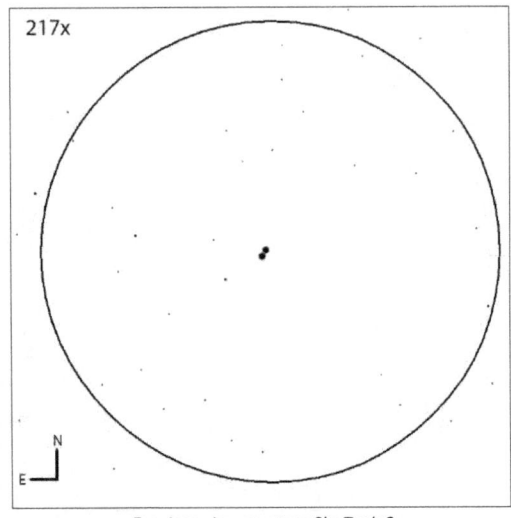

Eyepiece view courtesy *Sky Tools 3*

Achird is a relatively close multiple star; many sources will mention a gold primary with a reddish-purple companion, but I've noted other colors and a third, bluish star nearby.

Although it doesn't require much power to reveal the two main components, I've been unable to split it at a magnification of 26x but upping it a little to 35x or 40x should do the trick.

The first, and brightest, component appears a brilliant white to me (rather than the gold noted by others) but you'll need to focus carefully to reveal the much fainter companion.

At this magnification, it's difficult for me to notice any particular color but increasing the magnification again, to about 50x, helps to reveal the companion's coppery complexion.

While you have it at about 50x, try looking for the third component, which should appear as a tiny blue star on the opposite side of the white primary.

If not, increase the magnification once again. I've been able to spot it at 65x but 87x seemed to provide a particularly nice view with all three stars visible.

Designation(s):	NGC 457
Constellation:	Cassiopeia
R.A.:	01h 19m 33s
Declination:	+58° 17' 27"
Object Type:	Open Cluster
Location:	★
Rating:	★ ★ ★
Best Seen:	Autumn and Winter

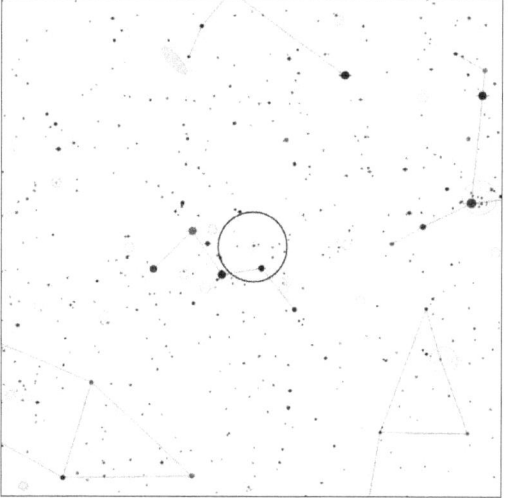

Map courtesy *Mobile Observatory*

A definite favorite, this cluster has a very distinctive shape that has given rise to a number of imaginative names over the years. Many see the stars forming the shape of an owl, while others see a kite. More recently, some have come to call it the E.T. Cluster.

Although the cluster can be seen with binoculars it doesn't truly shine until you turn your telescope toward it. You won't need a lot of magnification either – it can be seen with 26x and 35x provides a nice view, but somewhere between 50x and 90x is probably best. Once you get to about 100x, you'll have difficulty fitting the entire cluster into the field of view.

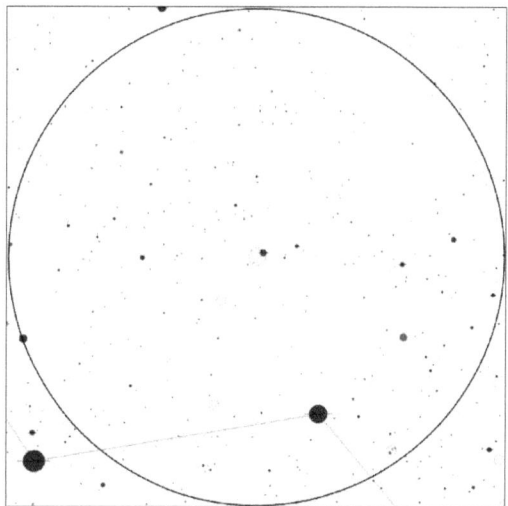

Finderscope view courtesy *Mobile Observatory*

The first time you see it, you're sure to be delighted as your eyes take in the sight and your imagination takes over. With the double star Phi Cassiopeiae marking the eyes, it's easy to see the cluster as either an owl with powerful wings or an alien with long, outstretched arms staring back at you.

The two stars of Phi Cassiopeiae are both white, with one being about 1½ times brighter than the other. It's a sparsely scattered cluster with the densest portion being around the chest area of the alien.

If the thought of an alien spooks you, look again. No owl? How about a goose or a swan in flight with Phi marking the tail?

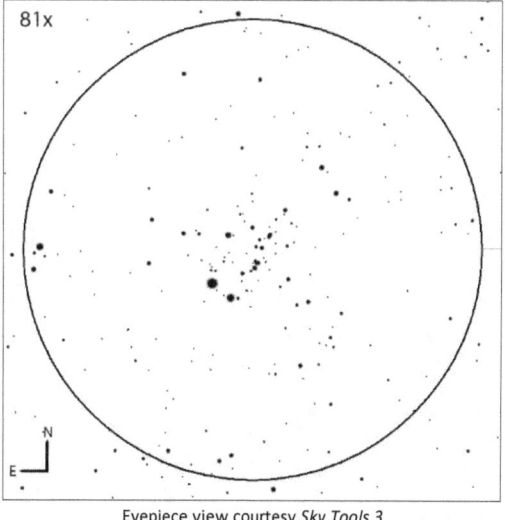

Eyepiece view courtesy *Sky Tools 3*

Messier 103

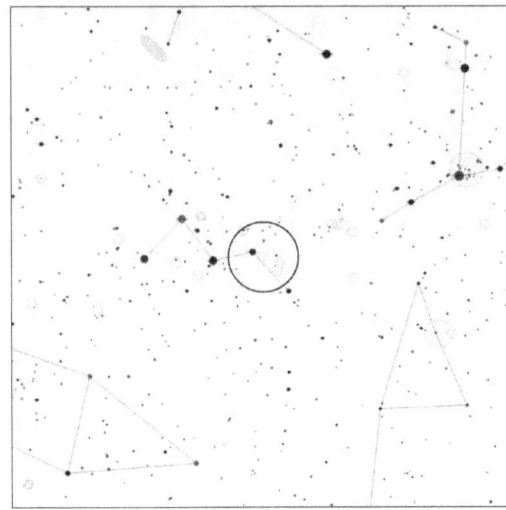

Map courtesy *Mobile Observatory*

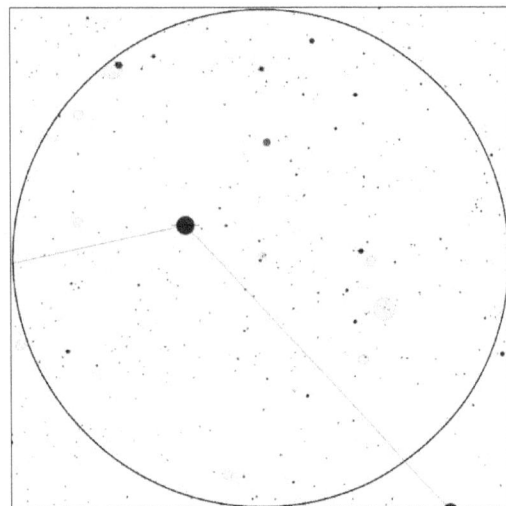

Finderscope view courtesy *Mobile Observatory*

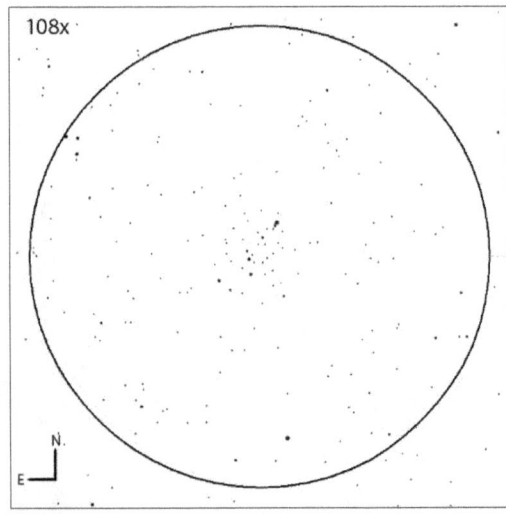

Eyepiece view courtesy *Sky Tools 3*

Designation(s):	Messier 103
Constellation:	Cassiopeia
R.A.:	01h 33m 22s
Declination:	+60° 39' 29"
Object Type:	Open Cluster
Location:	★
Rating:	★
Best Seen:	Autumn and Winter

Messier 103 is a small open cluster that's easily found but may be a little disappointing under city and suburban skies. It's barely visible as a tiny, compact elongated patch in binoculars but you may still be able to see the three stars that make up the slightly curved line across the center.

The telescopic view doesn't improve much from the city. I was only able to see about four or five stars at 26x – the central three for sure – with the center star appearing reddish.

The cluster fares better under suburban skies but it's best to observe it with low expectations. At 35x I could easily see the three bright stars across the center, like a mini Orion's belt. The middle star appeared to be golden with a blue companion close by while the other stars appeared to be white. I also noticed a triangle of stars to the north-east.

Upping the magnification to 91x definitely helped. The cluster still fitted within the field of view and I was able to see a lot more stars with averted vision.

There's one simple reason why the cluster appears so small: at a distance of about 9,000 light years, Messier 103 is one of the most distant open clusters known. There are thought to be over 150 member stars with the cluster being about 25 million years old.

Designation(s):	NGC 663
Constellation:	Cassiopeia
R.A.:	01h 46m 16s
Declination:	+61° 13' 06"
Object Type:	Open Cluster
Location:	★ ★
Rating:	★ ★ ★
Best Seen:	Autumn and Winter

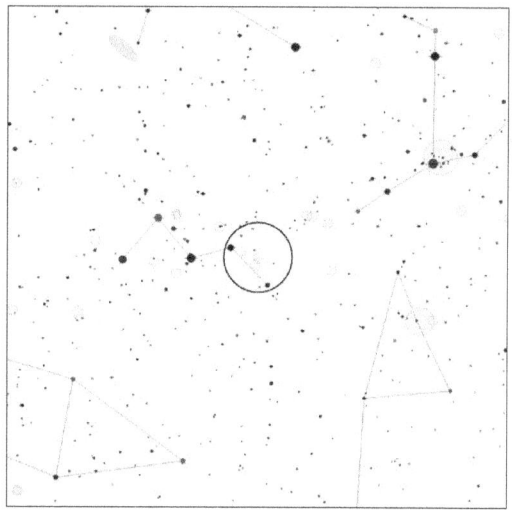

Map courtesy *Mobile Observatory*

NGC 663 is a little gem of a cluster. Like the Owl Cluster, it's easily found and is quite large and attractive, which leads you to wonder why Messier didn't include it in his famous list.

This cluster can be easily seen with binoculars as a large, faint, hazy circular patch about twice the size of Messier 103. It looks like a sparsely scattered globular cluster and you may be able to see individual stars with averted vision.

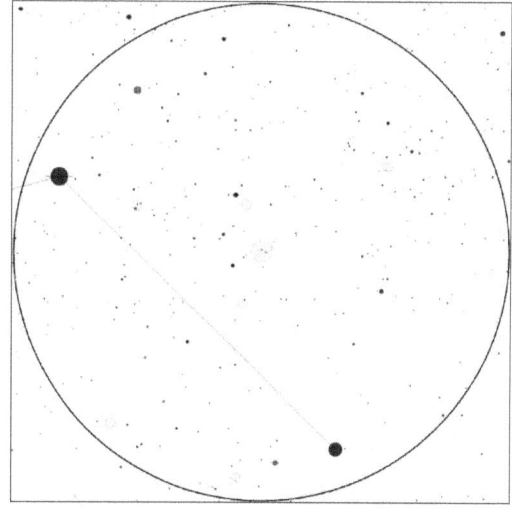

Finderscope view courtesy *Mobile Observatory*

However, it's best observed with a telescope at low to medium power, somewhere between 50x and 100x.

At 35x it appeared to be a sparsely scattered collection of stars that formed a diamond shape. The vast majority of stars were blue-white, but there were two pairings of stars – one just off center and the other near the edge – that appeared to be orange.

The two pairs of stars are actually a little spooky and make me think of two pairs of eyes. With the cluster high in the sky at Halloween, maybe they're ghosts. Or if NGC 457 is the E.T. Cluster, maybe E.T. is just the largest and friendliest of the aliens and NGC 663 are a smaller pair following close behind.

The cluster lies about 7,000 light years away and contains some 400 stars.

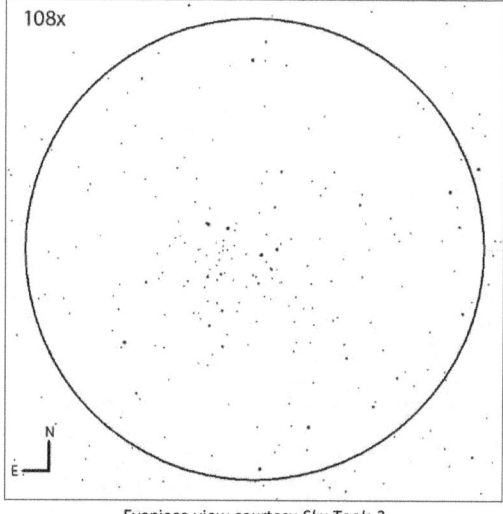

Eyepiece view courtesy *Sky Tools 3*

Mesarthim

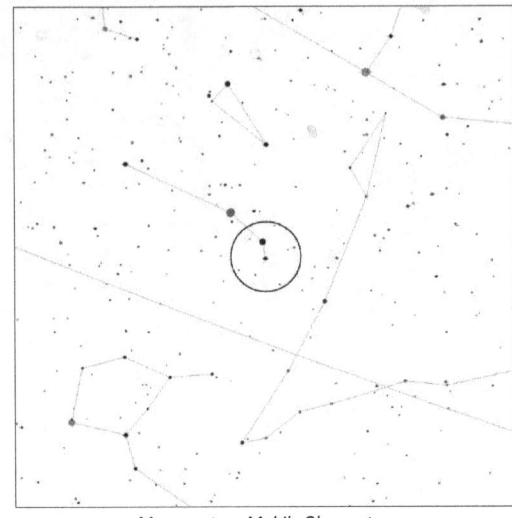

Map courtesy *Mobile Observatory*

Designation(s):	Gamma Arietis
Constellation:	Aries
R.A.:	01h 53m 32s
Declination:	+19° 17' 38"
Object Type:	Multiple Star
Location:	★ ★ ★
Rating:	★ ★ ★
Best Seen:	Autumn and Winter

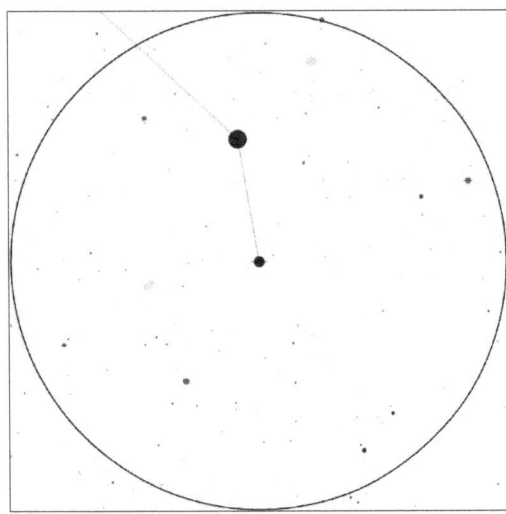

Finderscope view courtesy *Mobile Observatory*

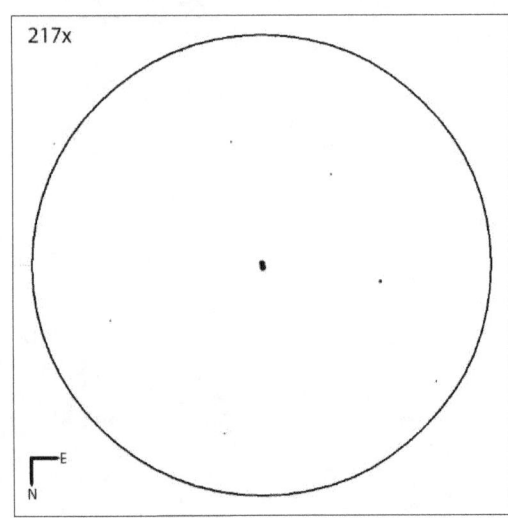

Eyepiece view courtesy *Sky Tools 3*

Mesarthim is a popular and relatively easy double star, known to many amateur astronomers across the world.

It's easily found and a good target if you're new to the hobby, especially as it can be used as a good benchmark for your equipment and the seeing conditions.

For example, you should be able to split the star at a fairly low magnification. From the city, I can barely split it at 26x with my 130mm 'scope but at 35x it's more apparent. I usually had no trouble splitting the star at that magnification from the suburbs with my 4 ½ inch XT.

How cleanly the star is split can depend upon the atmosphere; if the air isn't steady, it might appear to be barely split but otherwise it should be pretty clean.

You'll want to increase the magnification to get the best view; somewhere around 100x is probably best but even at lower magnification you should be able to see a third, unrelated fainter star nearby. What color is that star? To me it appears blue.

There's no mistaking the color of the Mesarthim stars themselves as both components are pure white and of equal brightness. Discovered by Robert Hooke in 1664, the two stars orbit one another every 5,000 years and lie about 165 light years away from us.

Designation(s):	Lambda Arietis
Constellation:	Aries
R.A.:	01h 57m 56s
Declination:	+23° 35' 46"
Object Type:	Multiple Star
Location:	★ ★ ★
Rating:	★ ★ ★
Best Seen:	Autumn and Winter

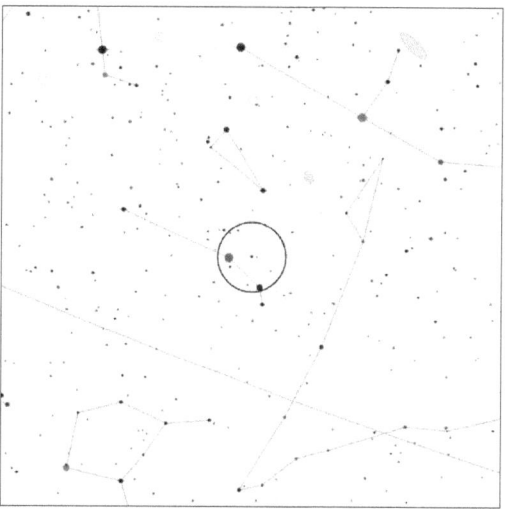

Map courtesy *Mobile Observatory*

Lambda Arietis is slightly off the beaten track but should still be quite easily found. It lies between two brighter stars – Hamal to the east and Sheratan to the west. Lambda appears slightly closer to Hamal in the map and finderscope views depicted here.

Some observers have reported this as being split with binoculars, but I've had no luck with my 10x50's.

However, a low magnification of 26x should easily reveal a brilliant white star that's about twice as bright as the bluish secondary. (On one occasion I noted hints of violet in the companion but the color seemed to disappear with a higher power eyepiece.)

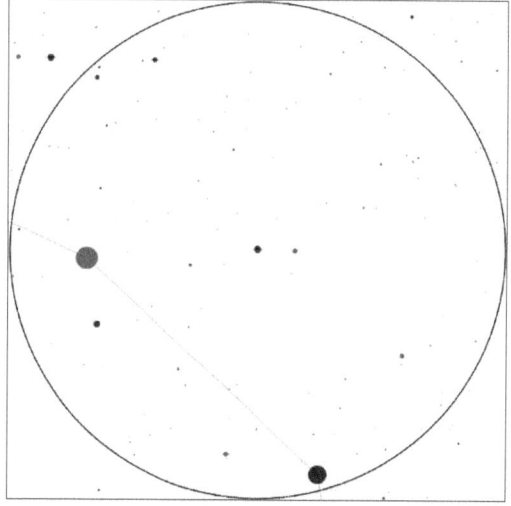

Finderscope view courtesy *Mobile Observatory*

A magnification of about 65x will provide a nice view. Look out for a third, slightly fainter companion nearby that forms an elongated triangle with the main pair.

Lambda is thought to be a true double star system as both stars appear to be moving together through space. In reality, both stars are actually yellow but the primary is hotter and its color is closer to yellow-white.

The reason the secondary appears blue is because your eyes are sensitive to contrasts in light and you see an exaggerated contrast effect as a result.

The stars lie at a distance of about 129 light years and take more than 33,000 years to orbit one another.

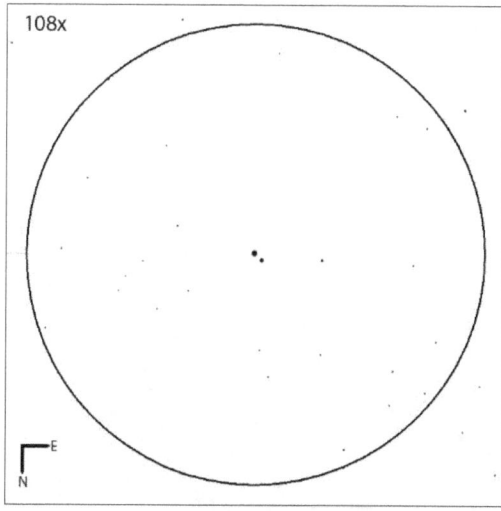

Eyepiece view courtesy *Sky Tools 3*

The Double Cluster

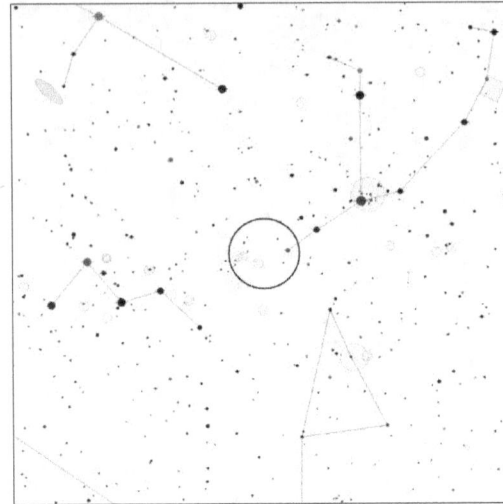

Map courtesy *Mobile Observatory*

Designation(s):	NGC 869 & NGC 884
Constellation:	Perseus
R.A.:	02h 22m 29s
Declination:	+57° 09' 00"
Object Type:	Open Cluster
Location:	★ ★
Rating:	★ ★ ★
Best Seen:	Autumn and Winter

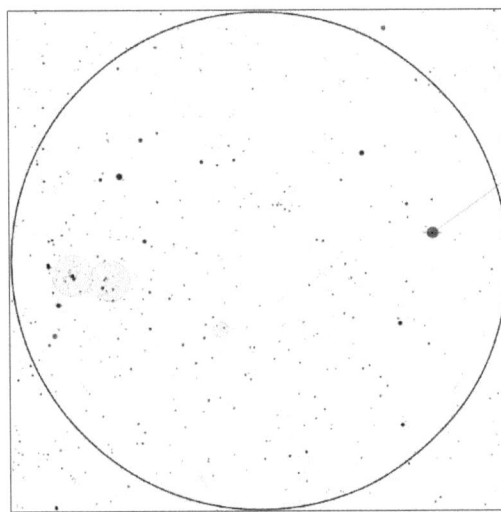

Finderscope view courtesy *Mobile Observatory*

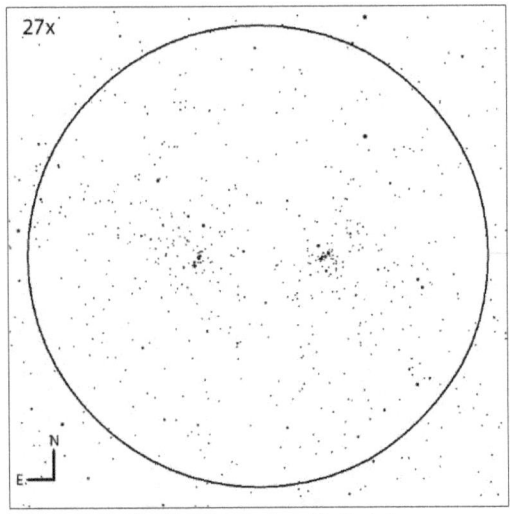

Eyepiece view courtesy *Sky Tools 3*

Known since antiquity, the Double Cluster is one of the highlights of the night sky and yet, somehow, like NGC 663, Messier never included it in his catalog. They are, quite simply, unmissable in every sense.

Depending on the quality of your sky, you may be able to glimpse this stellar pair with just your eyes. Look about midway between the top of Perseus, the Hero and Cassiopeia, the Queen and you might see a faint, elongated patch of hazy light.

If you don't have any luck, try scanning the area with binoculars or your finderscope as they should be easily seen in both. To me, in my small 8x30 binoculars, the clusters seemed to take on the form of a butterfly or bowtie.

These clusters appear fairly large in the sky and are consequently best viewed at low power but I've been able to fit them both into the same field of view up to about 70x or so.

NGC 884, the eastern cluster (to the left in the eyepiece depiction) is larger and more sparsely scattered than its neighbor. NGC 869 has a dense core with two bright blue-white stars, just off-center. Take your time with these clusters and allow your eyes to soak in the view. Most of the stars will be young and blue-white in color but you'll see a few older orange giants too.

Designation(s):	Messier 34
Constellation:	Perseus
R.A.:	02h 42m 07s
Declination:	+42° 44' 46"
Object Type:	Open Cluster
Location:	★ ★
Rating:	★ ★
Best Seen:	Autumn and Winter

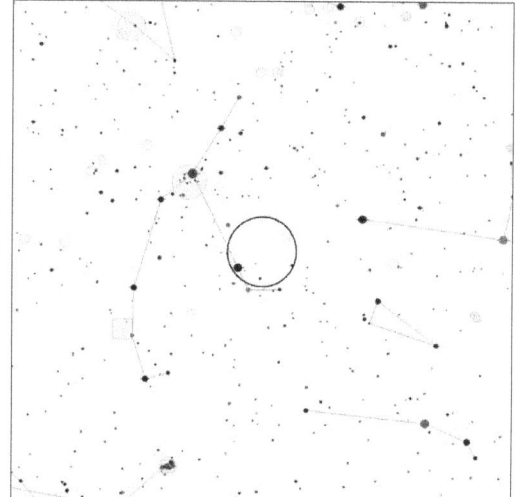

Map courtesy *Mobile Observatory*

Messier 34 is like a cousin to the Double Cluster and is often overlooked in favor of its flashier neighbors.

It can be found by pointing your finderscope or binoculars midway between Almach in Andromeda and Algol in Perseus. By placing Algol on the edge of the finderscope view, you should be able to see M34 on the opposite edge.

This is a fairly large, sparse cluster that reminds me of the Praesepe (M44) and is best viewed at low power. I found a magnification of about 50x to be best.

At 35x I noted that the cluster appeared to be comprised almost entirely of blue-white stars but with a few that show hints of amber. In particular, I've noticed a golden star on its western edge.

There are also a number of apparent double stars scattered throughout with a pair of equal brightness near a triangle at the center.

Do you notice any shapes? I saw an elongated X at 26x and I've also noted a pattern that's reminiscent of the Owl Cluster at the heart of the cluster.

The cluster lies about 1,500 light years away and is thought to contain close to 500 stars.

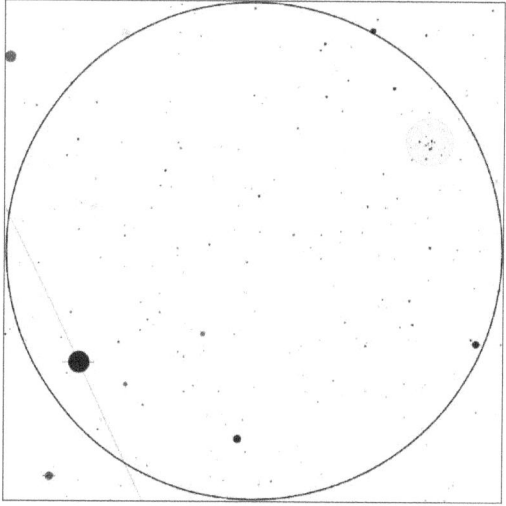

Finderscope view courtesy *Mobile Observatory*

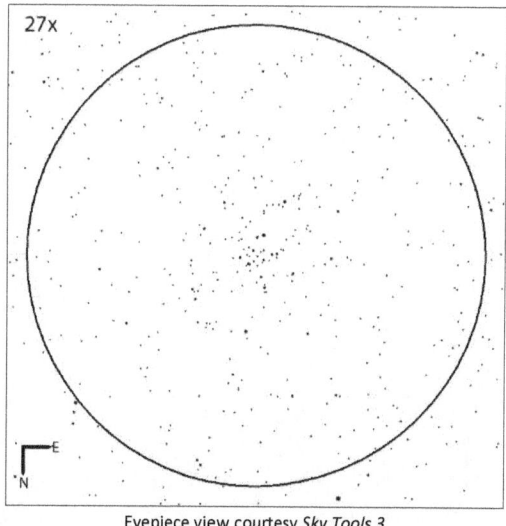

Eyepiece view courtesy *Sky Tools 3*

Polaris

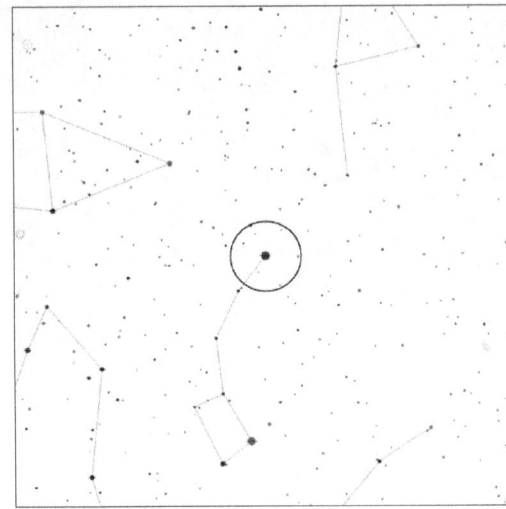

Map courtesy *Mobile Observatory*

Designation(s):	Alpha Ursa Minoris
Constellation:	Ursa Minor
R.A.:	02h 31m 49s
Declination:	+89° 15′ 51″
Object Type:	Multiple Star
Location:	★ ★ ★
Rating:	★
Best Seen:	All Year Round

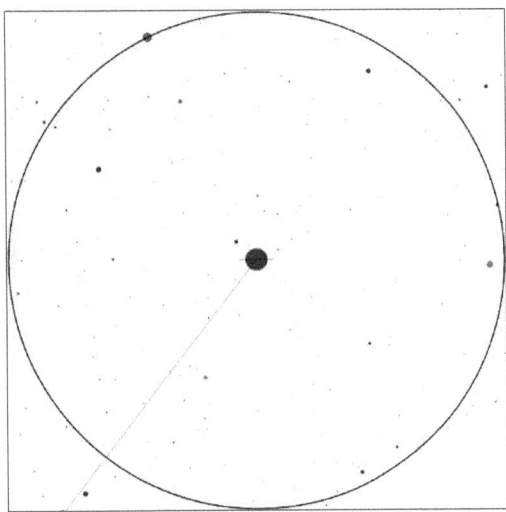

Finderscope view courtesy *Mobile Observatory*

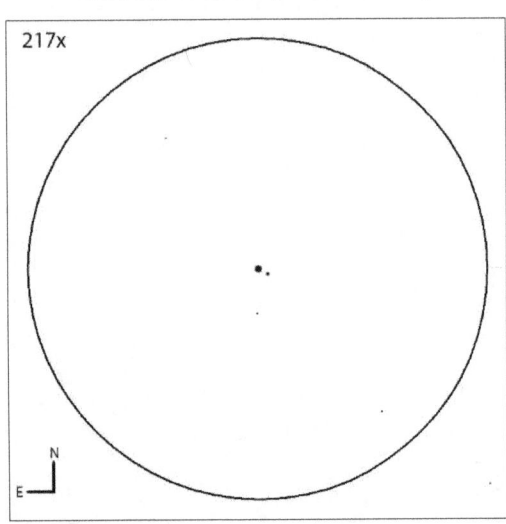

Eyepiece view courtesy *Sky Tools 3*

It's a common misconception among non-astronomers that Polaris, the pole star, is actually the brightest star in the sky. This has always baffled me as I can't imagine how this belief may have come about.

It's actually just a star of average brightness but it's very easily found and, since it's almost exactly at the north celestial pole, it never moves in the sky. (At least, not to the unaided eye.) It therefore has the convenience of being visible on every clear night of the year.

To be honest, it's not a very exciting multiple star but it's definitely one that should be on every newbie's bucket list. I haven't been able to split it at 26x and you may have to almost triple that power to about 75x to see the companion.

You should see a white primary with a very faint bluish secondary. It's an easy split at 130x but the secondary was difficult to see from the city and I needed to use averted vision to make it more apparent.

Once it's been split and you know where to look for the companion, gradually lower the magnification and see how low you can go before it disappears. Under suburban skies I've definitely seen it at 54x and have suspected the secondary at 35x.

Designation(s):	Messier 45
Constellation:	Taurus
R.A.:	03h 47m 58s
Declination:	+24° 09' 50"
Object Type:	Open Cluster
Location:	★ ★ ★
Rating:	★ ★ ★
Best Seen:	Autumn and Winter

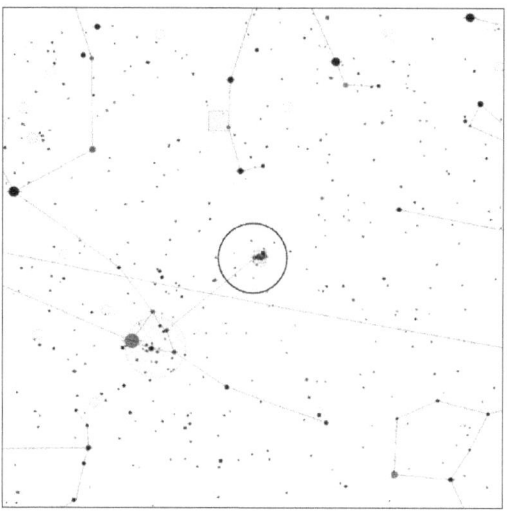

Map courtesy *Mobile Observatory*

Astronomically speaking, winter is a great time for observers in the northern hemisphere. Not only can we admire the beautiful Orion Nebula, but we're also treated to the Pleiades, a stunning open star cluster in the constellation of Taurus the Bull.

It's a famous cluster that's been known by numerous names across the world for many, many years. In Japan, it's known as the Subaru (hence, the car manufacturer) while elsewhere it's known as the Seven Sisters. And this is something of a mystery because although the cluster is easily seen with just your eyes, most observers can only count six stars at most. So what happened to the seventh?

It's a large cluster (it appears twice the size of the full Moon in the sky) and, consequently, some say it's best observed with binoculars where it can be appreciated against the background sky.

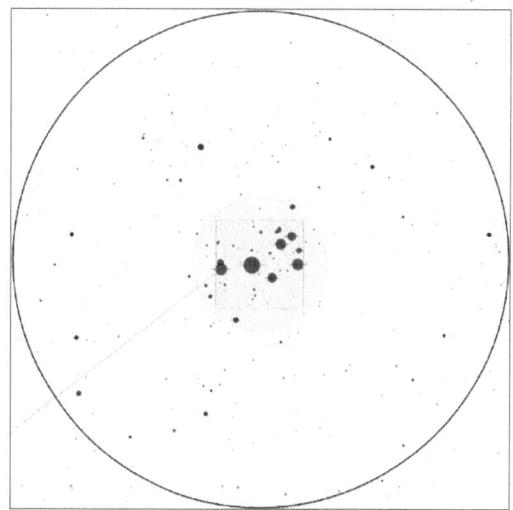

Finderscope view courtesy *Mobile Observatory*

However, if you turn a small telescope toward it with a low magnification eyepiece, you won't be disappointed. The entire cluster should fit into the field of view at about 35x with hundreds of blue-white stars scattered across the scene. Increasing the magnification might actually ruin the effect.

Lastly, if you get the opportunity, observe this cluster well away from any light pollution. You won't regret it.

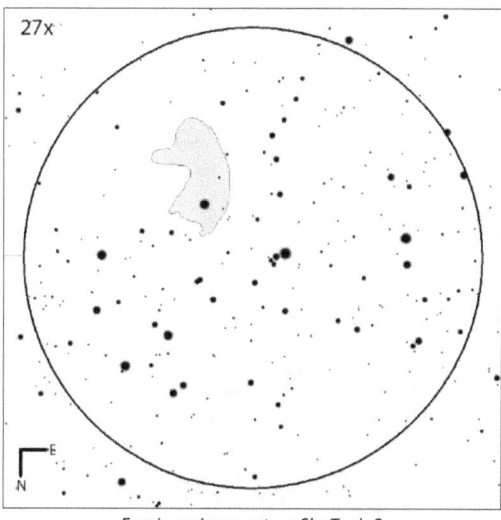

Eyepiece view courtesy *Sky Tools 3*

The Crab Nebula

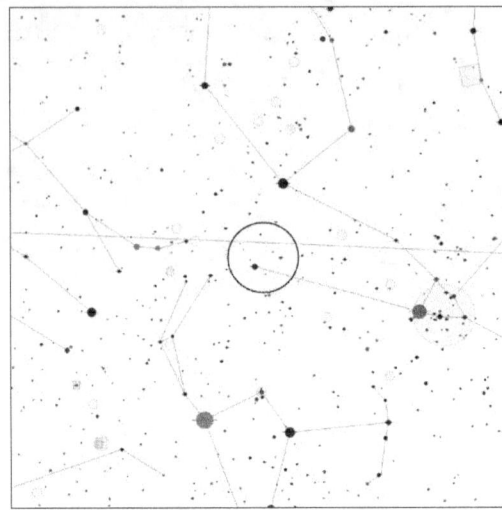
Map courtesy *Mobile Observatory*

Designation(s):	Messier 1
Constellation:	Taurus
R.A.:	05h 34m 32s
Declination:	+22° 00' 52"
Object Type:	Supernova Remnant
Location:	★
Rating:	★
Best Seen:	Autumn and Winter

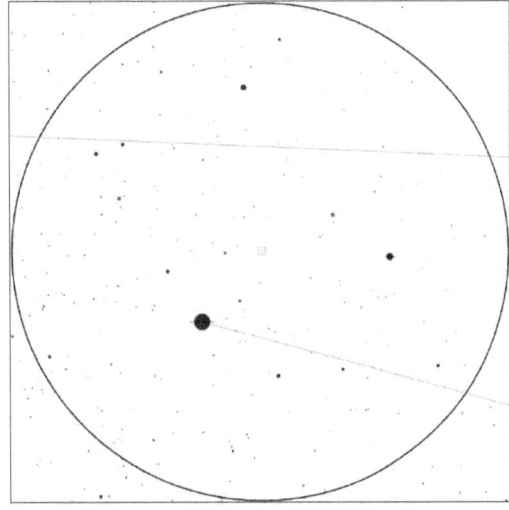
Finderscope view courtesy *Mobile Observatory*

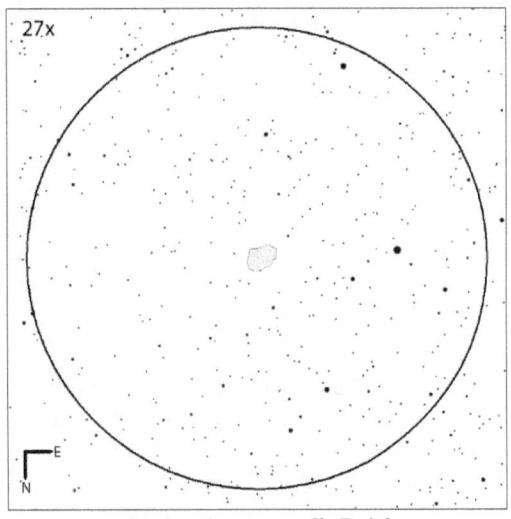
Eyepiece view courtesy *Sky Tools 3*

The Crab Nebula is thought to be the remains of a star that exploded almost a thousand years ago. On July 4th, 1054 Chinese astronomers recorded seeing a "guest star" in the sky. This supernova was so bright it was visible during daylight hours for about three weeks.

Nearly seven hundred years later, in 1731, the nebula was discovered by the astronomer John Bevis in the same area of the sky, but it wasn't linked to the supernova until the early 20th century.

To be honest, it can be disappointing when viewed in a small telescope, but it's worth remembering what you're looking at. There aren't too many supernovae remnants around and the Crab is the only example that can be readily observed by amateurs with inexpensive equipment.

Some can spot the nebula with binoculars but I've not had such luck and I suspect you'll need clear dark skies to be successful. However, it should be visible under suburban skies with a small telescope.

It's barely seen at about 35x but it's not an easy thing to see. Using averted vision, it appears as a very faint oval patch with little or no brightening near the center. I once described it as "a coal stain on a black carpet." Increasing the magnification may help, but if you live in the city, you'll probably be out of luck as I've never seen it from Los Angeles.

Designation(s):	Delta Orionis
Constellation:	Orion
R.A.:	05h 32m 00s
Declination:	-00° 17' 57"
Object Type:	Multiple Star
Location:	★ ★ ★
Rating:	★ ★
Best Seen:	Winter

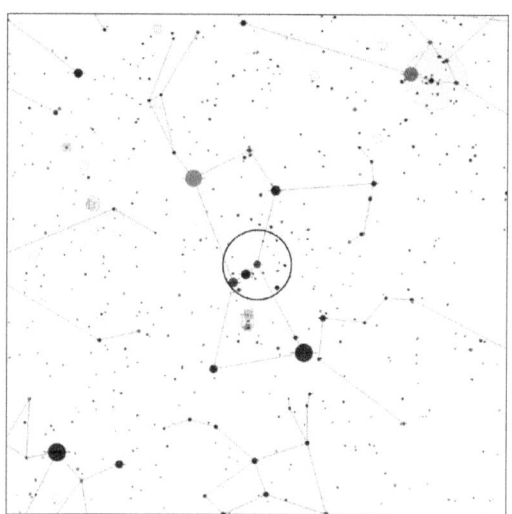

Map courtesy *Mobile Observatory*

Mintaka is a bright star, easily identified as being the most westerly star in Orion's belt. The belt has been a famous astronomical fixture for thousands of years with the three stars featuring prominently in cultures around the world.

But before you point your 'scope in its direction, scan the area with binoculars or a small telescope at low power. What can you see? The three belt stars are actually members of a very large open star cluster, known as the Orion OB1 Association and the area is studded with stars.

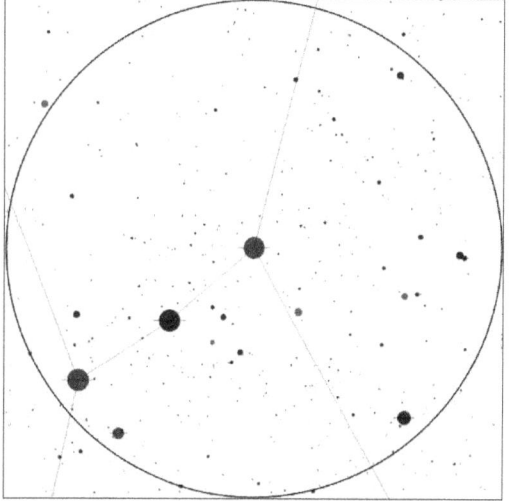

Finderscope view courtesy *Mobile Observatory*

Mintaka itself is a fairly wide double, but unfortunately not so wide that you can split it with binoculars. (Or at least I've been unable to.)

Having said that, you won't need a lot of power to split it as about 26x should do the trick. At this magnification the two components are easily seen, with the white primary appearing about three or four times brighter than the sky blue secondary.

It looks better at higher magnification with about 50x providing a good view. Something I've noticed is that the higher the power, the stronger the colors appeared to be. Unfortunately, the pair may be less visually appealing as the gap between them appears to increase.

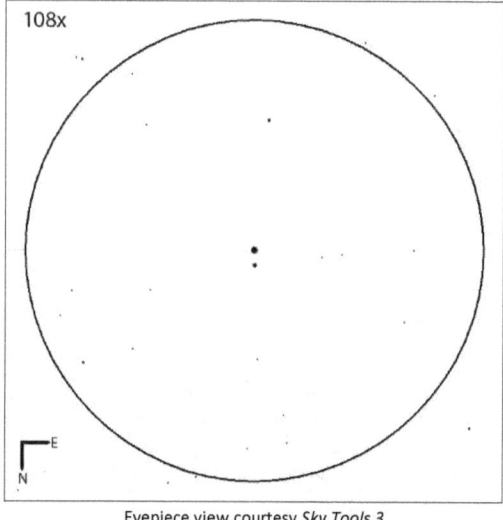

108x

Eyepiece view courtesy *Sky Tools 3*

Meissa

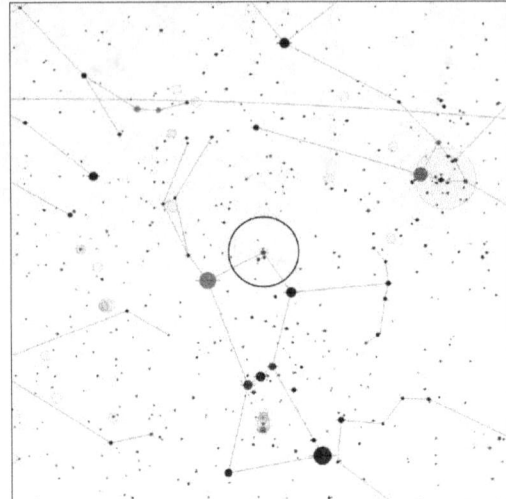

Map courtesy *Mobile Observatory*

Designation(s):	Lambda Orionis
Constellation:	Orion
R.A.:	05h 35m 08s
Declination:	+09° 56' 03"
Object Type:	Multiple Star
Location:	★ ★ ★
Rating:	★ ★ ★
Best Seen:	Winter

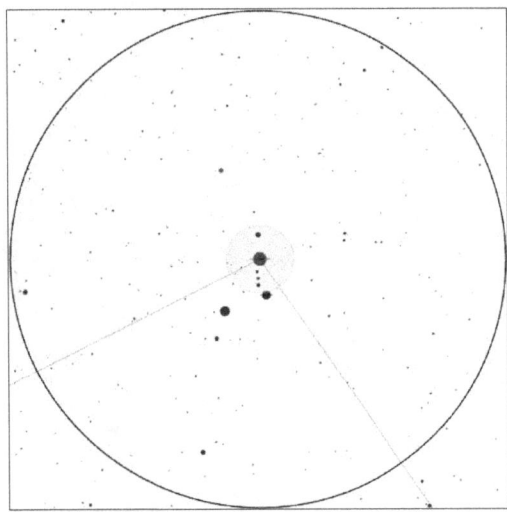

Finderscope view courtesy *Mobile Observatory*

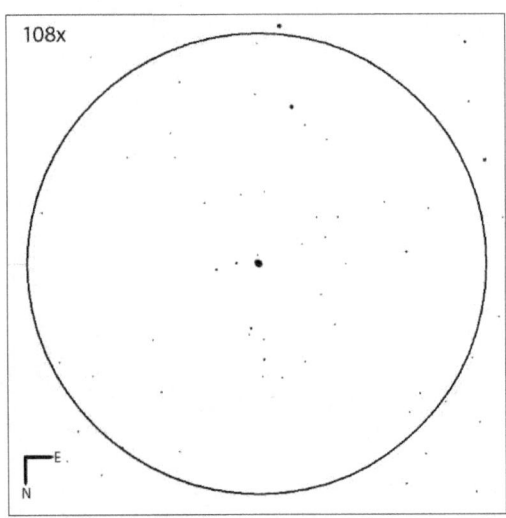

Eyepiece view courtesy *Sky Tools 3*

Meissa is the brightest member of a small cluster of stars that mark Orion's head. Through binoculars or a finderscope you should see a triangular asterism with a line of three faint stars close to Meissa itself.

At 26x, you'll see Meissa as a white star with a much fainter secondary nearby. If you look carefully with averted vision, you might be able to catch a third star between the two.

Up the power to about 35x and the third star should be easily seen, helping to form another line of three stars, equally spaced, with Meissa being at the end of the line and much brighter than the other two.

But that's not all. Increase the magnification again to about 70x and you might just be able to split Meissa itself. Stare for a few seconds and wait for the air to steady and a violet-blue companion will appear to be touching the star.

Meissa is one of those stars where a higher magnification will definitely benefit you. You'll need about 100x to cleanly split the pair (you might even see another, faint blue star nearby at this magnification) but the higher you go, the better the view.

Use your barlow!

Designation(s):	Messier 42
Constellation:	Orion
R.A.:	05h 35m 17s
Declination:	-05° 23' 28"
Object Type:	Nebula
Location:	★ ★ ★
Rating:	★ ★ ★
Best Seen:	Winter

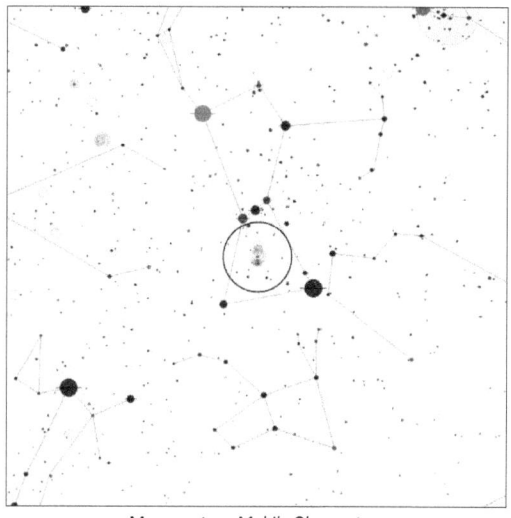

Map courtesy *Mobile Observatory*

By far the brightest and best nebula visible from the northern hemisphere, the Orion Nebula can be described with a single word: stunning.

Easily seen with just the unaided eye, it'll look good even through binoculars or a halfway decent finderscope. The cloud appears small and misty, but you should be able to pick out three or four bright stars in the heart of the nebula. These stars, known as the Trapezium, are young stars recently born from the nebula and still swaddled in the surrounding stellar nursery.

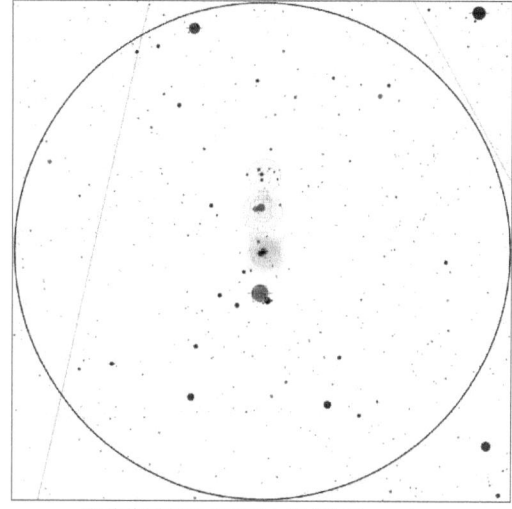

Finderscope view courtesy *Mobile Observatory*

Although best observed with low or medium magnification, the nebula stands up to quite a bit of power with greater magnifications revealing more details to the observer. Unlike some celestial sights, you'll notice some color here too – observers often report a greenish tint with the nebula appearing smoke-like. I've noted that it looks as though it's illuminated by moonlight.

While you're here, look out for a small, circular misty patch nearby with a lone star within it. This is Messier 43 (M43), a fragment that appears separated from the main nebula itself. You'll also see a darker, triangular shape on that edge, close to the Trapezium and M43. This is known as the Fish's Mouth and appears as a result of darker matter that obscures the light of the nebula beyond.

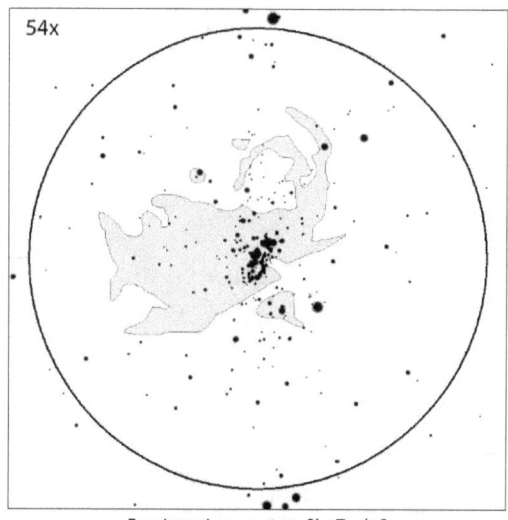

54x

Eyepiece view courtesy *Sky Tools 3*

Sigma Orionis

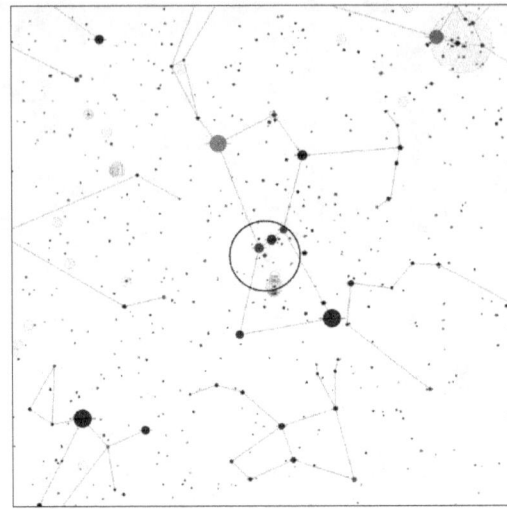

Map courtesy *Mobile Observatory*

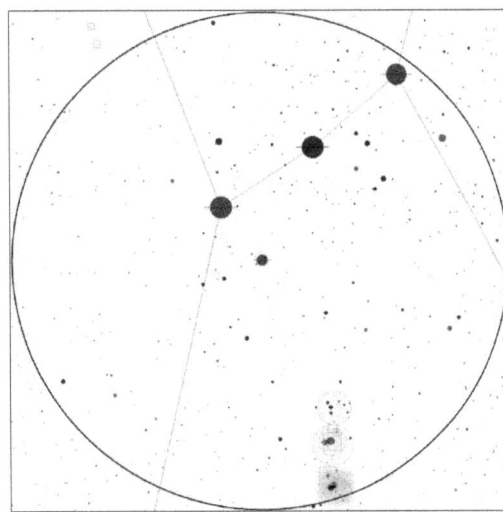

Finderscope view courtesy *Mobile Observatory*

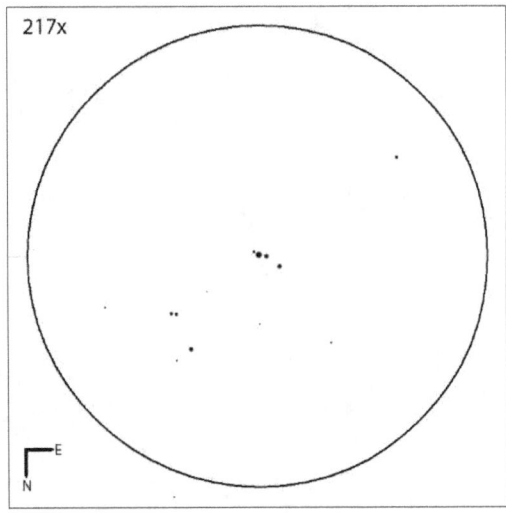

Eyepiece view courtesy *Sky Tools 3*

Designation(s):	Sigma Orionis
Constellation:	Orion
R.A.:	05h 39m 34s
Declination:	-02° 35' 36"
Object Type:	Multiple Star
Location:	★ ★ ★
Rating:	★ ★ ★
Best Seen:	Winter

Sigma is quite a fascinating multiple star that's easily found just below Orion's belt. What's particularly interesting is that it appears close to a mini belt at 26x.

I've also noted that it appears in the middle of an asterism that resembles the constellation Sagitta, with two stars to the west that form the feathers of the arrow. The southernmost star is itself a double with both components being of equal brightness.

Sigma itself has three components with the primary being a brilliant white star. There is a secondary, wide companion to the north-east that appears to be white and about 1½ times fainter than the primary. Look out for a slightly fainter third star, much closer than the second, which is also white and just east of the primary.

The system lies about 1,150 light years away and actually comprises of five stars. The primary you see in a telescope is a very close pair of young white dwarves that orbit one another once every 170 years.

The next star in the system is just to the west of the primary but will be too close and faint to be seen with small 'scopes from suburban or city skies.

The remaining two stars are also dwarves about seven times the mass of the Sun with one being a little-understood, helium-rich magnetic star.

Designation(s):	Gamma Leporis
Constellation:	Lepus
R.A.:	05h 44m 28s
Declination:	-22° 26' 54"
Object Type:	Multiple Star
Location:	★ ★ ★
Rating:	★ ★
Best Seen:	Winter

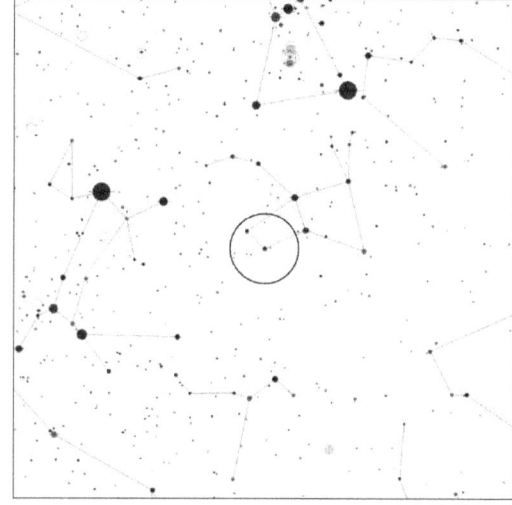

Map courtesy *Mobile Observatory*

Also split with a regular pair of 10x50 binoculars, this is a fairly wide double that should reveal both the two brightest components at a low magnification.

At 26x, the primary appears to be a creamy white and about twice as bright as the coppery-orange secondary. (On one occasion, I noted the companion appeared to have a blue-purple tint.)

Things get a little more interesting when you up the power. If you increase the magnification to about 90x, you might be able to see a third, fainter star, flickering in and out with averted vision. It appears about three times further away than secondary.

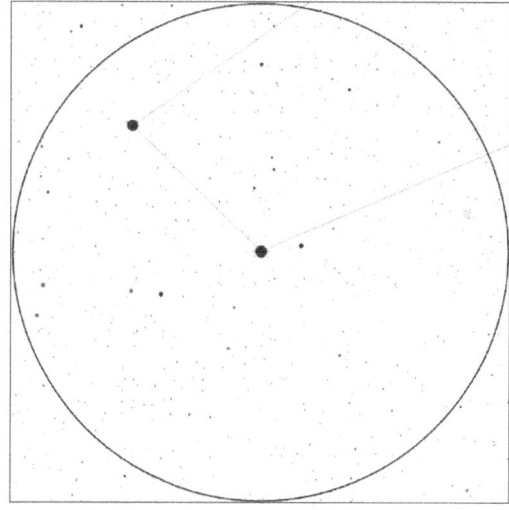

Finderscope view courtesy *Mobile Observatory*

Go a little further – about 100x or even a little higher – and you might catch a fourth star. It's very faint and appears on the opposite side of the primary from the second star. If you look at the distance between the primary and secondary stars, the fourth should appear a little further away.

At just under thirty light years, this system is relatively close to the Earth and is interesting for one other major reason: the primary star is Sun-like (although only about half the age) and may have Earth-sized planets orbiting it. Something to consider as you stare across space at this fascinating star.

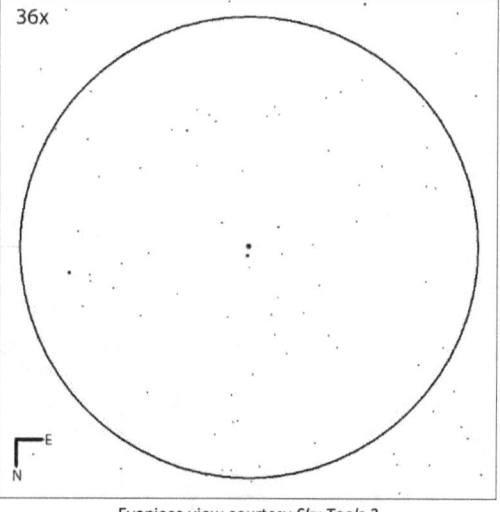

Eyepiece view courtesy *Sky Tools 3*

Messier 37

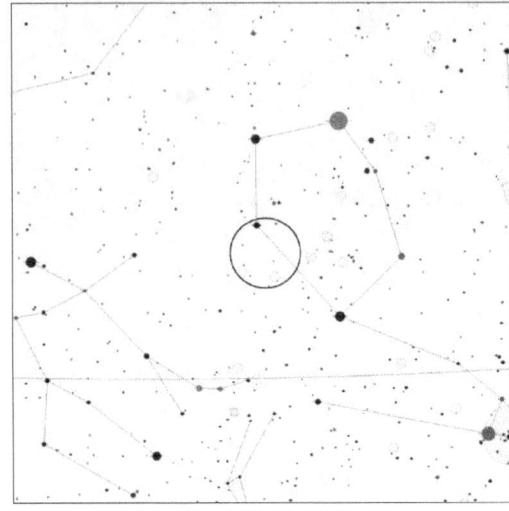

Map courtesy *Mobile Observatory*

Designation(s):	Messier 37
Constellation:	Auriga
R.A.:	05h 52m 18s
Declination:	+32° 33' 02"
Object Type:	Open Cluster
Location:	★
Rating:	★ ★ ★
Best Seen:	Winter

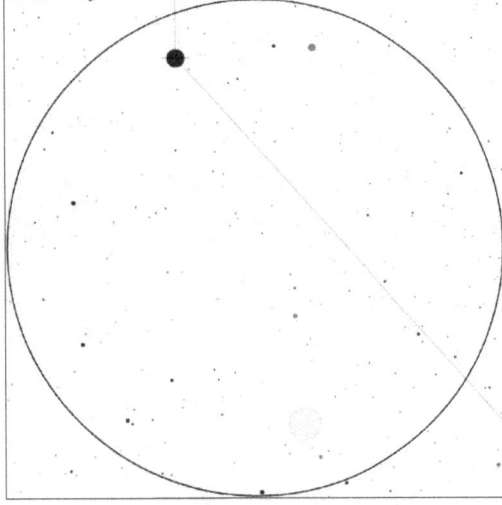

Finderscope view courtesy *Mobile Observatory*

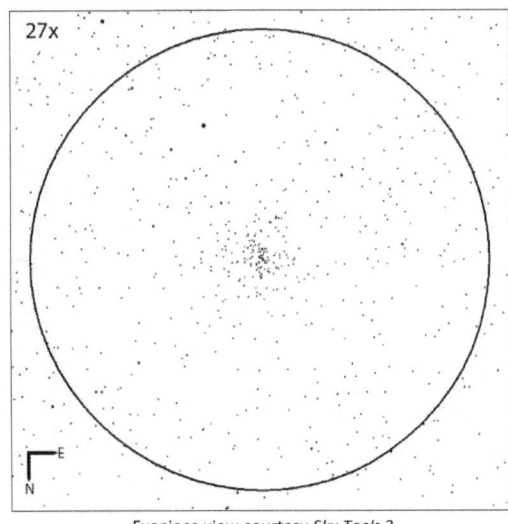

Eyepiece view courtesy *Sky Tools 3*

Messier 37 is the brightest of three Messier open star clusters in Auriga and, lying close to Theta Aurigae, is also the easiest found.

With Theta on the edge of your finderscope field of view, look for the cluster close to the opposite edge. It should appear as a very faint, misty patch under suburban skies.

The other two clusters, Messier 36 and Messier 38, lie to the west, in the south-eastern portion of Auriga's misshapen hexagon and are also worth seeking out.

It's a rich cluster (the richest of the three) and should be easily observed from suburban skies but may be more problematic from the city.

It presents a very nice view at 35x and I noted that there seemed to be a thousand or more stars in the cluster. Using averted vision, it seemed as though a thousand more would spring into view. I've also noted a number of triangular patterns within the cluster. You should be able to increase the magnification to about 50x and still fit the cluster within the field of view.

It's definitely a cluster that begs to be stared at. Just allow your eyes to take in the view, enjoy it and consider that it lies about 4,500 light years away and its 500 stars are scattered across some 25 light years of space.

Designation(s):	Messier 35
Constellation:	Gemini
R.A.:	06h 08m 56s
Declination:	+24° 21' 28"
Object Type:	Open Cluster
Location:	★ ★
Rating:	★ ★ ★
Best Seen:	Winter

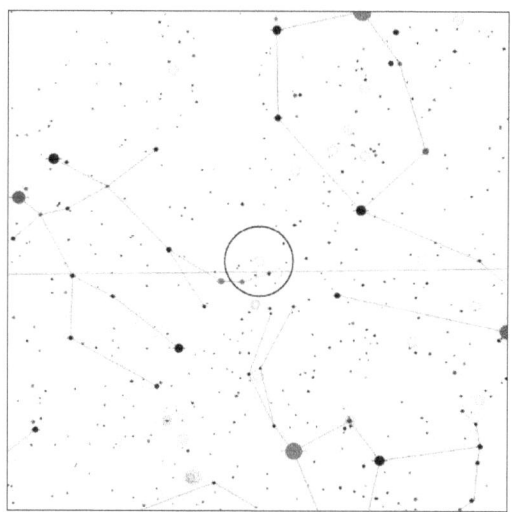
Map courtesy *Mobile Observatory*

Messier 35 is a large, bright open star cluster that should be easily found with your finderscope – or even with just your unaided eyes.

Through binoculars, it appears to be a faint, grey misty patch of uniform brightness but with an hourglass shape. If you look with averted vision, the patch may appear to be resolved into hundreds of tiny points of light – the cluster's individual stars.

Given its size, this is a cluster that's best observed at lower power and about 30x should do the trick. At this magnification, the cluster nicely fits into the field of view and appears to be surrounded by a background scattering of stars.

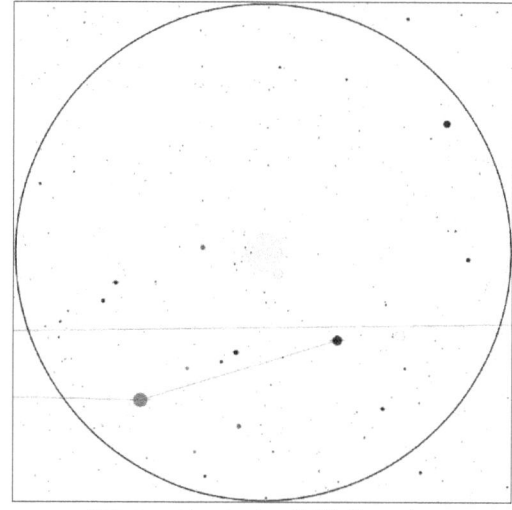
Finderscope view courtesy *Mobile Observatory*

It's fairly sparse but with plenty of faint stars. Look for a pale gold star on the edge of the cluster with a blue star close to it, giving it the impression of a double star. The golden star is about 1½ times brighter than its blue companion.

Of the others, the remaining stars appear to be either blue or blue-white.

The cluster was discovered in 1745 and lies at a distance of 2,800 light years from Earth.

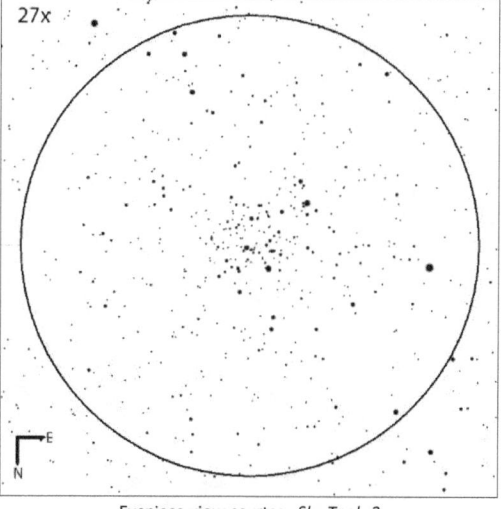
Eyepiece view courtesy *Sky Tools 3*

Castor

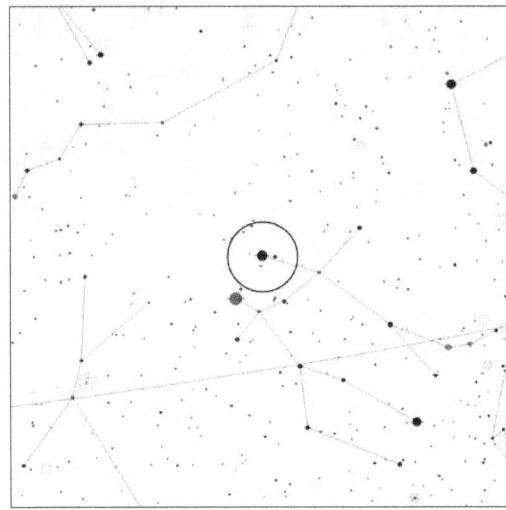

Map courtesy *Mobile Observatory*

Designation(s):	Alpha Geminorum
Constellation:	Gemini
R.A.:	07h 34m 36s
Declination:	+31° 53' 18"
Object Type:	Multiple Star
Location:	★ ★ ★
Rating:	★ ★ ★
Best Seen:	Winter

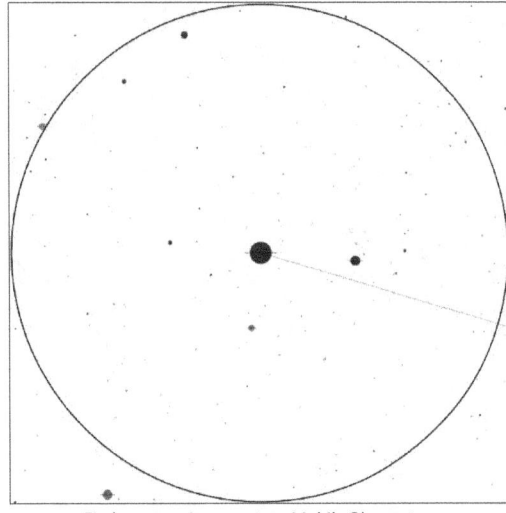

Finderscope view courtesy *Mobile Observatory*

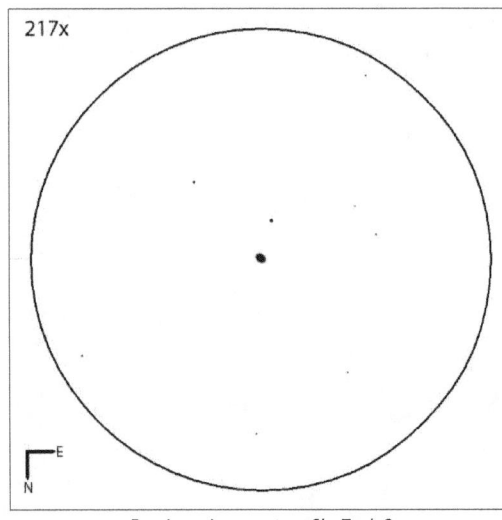

Eyepiece view courtesy *Sky Tools 3*

Castor, despite being designated as Alpha Geminorum, is actually the second brightest star in the constellation of Gemini, the Twins. To the unaided eye, it appears to be a solitary, bright white star but in reality it's a quadruple star system located some 51 light years away.

Unfortunately, only two of the four can be seen with amateur equipment and a magnification of about 50x is needed to split it.

At 26x you'll probably notice two very faint companions of almost equal magnitude within the same field of view. Although this is interesting in itself, this isn't what you're here to see.

Try upping the magnification to about 50x and take another look. Does Castor appear to be split? On some nights, I've barely split it at 54x. On other occasions, I can't say it's been resolved. A lot will depend upon your equipment and the sky conditions.

Double the magnification to about 100x and the view improves immensely. Castor should now be closely split into a pair of brilliant white stars, of almost equal magnitude. Through the 130mm 'scope, there also appeared three fainter stars visible nearby, equally spaced out like Orion's belt but with the middle star a little out of line from the others.

Don't miss this winter wonder!

Designation(s):	Beta Monocerotis
Constellation:	Monoceros
R.A.:	06h 28m 49s
Declination:	-07° 01' 59"
Object Type:	Multiple Star
Location:	★ ★
Rating:	★ ★
Best Seen:	Winter

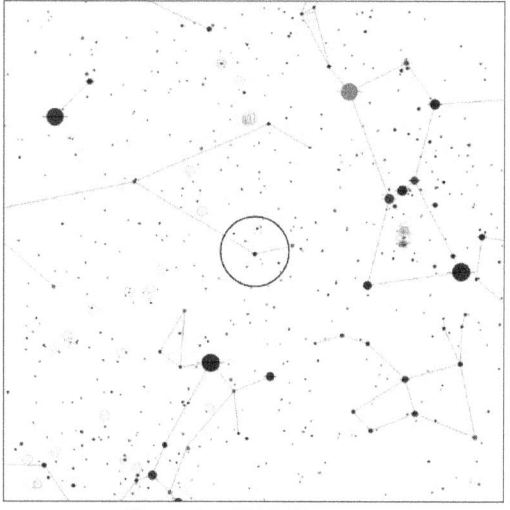

Map courtesy *Mobile Observatory*

Beta Monoceros is a true triple star system some 700 hundred light years away. Two of the stars should be a relatively easy target for a small 'scope and a low magnification eyepiece.

It's barely split at 26x but upping the magnification to 35x or more should make it cleaner and more obvious.

You'll see a pair of white stars of equal brightness with a third, fainter star nearby. (My estimates of the third star's brightness seem to vary somewhat.)

Increasing the magnification might reveal some color (pale yellow and pale blue,) but I only noticed this from the city so there's a real chance it had more to do with the quality of the air than the stars themselves.

You may need to push your 'scope to the limit to see all three stars. If you have a barlow, increase your magnification to as far as your 'scope can usefully go and try your luck.

At 217x, with the 150mm, I noted that one of the two brightest stars appeared to be elongated and, therefore, potentially multiple, but this was at the limit of the telescope's maximum magnification and I was unable to actually split them.

What do you see?

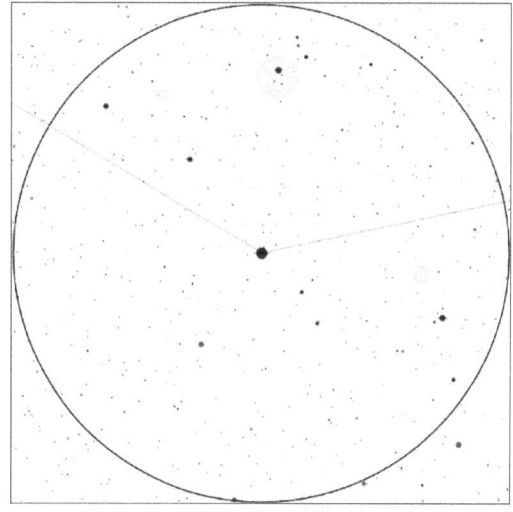

Finderscope view courtesy *Mobile Observatory*

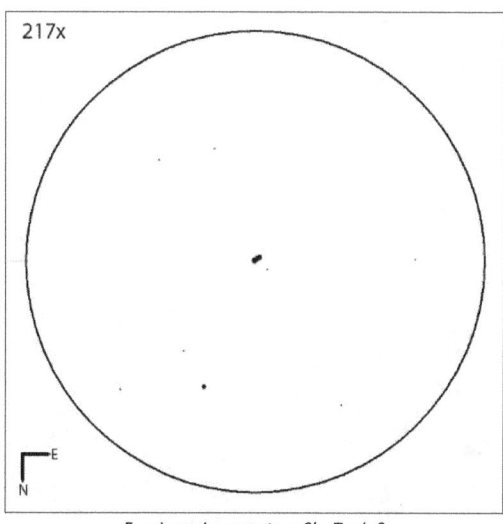

Eyepiece view courtesy *Sky Tools 3*

Sirius

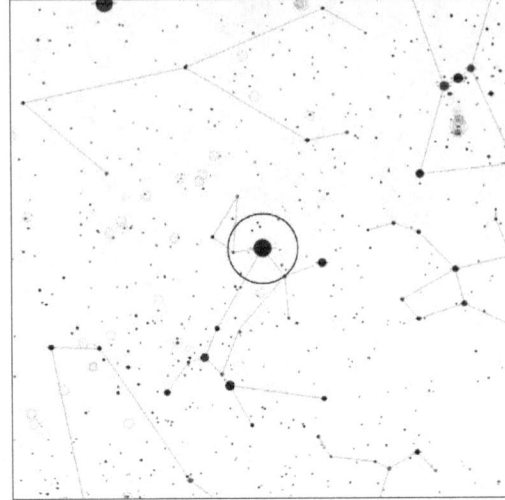

Map courtesy *Mobile Observatory*

Designation(s):	Alpha Canis Majoris
Constellation:	Canis Major
R.A.:	06h 45m 09s
Declination:	-16° 42' 58"
Object Type:	Multiple Star
Location:	★ ★ ★
Rating:	★
Best Seen:	Winter

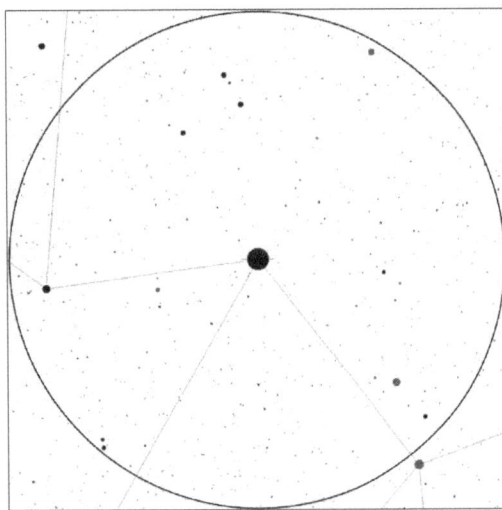

Finderscope view courtesy *Mobile Observatory*

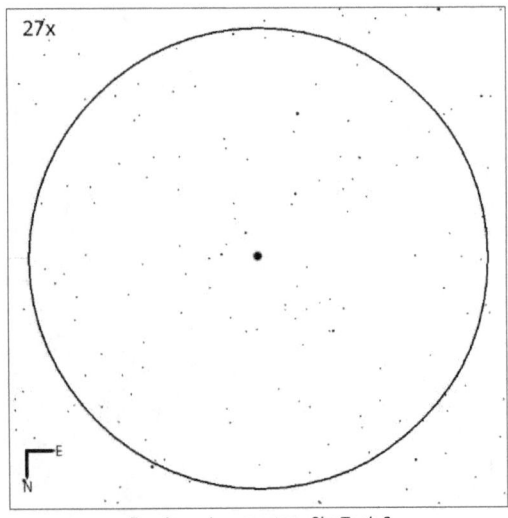

Eyepiece view courtesy *Sky Tools 3*

Sirius, as many people know, is the brightest star in the entire night sky and has been known throughout history by many different civilizations and cultures.

It's a brilliant blue-white star that will often sparkle and flash a myriad of colors when close to the horizon. In fact, its name is actually derived from the Greek for "glowing" and it's not unusual for unsuspecting folk to report it as a UFO.

Easily found by drawing a line through Orion's belt down and toward the east, it's a very easy target and you may have already turned your telescope toward it. What did you see? Personally, I was very surprised to find that it can be dazzling when observed, even at low power.

At 26x I've noticed that it forms a triangle with two much fainter stars and there's a fourth, coppery star between those two that's fainter than all of them.

Sirius is a little over eight and a half light years away and is one of the closest stars to the Earth. Besides the primary star itself, there's a smaller white dwarf companion that can be glimpsed with large telescopes.

This companion was once a red giant that slowly died and shrunk about 120 million years ago. As the primary, Sirius A, is often known as the "dog star" the white dwarf companion, Sirius B, is sometimes nicknamed "the pup."

Designation(s):	Messier 41
Constellation:	Canis Major
R.A.:	06h 46m 00s
Declination:	-20° 45' 15"
Object Type:	Open Cluster
Location:	★ ★
Rating:	★ ★ ★
Best Seen:	Winter

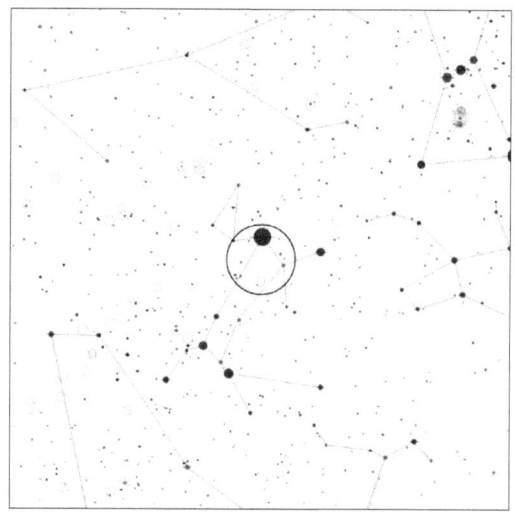

Map courtesy *Mobile Observatory*

Messier 41 is one of my favorite clusters, for a number of reasons. Firstly, it's very easily located, thanks to the proximity of Sirius. All you need to do is point your finder toward that bright star and with Sirius near the top, the cluster will appear near the bottom.

Through binoculars under suburban skies, it appears quite large and has a definite shape to it. It looks as though there's a circular hole on the western edge of the cluster that gives it the appearance of a lobster with its claws outstretched.

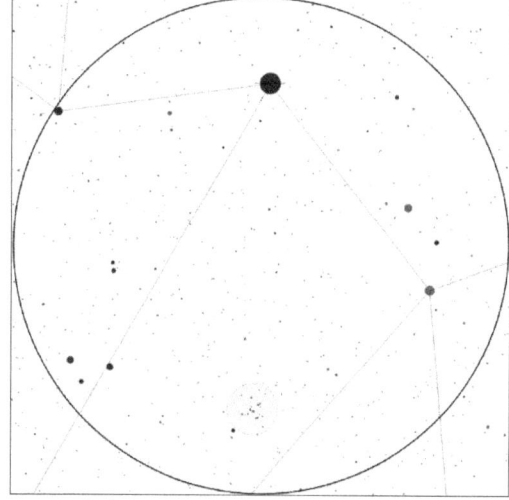

Finderscope view courtesy *Mobile Observatory*

That impression is reinforced when the cluster is observed at 35x. I feel it's one of the best (and under-rated) clusters in the sky, one that allows your imagination to run wild.

It's a nice, large cluster with sparsely scattered stars that reminded me of NGC 457 in Cassiopeia, only with a lot more stars. The vast majority are white or blue-white and many are of about the same brightness. However, there's a pair of stars, just a little brighter than the others and another with an orange hue. This reminds me of Phi Cassiopeiae.

The other stars in the cluster are grouped - to me - as though they're forming the shape of a man with his hands up, or the afore-mentioned lobster… or wings like an angel… or a dove… what do you see?

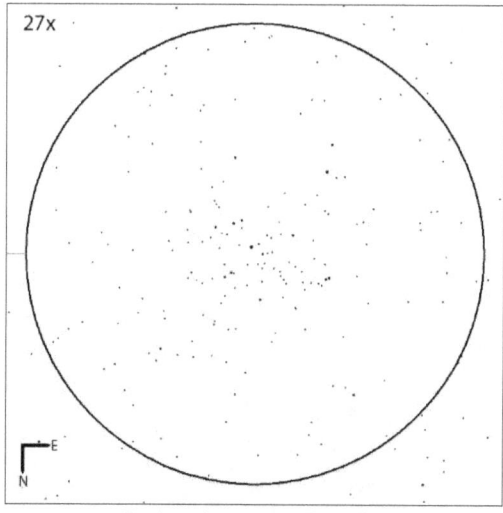

Eyepiece view courtesy *Sky Tools 3*

Procyon

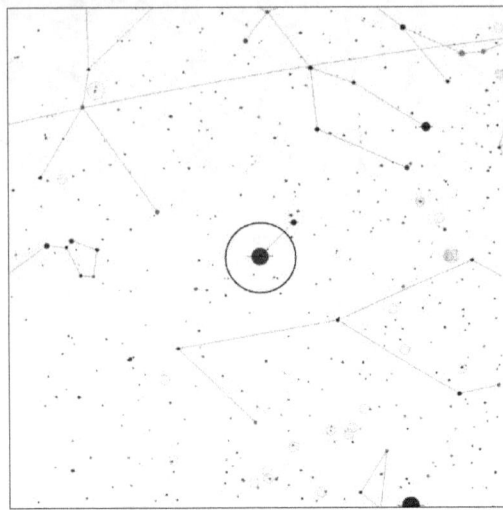

Map courtesy *Mobile Observatory*

Designation(s):	Alpha Canis Minoris
Constellation:	Canis Minor
R.A.:	07h 39m 18s
Declination:	+05° 13′ 29″
Object Type:	Multiple Star
Location:	★ ★ ★
Rating:	★
Best Seen:	Winter

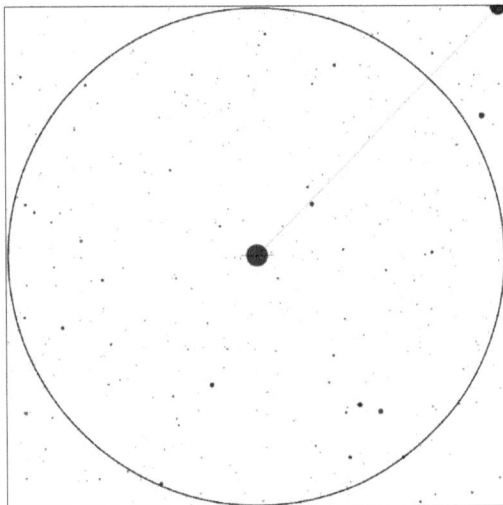

Finderscope view courtesy *Mobile Observatory*

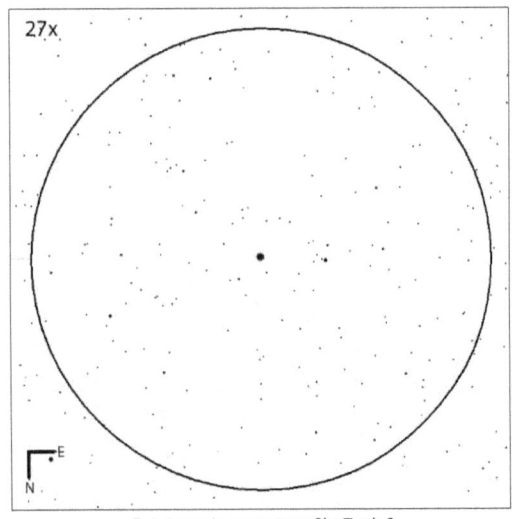

Eyepiece view courtesy *Sky Tools 3*

Procyon is a bright star, easily located to the east of Orion. Use the two stars of his shoulders - Betelgeuse and Bellatrix - to help identify it. (See page 49.)

Through the finderscope (or binoculars) it appears as a brilliant white star with a wide, faint blue-white companion nearby. This companion appears about ten times fainter than Procyon itself.

The pair will still appear within the same field of view at 35x but you'll also notice a scattering of many fainter stars in the background. In particular, look out for another faint star near Procyon and two others near the companion.

Procyon is the brightest of only two bright stars in the constellation of Canis Minor, the Little Dog. As with Sirius, the brightest star in Canis Major, the Great Dog, it follows Orion across the sky, just as the two dogs follow their master as he embarks on his winter hunting trips.

(They all seem totally oblivious to their prey, Lepus the Hare, which can be found below Orion. Maybe they've been distracted by Taurus the Bull, charging from the west.)

Again, like Sirius, Procyon is another bright, white star with a white dwarf companion that's just under eleven and a half light years away. The pair orbit one another every forty years.

Designation(s):	Messier 93
Constellation:	Puppis
R.A.:	07h 44m 29s
Declination:	-23° 51' 11"
Object Type:	Open Cluster
Location:	★
Rating:	★ ★
Best Seen:	Winter

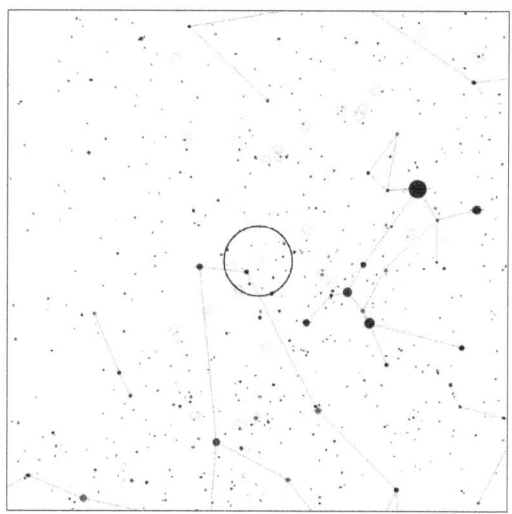

Map courtesy *Mobile Observatory*

This open star cluster is quite easily found, just to the west of Xi Puppis. It's a medium-sized cluster, quite dense with the vast majority of stars being blue-white with many other fainter stars visible with averted vision.

To get the best view, you'll want to get away from the city as I could barely see it at 26x. At that magnification, it appeared small but I was still able to see two bright stars at the tip of a group of much fainter stars. The group looked like a curved triangle.

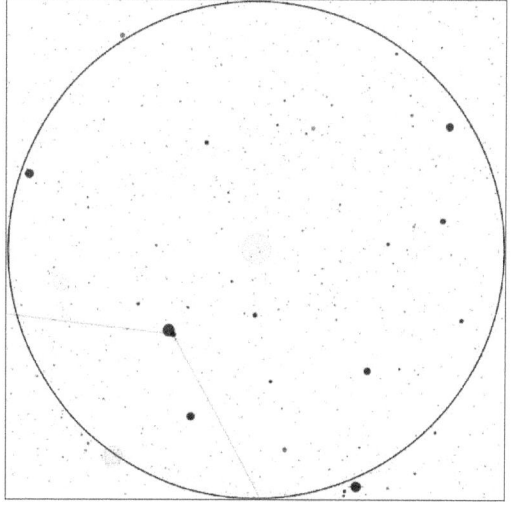

Finderscope view courtesy *Mobile Observatory*

The view from darkened skies is much better as many of the fainter stars are more easily visible. At 35x, the cluster made me think of a butterfly with the two bright, orange stars on the tip of the western wing.

Whether you're observing from the suburbs, the country or from the city itself, the cluster stands up well to higher magnification. I found the best view to be had at around 75x. Increasing the magnification to about 100x only made the cluster appear larger, rather than richer. (A larger telescope, with greater light gathering power, would almost certainly reveal more stars.)

The cluster was first recorded by Charles Messier in March 1781, just a week after William Herschel discovered the planet Uranus. The cluster lies about 3,600 light years away and is thought to contain about 80 stars and is about 25 light years in diameter.

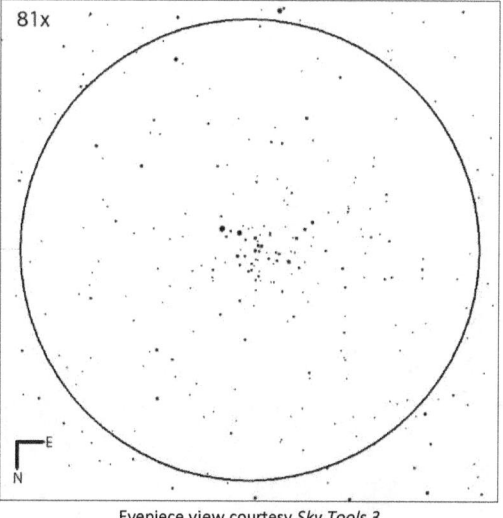

Eyepiece view courtesy *Sky Tools 3*

The Praesepe

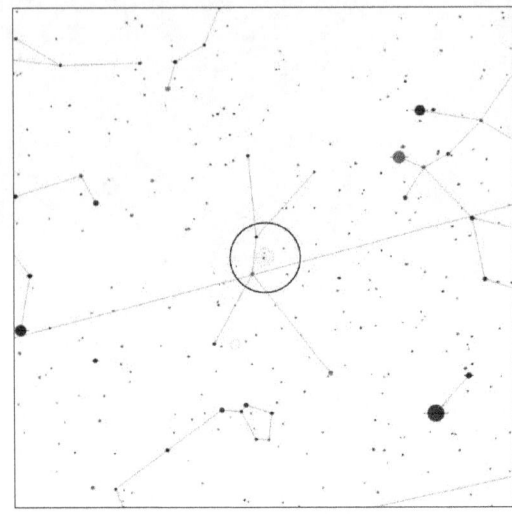

Map courtesy *Mobile Observatory*

Designation(s):	Messier 44
Constellation:	Cancer
R.A.:	08h 40m 22s
Declination:	+19° 40' 19"
Object Type:	Open Cluster
Location:	★ ★
Rating:	★ ★ ★
Best Seen:	Winter and Spring

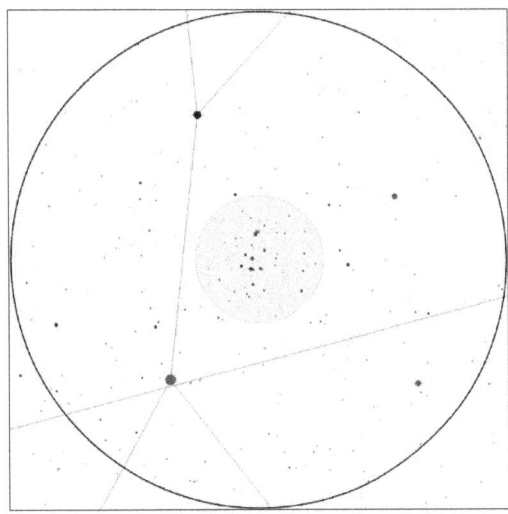

Finderscope view courtesy *Mobile Observatory*

Known since antiquity, the traditional name for the cluster is the Praesepe, which means "the manger." Another popular name is the Beehive Cluster. Personally, it always gives me the impression of a swarm of bees, but I also like the traditional name.

Whatever name you choose, you'll find spotting the cluster to be a good test of your environment. In theory, you should be able to see it with just your eyes but finding it can be a little problematic as it's actually brighter than any of the stars that form the constellation of Cancer itself.

Look midway between the mid-section of Gemini and Regulus, the brightest star in Leo, the lion. With luck, you should be able to see a tiny, misty patch. I've seen it from the suburbs but in the city I've always needed a pair of binoculars to help me.

It's a large cluster and, consequently, one that's best suited to lower magnifications. Even at 35x it'll barely fit within the field of view. But it's a very nice view, all the same with dozens of stars glinting against the night. The majority of stars appear blue-white, indicating a young age. Several others are older orange giants.

There are a lot of doubles to be found here, but look toward the center: can you see a mini Cepheus? It always leaps right out at me!

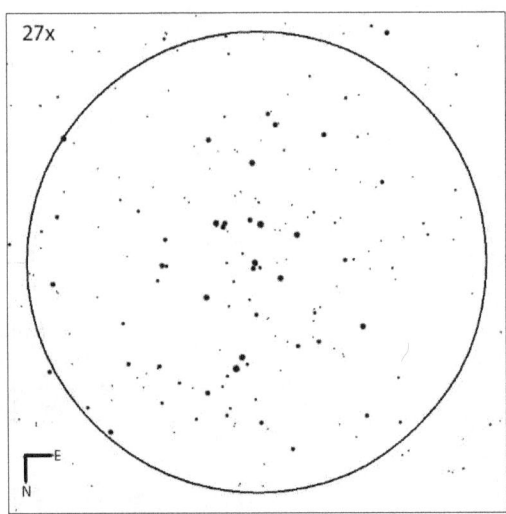

27x

Eyepiece view courtesy *Sky Tools 3*

Designation(s):	Iota Cancri
Constellation:	Cancer
R.A.:	08h 46m 42s
Declination:	+28° 45' 36"
Object Type:	Multiple Star
Location:	★ ★ ★
Rating:	★ ★ ★
Best Seen:	Winter and Spring

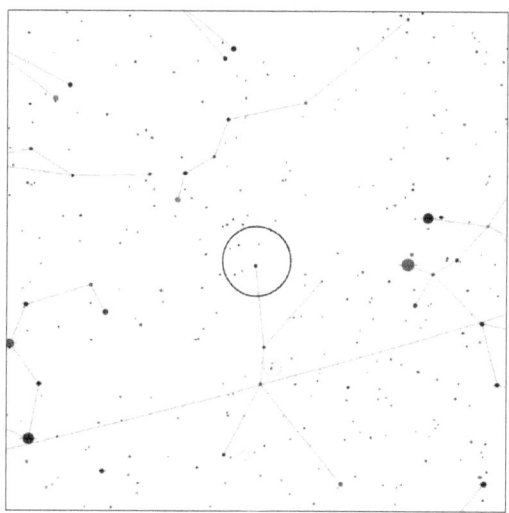

Map courtesy *Mobile Observatory*

Iota Cancri is like a pale Albireo for late winter and early spring. Unfortunately, it's not a prominent star, which means you might have difficulty finding it from the city. Scan the area about midway between Pollux and the top of the backwards question-mark that represents Leo's head.

Through a finderscope or binoculars, you'll see 53 and Rho Cancri - another, wider coppery double pair to the east but within the same field of view. This is a nice sight for binoculars and may be a consolation prize for anyone observing without a telescope.

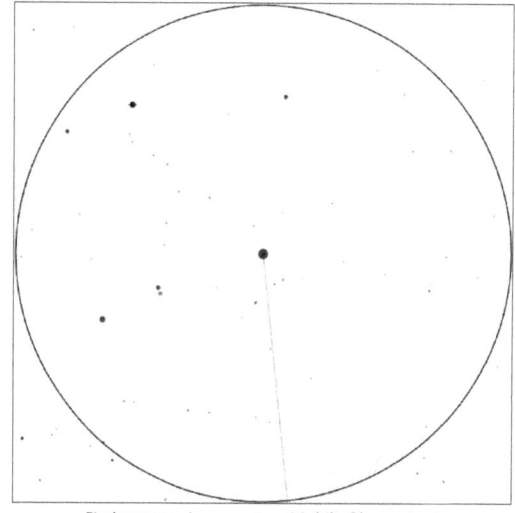

Finderscope view courtesy *Mobile Observatory*

Once you've found Iota, you'll find it easy to split with a low powered eyepiece. Just 26x will reveal both components. The primary is pale gold and appears to be about two or three times brighter than the very pale blue secondary.

Although it looks good at all magnifications, it's probably best at about 100x where a decent distance appears between the pair. As a bonus, I've noted that the secondary shows definite signs of purple and deep blue when observed at higher powers.

The pair is not a true multiple star system – it's what's known as an *optical* binary. In other words, the stars only appear close to another due to chance alignment.

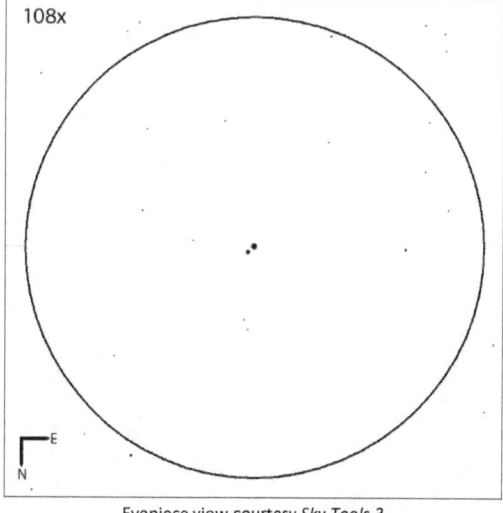

Eyepiece view courtesy *Sky Tools 3*

Messier 67

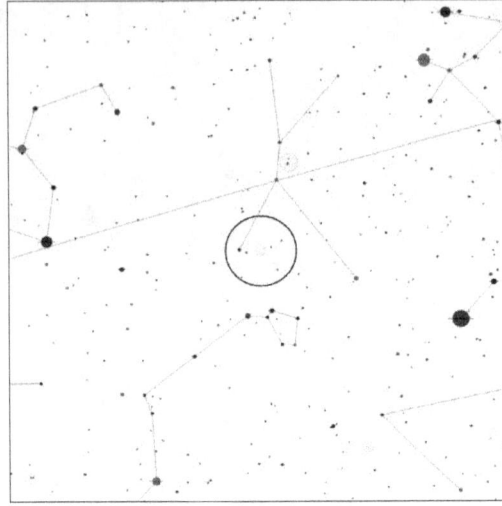

Map courtesy *Mobile Observatory*

Designation(s):	Messier 67
Constellation:	Cancer
R.A.:	08h 51m 20s
Declination:	+11° 48' 43"
Object Type:	Open Cluster
Location:	★
Rating:	★ ★
Best Seen:	Winter and Spring

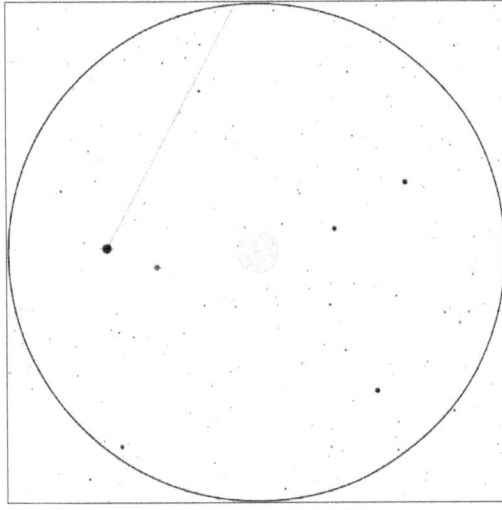

Finderscope view courtesy *Mobile Observatory*

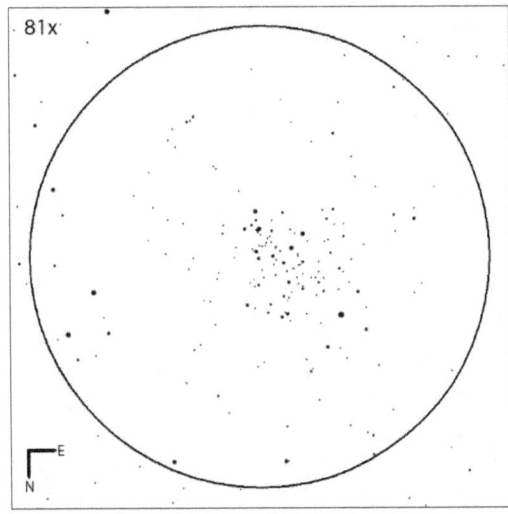

Eyepiece view courtesy *Sky Tools 3*

This late Winter / early Spring cluster is reminiscent of Messier 37 in Auriga. It's a tightly compacted cluster of faint blue-white stars that mostly appear to be about the same magnitude. There are a couple of notable exceptions, however.

It's barely visible from the city but can be glimpsed at 26x and using averted vision will help to bring out the faintest stars. In particular, look for a bright orange star that appears on the edge of the shield-shaped cluster. You should also see a pair of faint stars on the opposite edge but the remaining stars might be difficult.

Increasing the magnification will help – it will still fit within the field of view at about 90x - but the best views are reserved for those observing from a dark location.

Even at 35x, you'll see a glimmering diamond shape that appears to contain hundreds of tiny stars. To me, again, like Messier 37, it gave me the impression of sugar or even diamond dust on velvet but besides the overall shape of the cluster itself I didn't notice any particular patterns.

At more than three billion years old the cluster is one of the oldest known and lies at a distance of over 2,500 light years. Containing more than five hundred stars, it's thought to be about ten light years in diameter.

Designation(s):	Alpha Leonis
Constellation:	Leo
R.A.:	10h 08m 22s
Declination:	+11° 58' 02"
Object Type:	Multiple Star
Location:	★ ★ ★
Rating:	★
Best Seen:	Spring

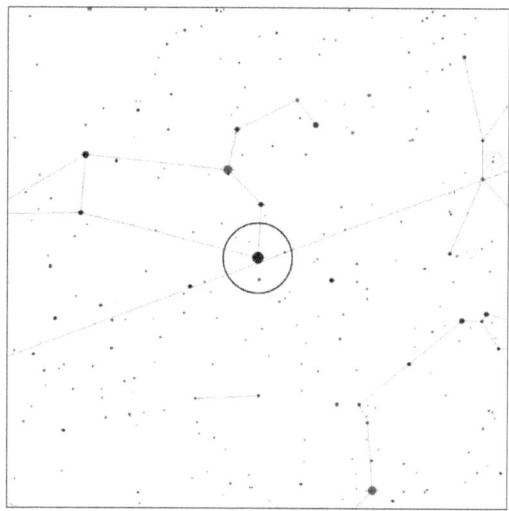

Map courtesy *Mobile Observatory*

Regulus is one of the brightest stars in the sky and is also one of the most overlooked – at least by experienced amateurs.

Through binoculars or a finderscope, you'll see the reasonably bright stars Nu and 31 Leonis within the same field of view, but look carefully with binoculars and you might see something else.

You'll need to hold your binoculars steady and you may have to wait for the air to be still but you might just see a very faint companion to the north-west. (If your finder magnifies the view, try your luck with that too.)

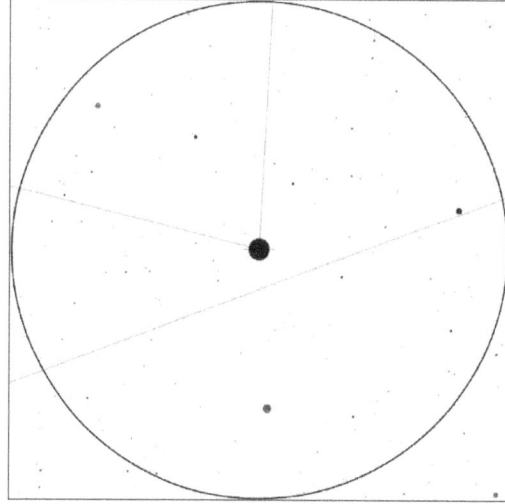

Finderscope view courtesy *Mobile Observatory*

The pair are much more apparent at low power in a small telescope. At 27x, you'll see Regulus as a brilliant white star, much brighter than the bluish secondary. It'll be an easy and wide split but if you want the secondary to be more easily seen, you'll need to up the power some more. I found that 50x provided quite a nice view.

Regulus is the brightest star in the constellation of Leo the Lion and, at 79 light years away, is relatively close to Earth. The pair are a part of a true multiple star system with both components orbiting one another once every two million years.

Each component is itself double, making this a four star system in all.

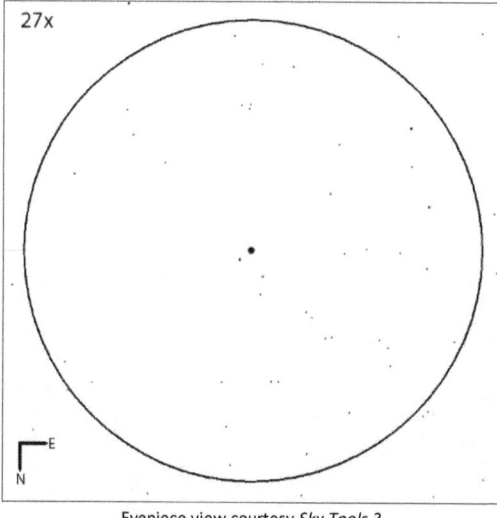

Eyepiece view courtesy *Sky Tools 3*

Adhafera

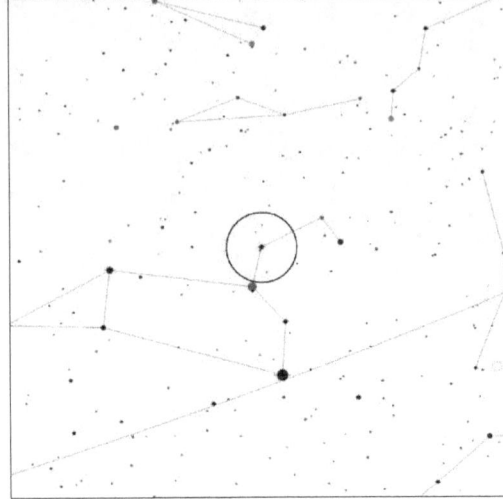

Map courtesy *Mobile Observatory*

Designation(s):	Zeta Leonis
Constellation:	Leo
R.A.:	10h 16m 41s
Declination:	+23° 25' 02"
Object Type:	Multiple Star
Location:	★ ★ ★
Rating:	★ ★
Best Seen:	Spring

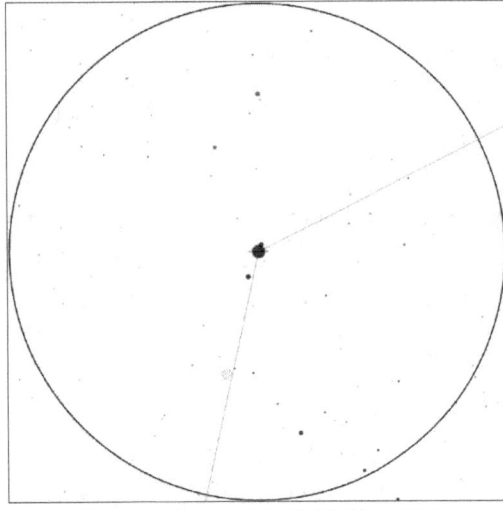

Finderscope view courtesy *Mobile Observatory*

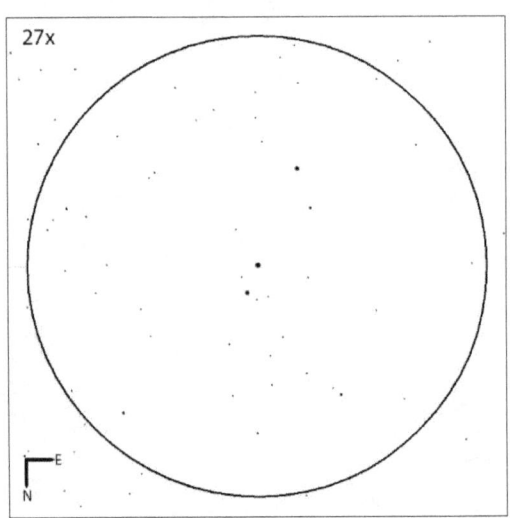

Eyepiece view courtesy *Sky Tools 3*

Adhafera is another overlooked star in Leo, the Lion. Through the finder or binoculars you'll see a fainter star, 35 Leonis, to the south. This star is not associated with Adhafera and its position is purely due to chance.

Specifically, Adhafera is about 260 light years away while 35 Leonis is about 100 and both stars are moving in different directions.

Unfortunately, I don't have any notes from binocular observations, but you may also be able to spot another, fainter companion to the north-east. It'll appear very close to Adhafera itself with the pair best observed through a small telescope.

At 27x you'll see a wide pair of stars comprising of a lemony-white primary that's about 1½ times brighter than its white companion.

As with many double stars, I've noted some variance in the colors; sometimes the primary appears to be purely white while the secondary can appear peachy.

Adhafera is notable for another important reason – the Leonid meteors appear to radiate from a point just to the west of the star. This shower peaks around November 18th and usually produces about ten or twenty shooting stars per hour. However, on occasion, the shower has been known to exceed a hundred.

Designation(s):	Gamma Leonis
Constellation:	Leo
R.A.:	10h 19m 58s
Declination:	+19° 50' 30"
Object Type:	Multiple Star
Location:	★ ★ ★
Rating:	★ ★ ★
Best Seen:	Spring

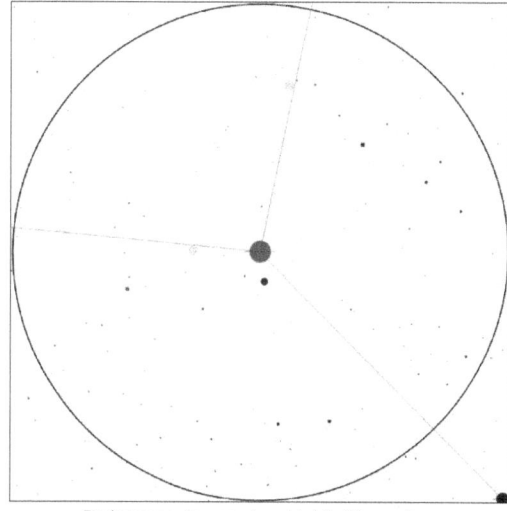

Map courtesy *Mobile Observatory*

Algieba is an easy, wide double star that can be easily seen with either binoculars or even your own unaided eyes. The primary appears gold and about two or three times as bright as the blue-ish companion. You might also notice a third, slightly fainter star that forms a triangle with the two brighter stars.

The real magic happens when you turn a telescope toward the trio. You won't notice anything different until you get to about 50x or higher. The gold star may be barely split at this magnification with a pale white gold companion appearing close beside it.

If you're having difficulty with the pair, it might be due to your location and/or your observing conditions. I've been able to comfortably split it at under 100x from the suburbs but from the city it's barely split at even 162x. (This, no doubt, is also due to the higher humidity of my location and the fact that I was observing the star in June, when temperatures were still 70°F at night.)

What's the lowest magnification you can use to split the star and see the secondary? Find a magnification where the secondary is clearly seen and then trying working your way back down until it disappears again. At what point does Algieba go from being a clean split to elongated and then just a single point of light?

Finderscope view courtesy *Mobile Observatory*

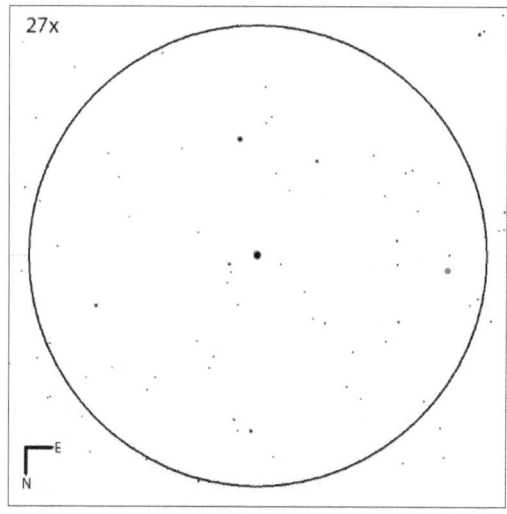

Eyepiece view courtesy *Sky Tools 3*

Denebola

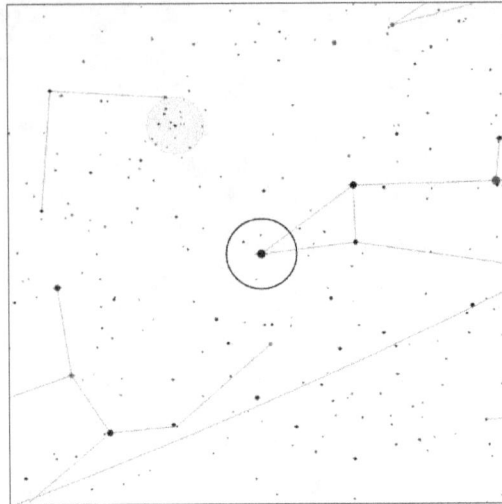

Map courtesy *Mobile Observatory*

Designation(s):	Beta Leonis
Constellation:	Leo
R.A.:	11h 49m 04s
Declination:	+14° 34' 19"
Object Type:	Multiple Star
Location:	★ ★ ★
Rating:	★ ★
Best Seen:	Spring

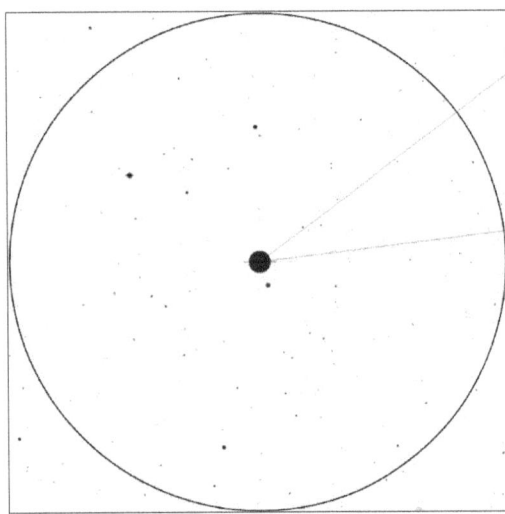

Finderscope view courtesy *Mobile Observatory*

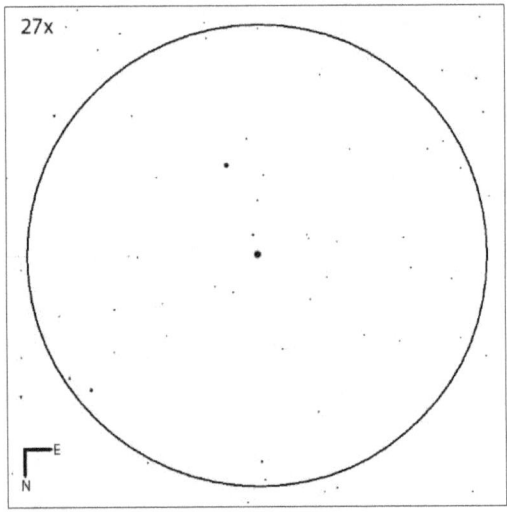

Eyepiece view courtesy *Sky Tools 3*

Denebola is a bright star that marks the tail of Leo the Lion – and it's not one you usually find highlighted in an astronomy book.

In fact, I really didn't pay any attention to it until I bought a small 70mm reflector for my son and was looking for easy objects for him to observe. In essence, it was the beginnings of the very book you're reading now and demonstrates the benefits of going off the beaten track and sometimes just randomly exploring the night sky.

Through binoculars or a finderscope, you'll see a wide pair of stars, similar to Algieba. Both are white with the primary about four times brighter than the secondary.

Similarly, if you observe the star through a small telescope – even at a low magnification of 15x – you'll easily see the pair but a third, much fainter star will become apparent. It's a lot fainter than the primary and can be seen about a quarter of the way between the first two stars. You may need to use averted vision to see it properly.

I then decided to take a look through my 130mm 'scope and increased the magnification even further. Once I got to 108x, I got another surprise. While all three stars could still be seen within the field of view, a fourth star now appeared, fainter than the third. Look for it about midway between the first two and a little out of line with the others.

Designation(s):	Delta Corvi
Constellation:	Corvus
R.A.:	12h 29m 52s
Declination:	-16° 30' 56"
Object Type:	Multiple Star
Location:	★ ★ ★
Rating:	★ ★
Best Seen:	Spring

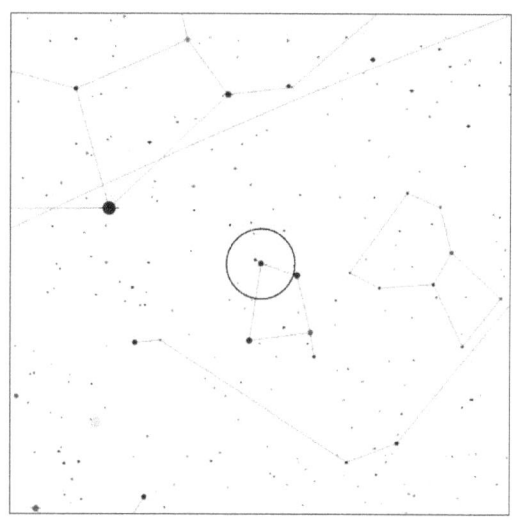

Map courtesy *Mobile Observatory*

Algorab is the third brightest star in the constellation of Corvus the Crow. In Greek mythology, the god Apollo sent the crow to fetch him a cup of water but the bird stopped to eat figs and was slow to return. To punish the bird, Apollo threw him and the cup into the sky. Consequently, Crater, the Cup, appears to the west of Corvus and will forever be out of reach of the bird.

Through a finder or binoculars, you'll see a wide, unrelated companion star, just to the north-west. You may even be able to see it with just your eyes. However, you'll need a telescope to see both components of Algorab itself.

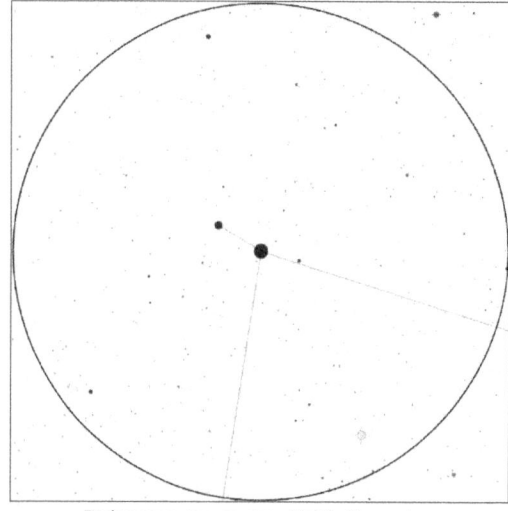

Finderscope view courtesy *Mobile Observatory*

The star was usually split at 26x but it was dependent upon the quality of the atmosphere, especially from the city. It can be a clean split but the secondary is very faint and may not be seen unless you use a higher magnification first.

It was a lot easier under suburban skies at 35x. The primary appeared white with a very pale and faint rusty colored secondary close beside it. Increasing the magnification might improve the view but I've noticed the colors appear to fade in doing so.

In reality, the two stars are moving along a similar path through space but they're not thought to be a true multiple star system.

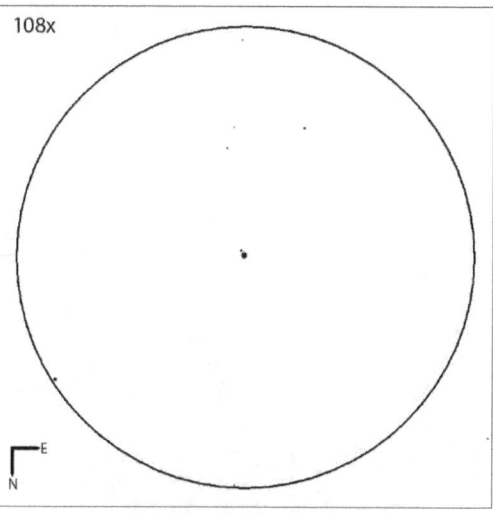

Eyepiece view courtesy *Sky Tools 3*

Cor Caroli

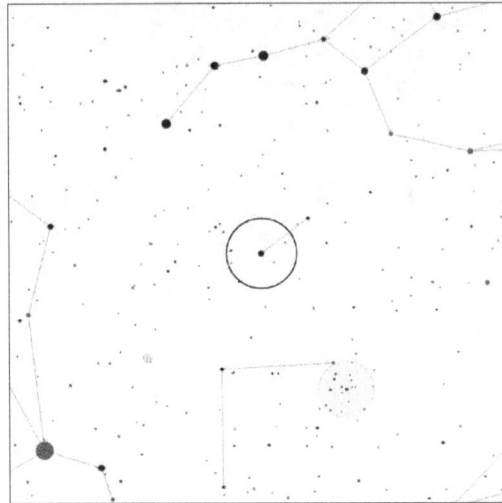

Map courtesy *Mobile Observatory*

Designation(s):	Alpha Canum Venaticorum
Constellation:	Canes Venatici
R.A.:	12h 56m 02s
Declination:	+38° 19' 06"
Object Type:	Multiple Star
Location:	★ ★ ★
Rating:	★ ★ ★
Best Seen:	Spring and Summer

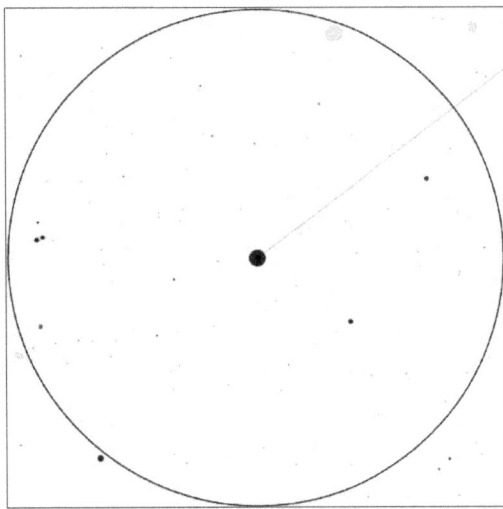

Finderscope view courtesy *Mobile Observatory*

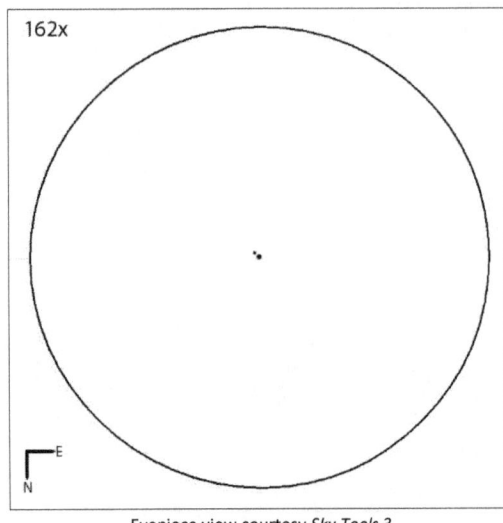

Eyepiece view courtesy *Sky Tools 3*

The name Cor Caroli actually means "Charles's Heart" and it originates from the 17th century when it was said the star shone particularly brightly upon the King's return to England.

It's a showpiece multiple star for the Spring and Summer skies and another favorite with astronomers throughout the northern hemisphere. Even at a low magnification of just 26x the star is clearly split and provides a beautiful view. Increasing the power to about 50x or 60x will probably give you the best view.

The primary is white and appears about three times brighter than the secondary. However, I've noted slightly different colors for the secondary; more often than not, I've noted that it appears white too, but on occasion it seems to hold a creamy or pale gold color. (I've even noted a grey-greenish color at a magnification of 182x.)

What colors do you see?

(Most other sources describe the pair as both being either white or blue-white in color but some have also noted the different colors reported by observers.)

Cor Caroli is a true binary star system some 110 light years away with the pair orbiting one another roughly every 8,300 years. Curiously, the primary component is highly magnetic and has a magnetic field thousands of times stronger than the Earth's.

Designation(s):	Zeta and 80 Ursae Majoris
Constellation:	Ursa Major
R.A.:	13h 23m 56s
Declination:	+54° 55' 31"
Object Type:	Multiple Star
Location:	★ ★ ★
Rating:	★ ★ ★
Best Seen:	Spring and Summer

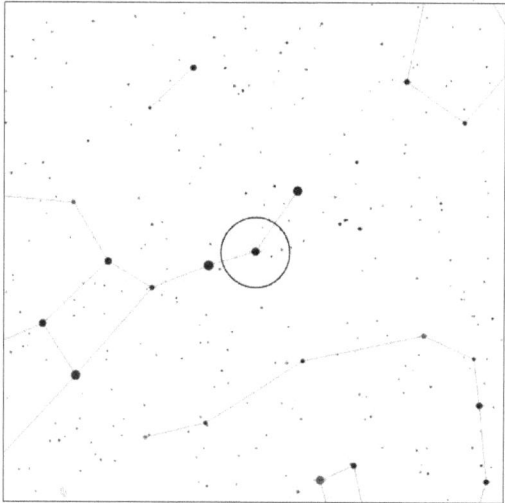

Map courtesy *Mobile Observatory*

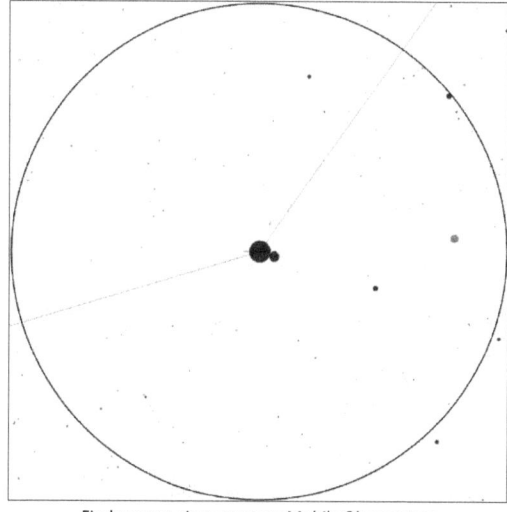

Finderscope view courtesy *Mobile Observatory*

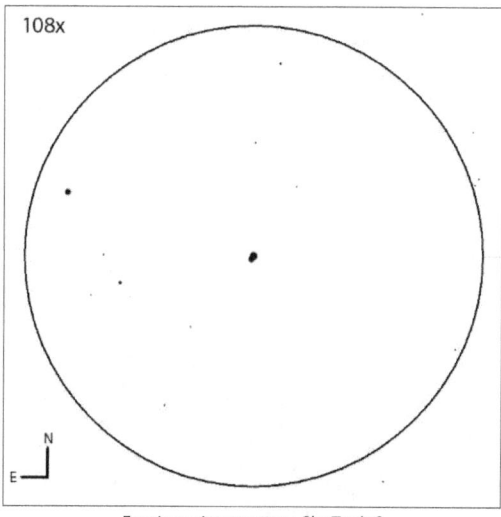

Eyepiece view courtesy *Sky Tools 3*

Mizar and Alcor are a wide pair of stars that have been considered a test of eyesight since ancient times. As both stars are well within the limits of naked eye visibility, I've never had any trouble seeing them both – even from the city.

If you're familiar with the seven stars of the Dipper (or the Plough, as it's known in the United Kingdom) you'll find the pair in the middle of the curved line of three that make up the handle. Mizar is, of course, immediately visible but you might need to stare for a few moments to spot Alcor.

The pair make for an attractive sight in binoculars with Mizar being about twice as bright as Alcor and both stars appearing blue-white.

However, like many other doubles, it's only when you turn a telescope toward the stars that everything comes into view. At a low power of just 27x, Mizar is clearly and easily split with the secondary also blue-white and about twice as faint as the primary.

While you're in the area, look out for a third star that forms a triangle with Alcor and the two components of Mizar.

Mizar and Alcor is not a true double star system but Mizar itself is a quadruple star some 85 light years away.

Arcturus

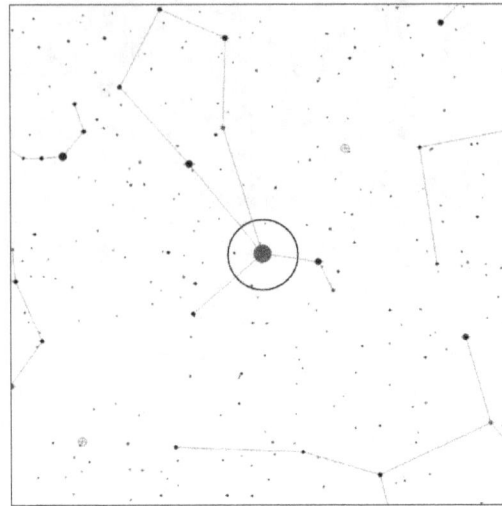

Map courtesy *Mobile Observatory*

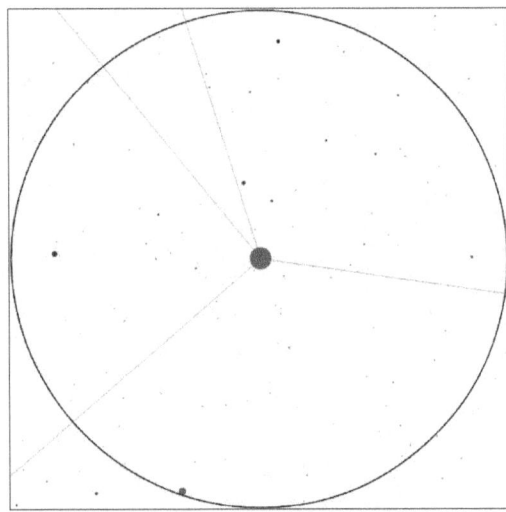

Finderscope view courtesy *Mobile Observatory*

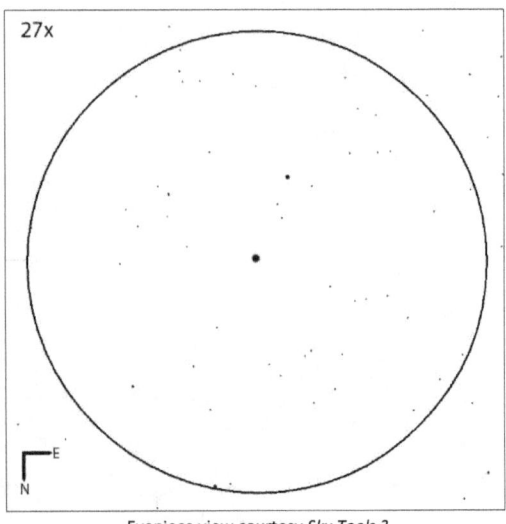

Eyepiece view courtesy *Sky Tools 3*

Designation(s):	Alpha Boötis
Constellation:	Boötes
R.A.:	14h 15m 40s
Declination:	+19° 10' 56"
Object Type:	Multiple Star
Location:	★ ★ ★
Rating:	★
Best Seen:	Spring and Summer

Arcturus is the fourth brightest star in the sky and, at just under 37 light years away, one of the closest to the Earth.

It's a red giant star that appears orange to the unaided eye but despite this, at 27x I've noted that it shows a brilliant, white gold hue. Other observers have noted a range of colors including yellow, peach and topaz.

(If you want to see the star truly sparkle, wait until it's low on the horizon. Like Sirius, you'll find it flashing a multitude of colors when it's close to setting in the west.)

This has always been something of a favorite star for me. It's easily found by following the curved tail of Ursa Major, the Great Bear, down towards the south. ("Arc down to Arcturus.") It appears at the bottom of kite-shaped Boötes, a large constellation representing a herdsman of the same name. (The name Arcturus actually means "guardian of the bear.")

Arcturus is an old star, maybe seven billion years old, but is still over a hundred times more luminous than the Sun. As one of the fastest moving stars, it's currently drawing nearer and will be at its closest in about 4,000 years. In the far more distant future, it will eventually shed its outer layers and shrink down to a white dwarf, forming a planetary nebula in the process. Unfortunately, that sight will only be seen by our descendants as you and I will both be long gone!

Designation(s):	Delta Boötis
Constellation:	Boötes
R.A.:	15h 15m 30s
Declination:	+33° 18' 53"
Object Type:	Multiple Star
Location:	★ ★ ★
Rating:	★ ★ ★
Best Seen:	Spring and Summer

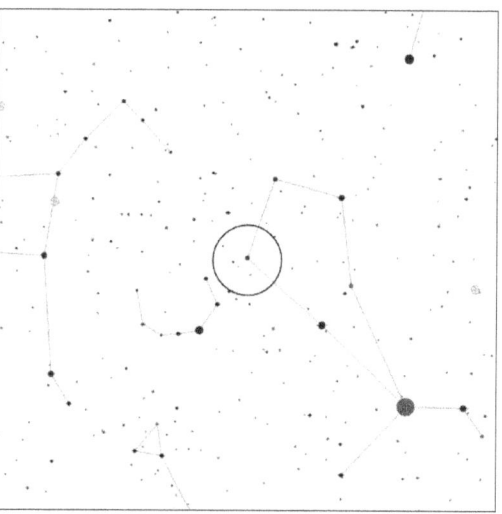

Map courtesy *Mobile Observatory*

There are a number of multiple stars to be found in Boötes but Delta Boötis is one of the easiest to find. It's a relatively bright star that marks the upper-left corner of the kite-shaped constellation.

Although a fairly wide double, I haven't been able to split the pair with binoculars but others have noted that it's been done so it's definitely worth a try. (I was using a smaller pair of 8x30's so a standard pair of 10x50's might be just enough to split them.)

If you're using a small telescope, a low magnification of about 25x should easily do the trick.

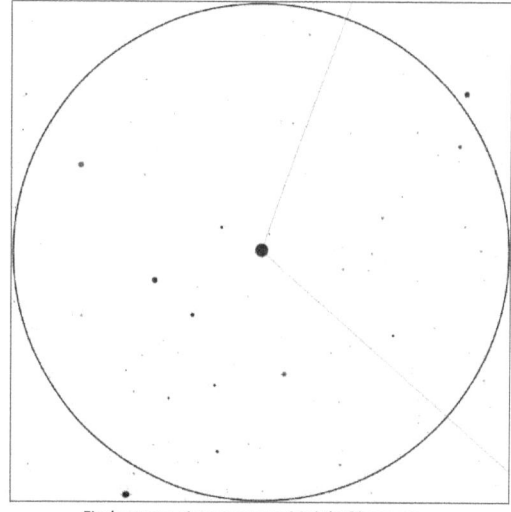

Finderscope view courtesy *Mobile Observatory*

At that magnification, it appears to be a wide and pretty pair with a nice color contrast between the two stars. The pale yellow-white primary appears about five times brighter than pale blue secondary. Add to that a number of other stars that appear within the same field of view and you have a sight worth looking out for.

Delta is thought to be a true double star system, lying some 121 light years from Earth. The primary is a giant star nearing the end of its life and some ten times larger than our own Sun while the secondary is a dwarf star a little smaller than the Sun. The pair are thought to orbit one another once every 120,000 years.

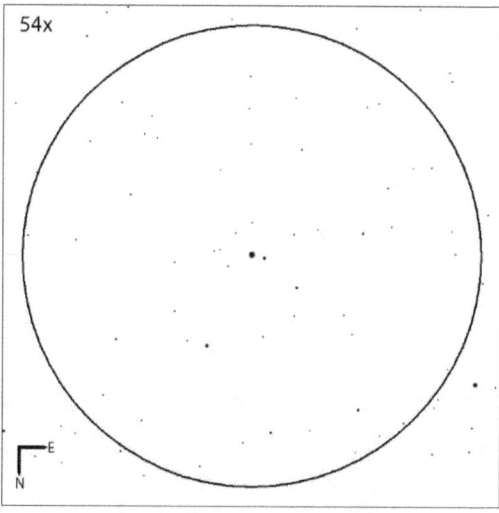

Eyepiece view courtesy *Sky Tools 3*

Zuben Elgenubi

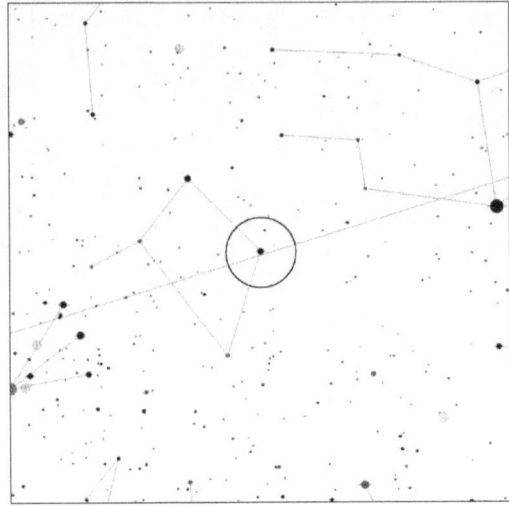

Map courtesy *Mobile Observatory*

Designation(s):	Alpha Librae
Constellation:	Libra
R.A.:	14h 50m 41s
Declination:	-15° 59' 50"
Object Type:	Multiple Star
Location:	★ ★ ★
Rating:	★ ★
Best Seen:	Spring

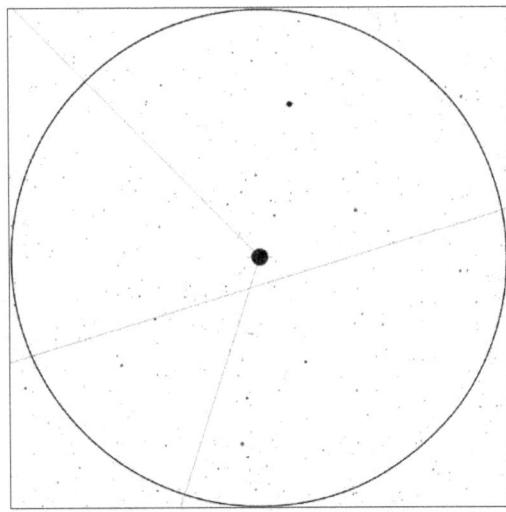

Finderscope view courtesy *Mobile Observatory*

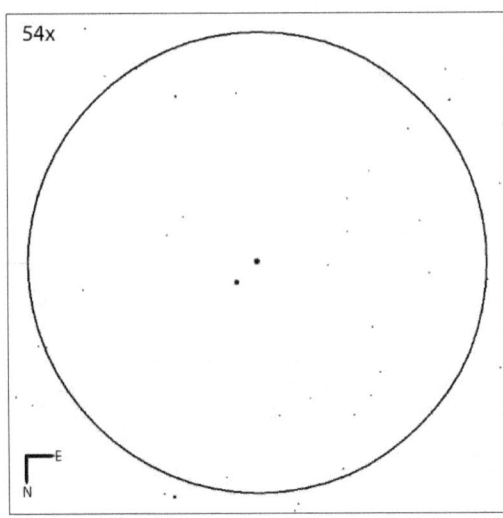

Eyepiece view courtesy *Sky Tools 3*

Zuben Elgenubi – or Alpha Librae, if you prefer – is a wide pair of stars easily seen with both binoculars and telescopes at low power. In fact, this is one example where the object is best at low power as increasing the magnification too much reduces the aesthetic appeal of the stars.

Both stars are white but the primary is about 1½ times brighter than the secondary.

The name Zuben Elgenubi is Arabic and means "the southern claw." In case you're wondering how that relates to a constellation that represents scales, it originates from a time when the stars of Libra were still considered part of nearby Scorpius, the Scorpion.

Despite it having the designation of Alpha Librae, it is in fact the second brightest star in the constellation. It's thought to be a true binary system (as opposed to being just a chance alignment of two stars) that lies about 77 light years away. The pair are estimated to obit one another once every 200,000 years.

Each star is itself a binary system, making four components in all with a suspected fifth member – KU Librae – appearing nearby.

Designation(s):	Beta Scorpii
Constellation:	Scorpius
R.A.:	16h 05m 26s
Declination:	-19° 48' 20"
Object Type:	Multiple Star
Location:	★ ★ ★
Rating:	★ ★ ★
Best Seen:	Spring and Summer

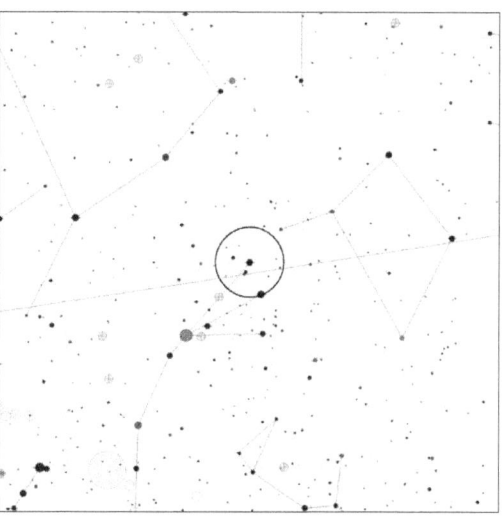

Map courtesy *Mobile Observatory*

Graffias is an interesting star – and also slightly confusing as the name applies to both Beta and Xi Scorpii. To clarify, Beta appears in the center of the finderscope depiction to the left and is the star we're interested in.

If you're using binoculars, you should easily see Omega[1] and Omega[2], a fairly wide pair of white stars nearby. These are the two stars that can be seen off center, just to the lower left of Beta in the finderscope view to the left. You might also notice a couple of other fainter stars in the field of view.

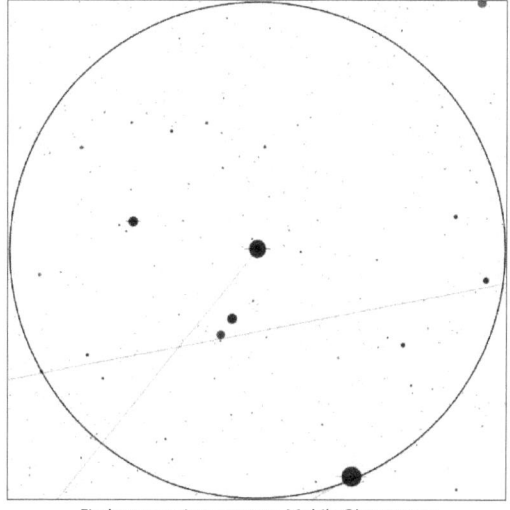

Finderscope view courtesy *Mobile Observatory*

This in itself is a nice enough view, but if you turn your telescope toward Beta you should be able to split it at 27x without any trouble.

Both components of the primary appear white (or sometimes blue-white) with the secondary being about three times fainter than the primary. It's a pretty nice view, even at this magnification.

Look carefully at Beta's secondary as I've noticed it showed signs of a violet color at 81x. However, this was while observing from the city, so it's possible the atmospherics were not great at the time.

There are, in all, six components to Beta Scorpii and the entire system lies some 400 light years away.

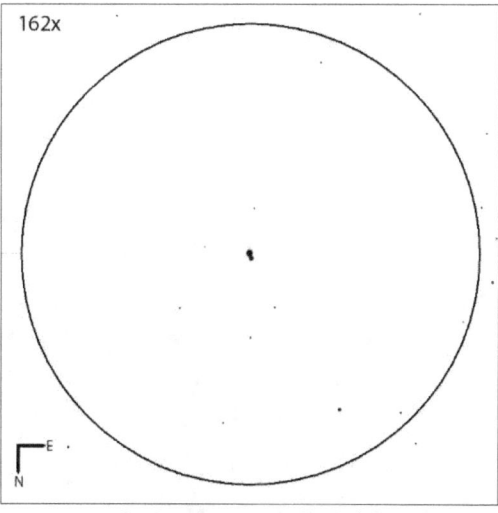

Eyepiece view courtesy *Sky Tools 3*

Jabbah

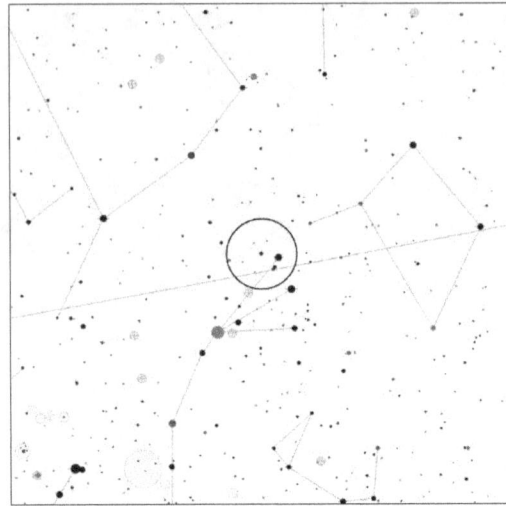

Map courtesy *Mobile Observatory*

Designation(s):	Nu Scorpii
Constellation:	Scorpius
R.A.:	16h 12m 00s
Declination:	-19° 27' 39"
Object Type:	Multiple Star
Location:	★ ★
Rating:	★ ★ ★
Best Seen:	Spring and Summer

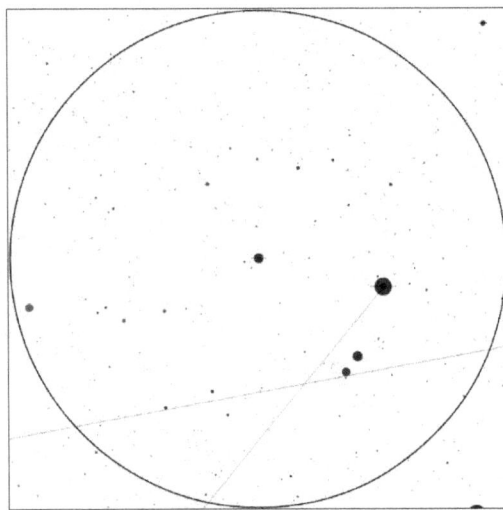

Finderscope view courtesy *Mobile Observatory*

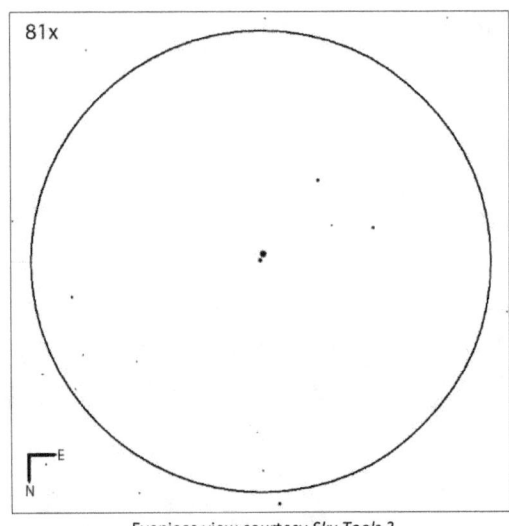

Eyepiece view courtesy *Sky Tools 3*

If you're able to locate Graffias, you should have no difficulty spotting Jabbah as it lies to the east and within the same finderscope field of view.

Unfortunately, despite what you might think (or hope) Jabbah has nothing to do with *Star Wars* - but it's still a pretty interesting sight for telescope observers at both low and medium magnification.

At about 30x you'll see a pair of white stars of similar brightness at one corner of a triangle of three. But increase the magnification and what do you see?

At 81x one of the stars appeared blue-ish, with hints of violet but increasing the magnification again to 107x reveals a third companion.

Look carefully at the fainter component and you might see that it's split into two white stars of almost equal magnitude.

Like many other multiple stars, there are other components to this system that can't be seen with amateur equipment. In fact, it's thought there may be up to seven stars in this system, which would make for an amazing (and possibly blinding) sight if you could ever get there. And you thought a twin sunset over the deserts of Tatooine was cool!

Designation(s):	Messier 80
Constellation:	Scorpius
R.A.:	16h 17m 02s
Declination:	-22° 58' 34"
Object Type:	Globular Cluster
Location:	★
Rating:	★
Best Seen:	Spring and Summer

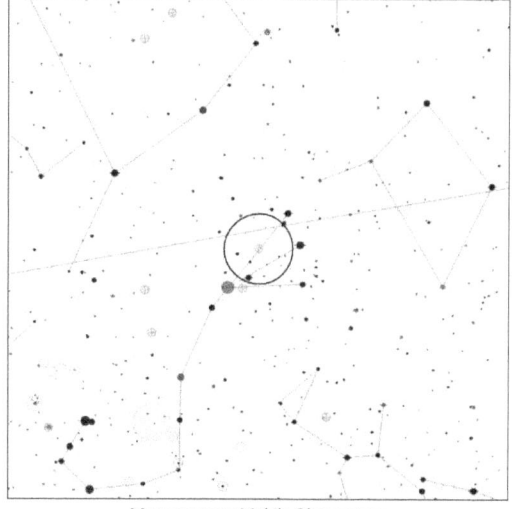

Map courtesy *Mobile Observatory*

Messier 80 is the *other* globular cluster that's easily found within the constellation of Scorpius, the Scorpion. It's often overlooked in favor of it's bigger and brighter neighbor, Messier 4, but that's not to say it's not worth seeking out.

From the suburbs, I was able to easily find the cluster at 35x. It appeared as a grey smudge close to a brighter star to the west and a group of stars to the east.

Although it's smaller than Messier 4, it appeared a little brighter to me with a bright core. Unfortunately, at both 70x and 182x there didn't seem to be much change and I wasn't able to resolve the cluster into individual stars.

Finderscope view courtesy *Mobile Observatory*

It was another ten years before I tried again, but this time it was from the brighter skies of the city and I had less success than before. I was able to spot it at 27x, when it appeared as a very small, faint, circular grey sphere close to a star, but again, I wasn't able to improve the view.

So why hunt it down at all? Well, think about what you're seeing here. Messier 80 is over 32,000 light years away and is thought to contain about 200,000 stars, all crammed into an area some 95 light years across. In comparison, there are only about 500 stars within 100 light years of the Earth, which means it contains about 400 times as many stars as our local stellar neighborhood.

Are you still a little disappointed?

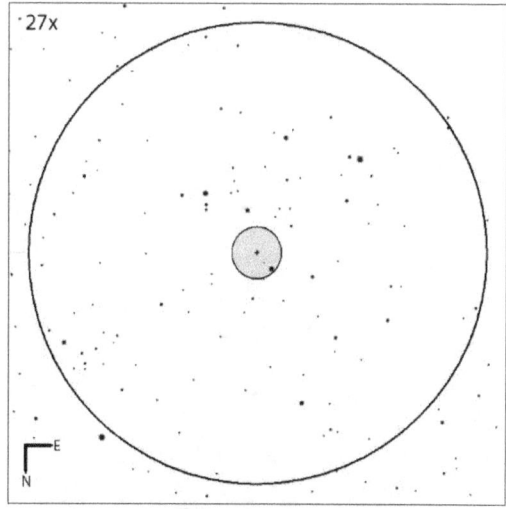

Eyepiece view courtesy *Sky Tools 3*

Messier 4

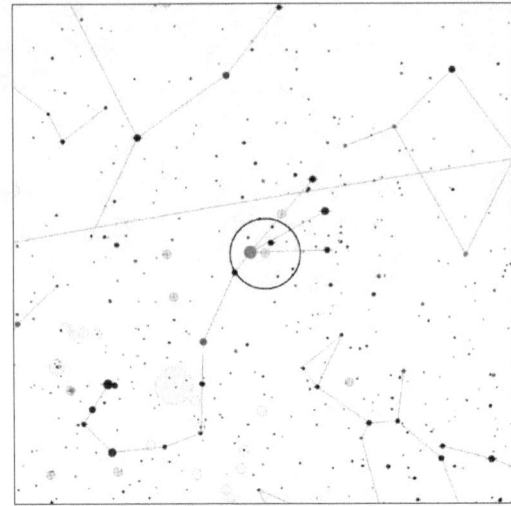

Map courtesy *Mobile Observatory*

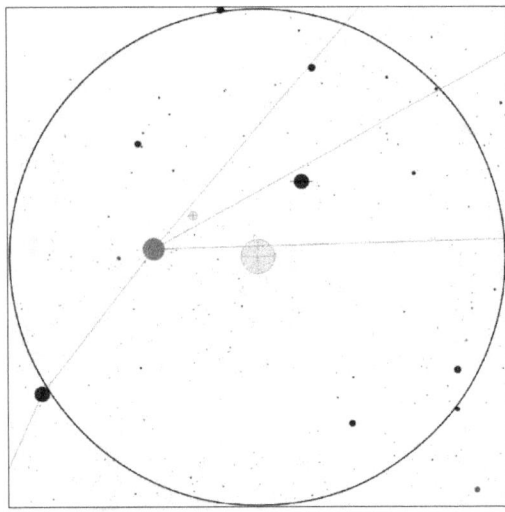

Finderscope view courtesy *Mobile Observatory*

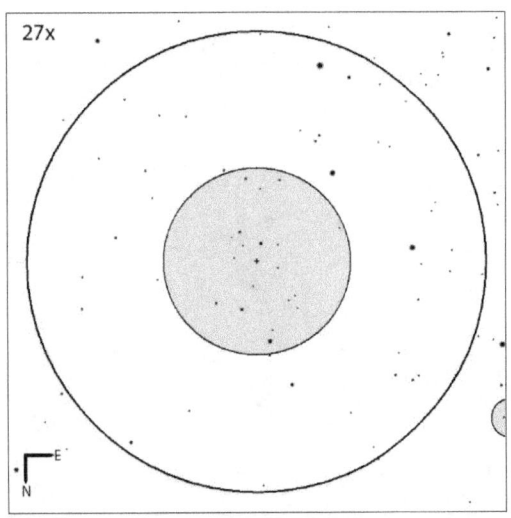

27x

Eyepiece view courtesy *Sky Tools 3*

Designation(s):	Messier 4
Constellation:	Scorpius
R.A.:	16h 23m 35s
Declination:	-26° 31' 33"
Object Type:	Globular Cluster
Location:	★
Rating:	★ ★
Best Seen:	Spring and Summer

If Messier 80 failed to impress, try your hand at Messier 4. It appears very close to Antares, the reddish star that marks the beating heart of Scorpius the Scorpion.

In fact, it's fair to say the cluster is almost unmissable – unless you're trying to find it with binoculars in the city. I've never had any luck there, but if you live under dark skies you shouldn't have much problem at all.

As with almost every deep sky object (such as star clusters, nebulae and galaxies) you'll see more under darker skies and Messier 4 is no exception.

I've been able to pick it out at low power from both the suburbs and the city, but a magnification of about 30x is only enough to show it as a very faint and fuzzy sphere.

At around 50x the cluster appeared elongated, with a band of bright stars stretching across the center and some resolution around the edges with averted vision.

City dwellers may not be so lucky but I've noted some resolution at 81x – again, with averted vision.

In comparison, at around 80x or 90x the view from suburban skies improves quite considerably and the cluster appears quite spectacular. I've noted that thousands of stars could be seen with averted vision and the cluster looked like sugar or salt spilt on black satin.

The Butterfly Cluster

Designation(s):	Messier 6
Constellation:	Scorpius
R.A.:	17h 40m 21s
Declination:	-32° 15' 15"
Object Type:	Open Cluster
Location:	★ ★
Rating:	★ ★ ★
Best Seen:	Summer

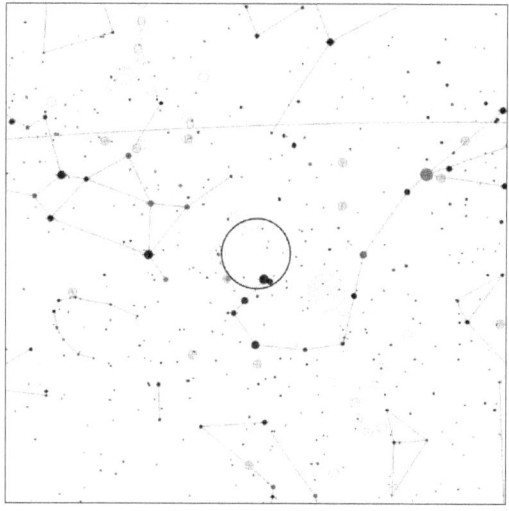

Map courtesy *Mobile Observatory*

The Butterfly cluster is one of two (the other being Messier 7) that won't be visible from the United Kingdom and skims the horizon from North America. If you live in the United States, you shouldn't have much problem as long as the whole of Scorpius can be seen from your location.

To find it, place the two stars representing the sting of the Scorpion's tail on the edge of your field of view in about the five o'clock position. (Messier 7 may be just outside the view at about the nine o'clock position.) The Butterfly cluster may appear near to the eleven o'clock position. I've noted that it's barely visible from suburban skies and is the middle of the three circles in the finderscope depiction.

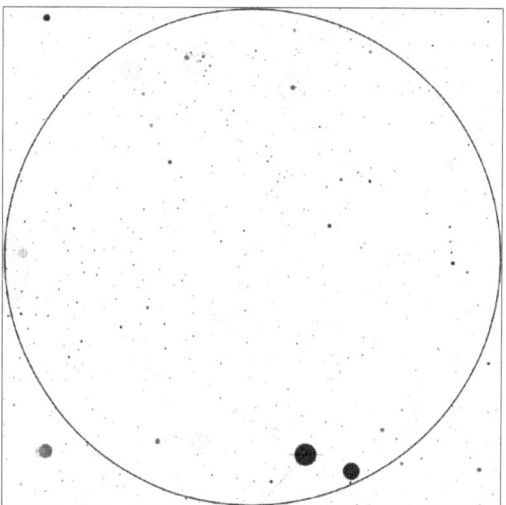

Finderscope view courtesy *Mobile Observatory*

Through binoculars it's about half the size of Messier 7 and looks like tiny, very fine granules of star dust in a triangular shape that points toward the east.

Denser and more compact than Messier 7, the cluster provides a beautiful view at 35x with its butterfly shape being quite apparent. The brightest five stars form a shape like a flattened Pleiades. Four of those are blue-white with the fifth, on the south-western side, being white, possibly orange-ish.

The majority of the stars appear much fainter, look like diamond dust and must number in the hundreds.

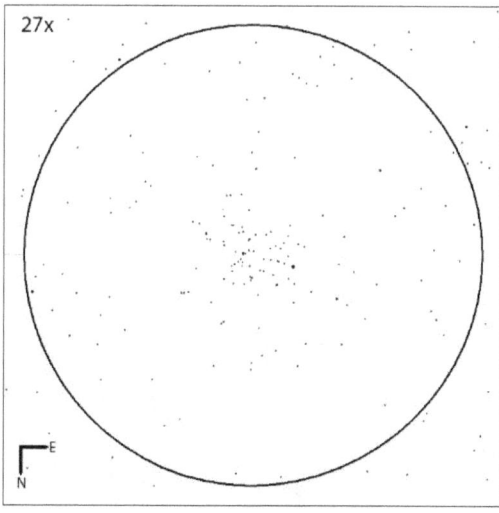

Eyepiece view courtesy *Sky Tools 3*

Messier 7

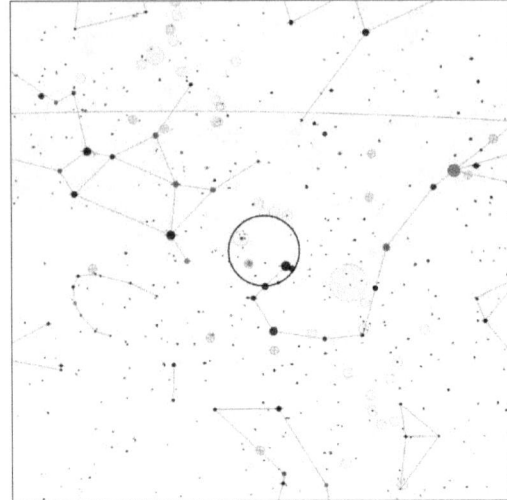

Map courtesy *Mobile Observatory*

Designation(s):	Messier 7
Constellation:	Scorpius
R.A.:	17h 53m 51s
Declination:	-34° 47' 34"
Object Type:	Open Cluster
Location:	★ ★
Rating:	★ ★
Best Seen:	Summer

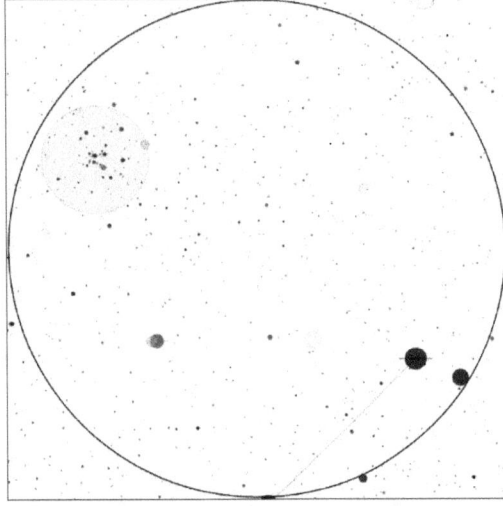

Finderscope view courtesy *Mobile Observatory*

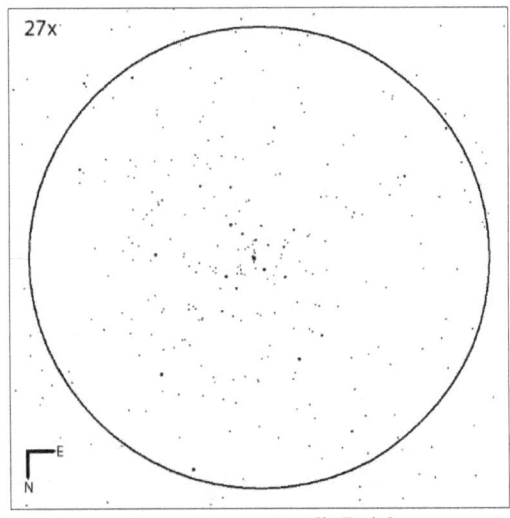

Eyepiece view courtesy *Sky Tools 3*

Close to the Butterfly Cluster in the sky, Messier 7 is larger, brighter and is consequently a little easier to see. You may be able to spot it with just your eyes under a clear, dark sky.

With the two stars of the scorpion's stinger at roughly the four o'clock position in your finder, it should be quite apparent on the opposite edge of the field of view. In fact, the two stars will point you right to it.

A low power magnification is all you'll need to enjoy this cluster. It's not as dense as the Butterfly and, in my opinion, is not as attractive. At 35x I've noted plenty of sparsely scattered stars that appear to form the shape of a bow tie (bow ties are cool, by the way.)

You may also see some gold stars among the many blue-white members; in particular, look out for an apparent double at the very heart of the cluster. Both stars are of equal brightness but with one being pale gold and the other blue-white.

Increasing the magnification to 54x may also reveal a small trio of stars forming an equilateral triangle on the eastern edge.

Messier 7 lies about 980 light years away and is thought to be about 25 light years in diameter.

The Keystone Cluster

Designation(s):	Messier 13
Constellation:	Hercules
R.A.:	16h 41m 41s
Declination:	+36° 27' 37"
Object Type:	Globular Cluster
Location:	★ ★
Rating:	★ ★ ★
Best Seen:	Spring and Summer

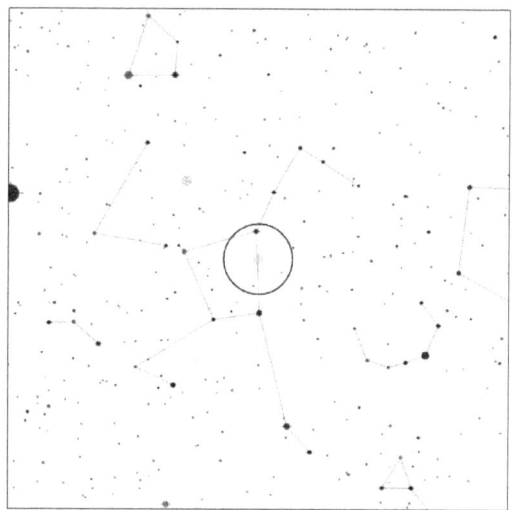

Map courtesy *Mobile Observatory*

The Keystone Cluster is by far the best globular cluster visible from the northern hemisphere.

If you have sharp eyesight and you live under clear, dark skies, you may be able to spot the cluster with just your eyes, but (more likely) you'll probably need binoculars to see it.

Through my small 8x30's from suburbia it appeared very small and looked like a faint, fuzzy star close to two other stars. It had a star-like core and appeared comet-like.

The darker your sky, the more you'll see and you can certainly reveal a lot of detail under the right conditions. Unfortunately, it's a disappointment from the city and a shadow of its true self. Just 27x will allow you to see the cluster but it's not much better than using binoculars and increasing the magnification doesn't improve the view.

From the suburbs it's another story. At 35x I noted the bright core extending some way toward the edge, where it then faded rapidly. I also noted some resolution and – best of all – chains of stars extending out from the core.

I've found a magnification of about 100x to be best for small 'scopes. I've often thought it like some strange sea creature, with tentacles reaching out from the depths of space.

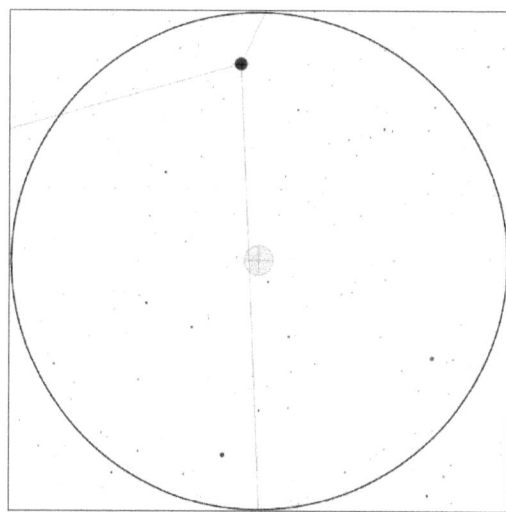

Finderscope view courtesy *Mobile Observatory*

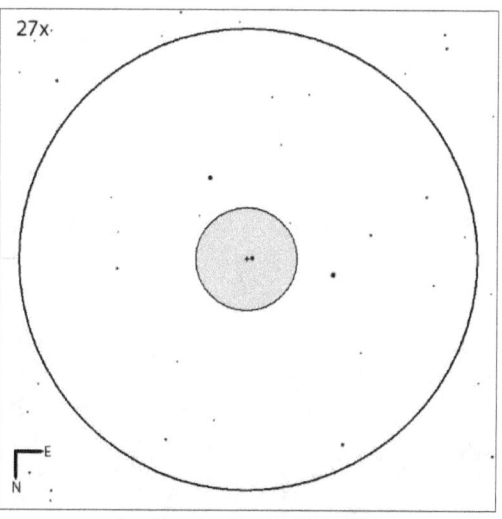

Eyepiece view courtesy *Sky Tools 3*

Rho Herculis

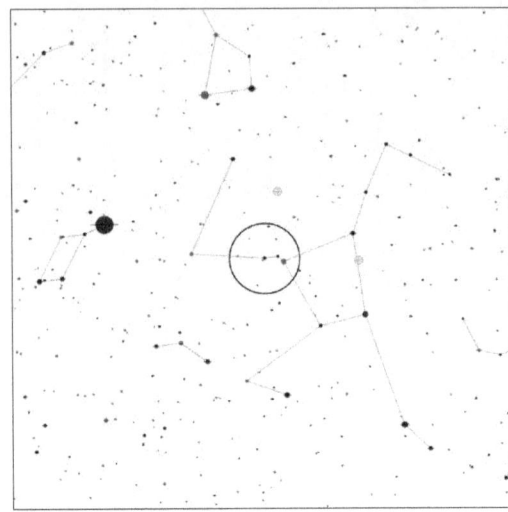

Map courtesy *Mobile Observatory*

Designation(s):	Rho Herculis
Constellation:	Hercules
R.A.:	17h 24m 12s
Declination:	+37° 08' 16"
Object Type:	Multiple Star
Location:	★ ★
Rating:	★ ★
Best Seen:	Spring and Summer

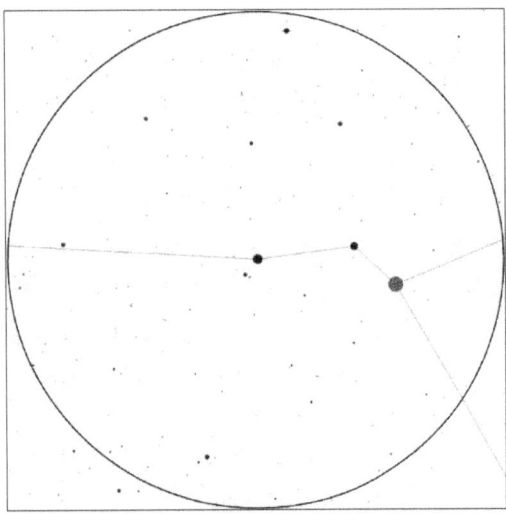

Finderscope view courtesy *Mobile Observatory*

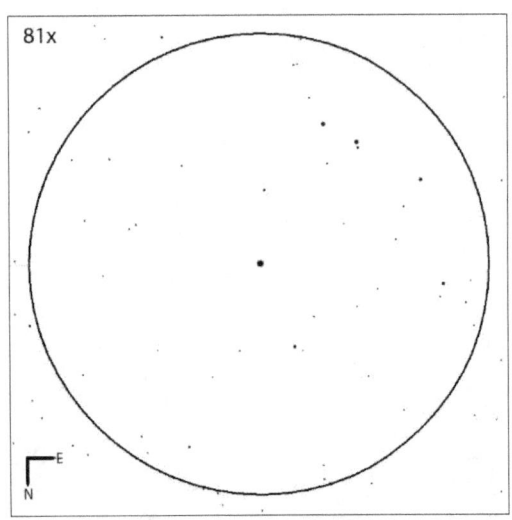

Eyepiece view courtesy *Sky Tools 3*

Rho Herculi is a white star with a nearby blue companion that's easily seen with binoculars. You might also see a fainter, third companion close to that secondary.

A telescope won't reveal much else until you get to about 50x when the primary can be barely split. Both components appear white with the primary being slightly brighter than the secondary.

Increasing the power helps to reveal some color in the pair. A magnification of about 100x shows a clear split with the primary being a pale gold. The secondary has shown some interesting colors – I've noted olive green with flecks of mustard at 107x.

The system lies just over 400 light years away and comprises of a subgiant primary and a dwarf secondary. The system is thought to be about 300 million years old with a third, very faint suspected companion appearing some distance away.

Rho marks the shoulder of Hercules, the Hero. In ancient mythology, he was known as Heracles and was the mortal son of Zeus. He's famous for performing twelve labors, by order of King Eurystheus; his first was to slay a lion and his second was to slay a sea serpent. Both these creatures can now be found toward the west in the night sky, some way from Hercules, in the form of the constellations Leo and Hydra.

Designation(s):	Nu Draconis
Constellation:	Draco
R.A.:	17h 32m 11s
Declination:	+55° 11' 03"
Object Type:	Multiple Star
Location:	★ ★ ★
Rating:	★ ★ ★
Best Seen:	Spring and Summer

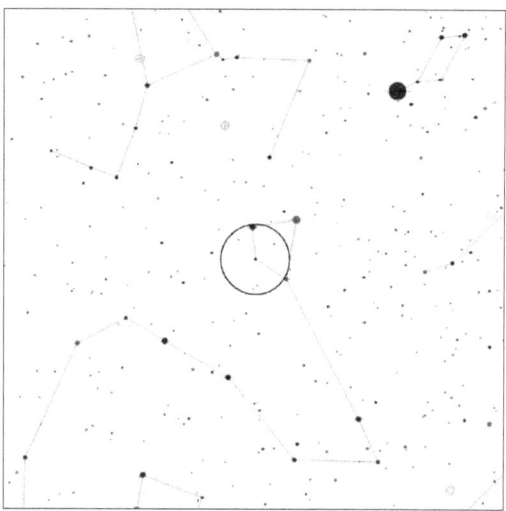

Map courtesy *Mobile Observatory*

Kuma is a gem and a very easy target for small telescopes in the northern hemisphere. I was able to split the pair with both my finderscope and my small 8x30 binoculars (although I had to hold them steadily to do it. Leaning against a wall will definitely help!)

This is a double that's best observed at low power. It's an easy and clean split at just 27x with both stars appearing to be brilliant white and of equal brightness. They've always made me think of a pair of eyes staring back at me from space – in fact, some observers have taken to calling them "the eyes of the dragon" for that reason.

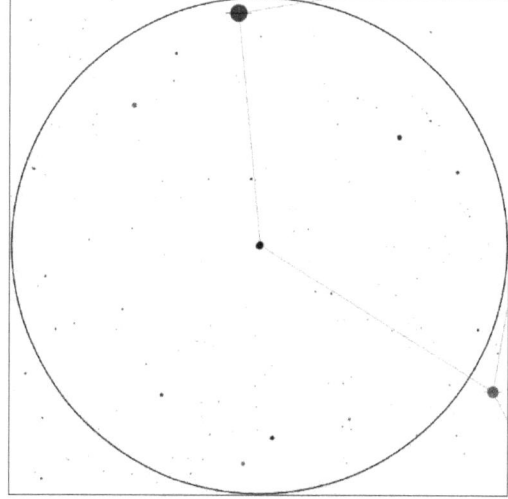

Finderscope view courtesy *Mobile Observatory*

A true multiple star system, the pair lie just under a hundred light years away with one component, Nu Draconis[2], having its own very close companion.

The constellation of Draco, the Dragon, snakes around Polaris, the Pole Star, and is actually circumpolar. This means the constellation never sets when seen from most of the northern hemisphere but appears to spin about the pole star instead.

Its head appears close to Hercules, the Hero and appears to chase him about the pole. It's at its highest in the Summer when the constellation can be seen snaking over much of the northern horizon.

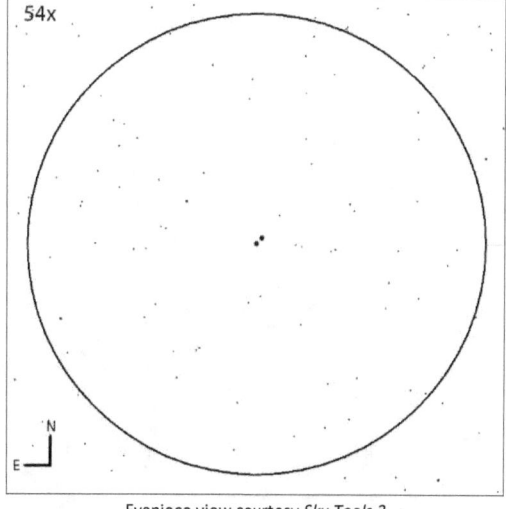

Eyepiece view courtesy *Sky Tools 3*

Messier 22

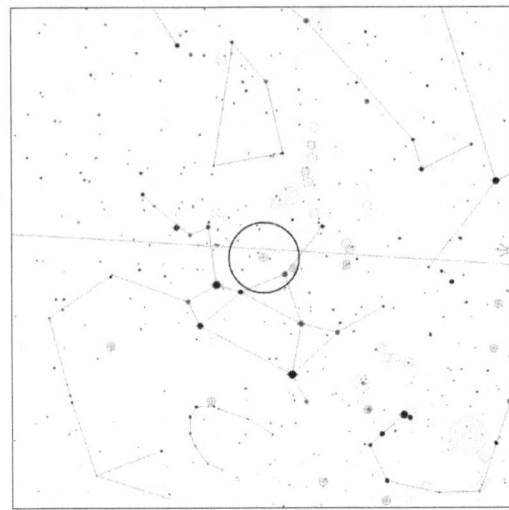

Map courtesy *Mobile Observatory*

Designation(s):	Messier 22
Constellation:	Sagittarius
R.A.:	18h 36m 24s
Declination:	-23° 54' 17"
Object Type:	Globular Cluster
Location:	★ ★
Rating:	★ ★
Best Seen:	Summer

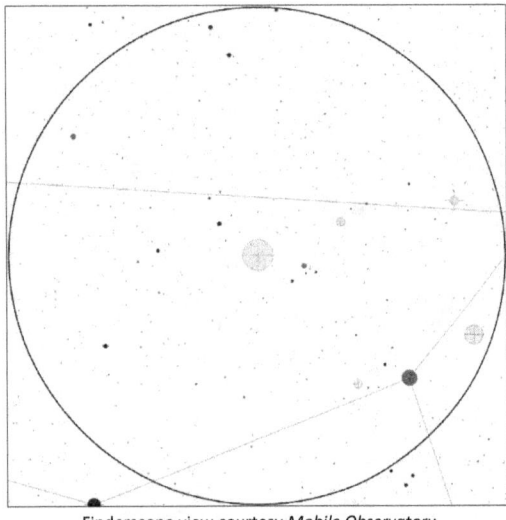

Finderscope view courtesy *Mobile Observatory*

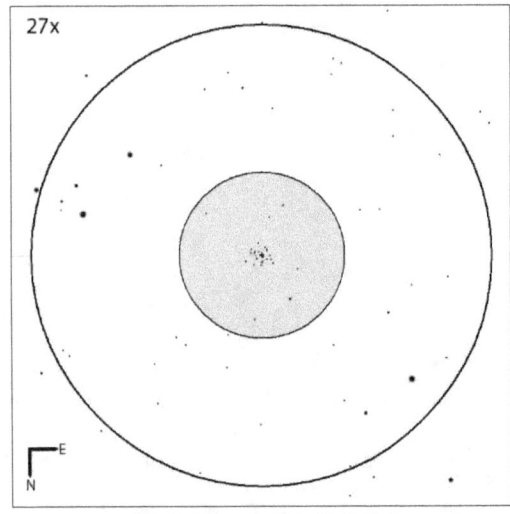

Eyepiece view courtesy *Sky Tools 3*

Messier 22, like the Keystone Cluster, is one of the best globulars in the entire night sky but unlike its cousin in Hercules, it's not well seen from the northern hemisphere.

This is a shame as it would certainly would be much better known if it were more easily seen. However, if you live in the United States it shouldn't be too much of a problem and even observers in the United Kingdom and Canada may still stand a chance.

It's conveniently located close to Kaus Borealis, the star that marks the top of the famous teapot of Sagittarius. It's easily seen with binoculars from suburban skies and appears as a small, nebulous sphere with no bright center and a uniform grey color.

A low magnification of 35x won't improve the view too much; it appears circular and could be easily mistaken for a nebula.

Increasing the magnification to 50x or higher and using averted vision might help reveal some resolution but take a look at its shape; does it still appear circular or has its appearance altered slightly? Does it look a little elongated to you?

Messier 22 is only 10,600 light years away (quite close for a globular) and contains about 70,000 stars.

The Double Double

Designation(s):	Epsilon Lyrae
Constellation:	Lyra
R.A.:	18h 44m 20s
Declination:	+39° 40' 12"
Object Type:	Multiple Star
Location:	★ ★ ★
Rating:	★ ★ ★
Best Seen:	Summer

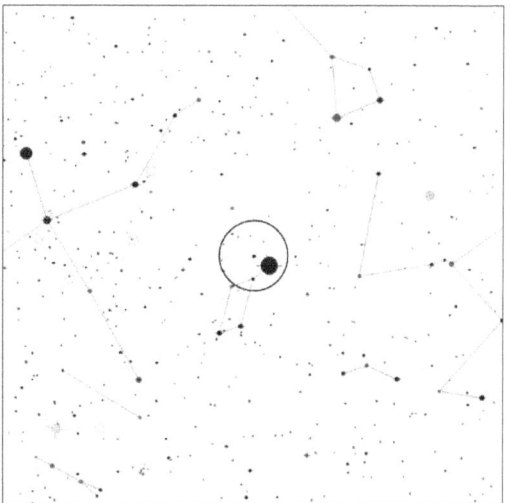

Map courtesy *Mobile Observatory*

Epsilon Lyrae, commonly known as the "Double Double," is one of the most famous and popular multiple stars in the sky – and for good reason.

First of all, turn a pair of binoculars toward it – what do you see? You might need to steady yourself against a wall, but you should easily be able to split the star into two close white components of equal brightness. It's a nice view with both Vega and Zeta Lyrae appearing within the same field of view.

In a small 'scope at low power (say, about 30x) you might notice something odd. I've seen both stars appear elongated, as though they were almost but not quite split.

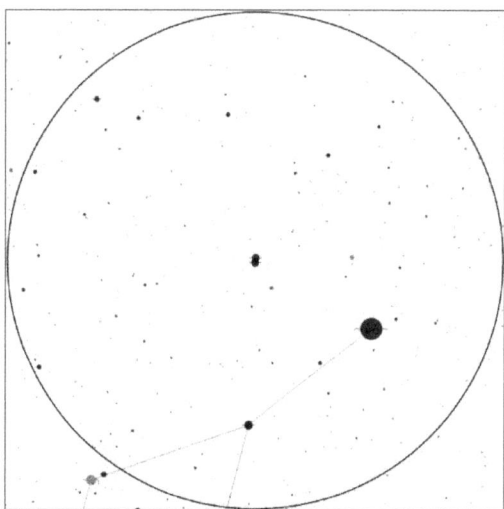

Finderscope view courtesy *Mobile Observatory*

Try increasing the magnification – first to about 50x, then about 75x. Are the stars split yet? Try again at about 100x and you should be able to barely split both stars into their components and the view is quite fascinating. Two pairs of bright white stars, with all four being about the same brightness. One pair might appear slightly brighter and wider than the other.

You're looking at a fascinating multiple star system some 162 light years away. One pair orbit each other once every 1,800 years while the other pair circle one another every 700 years. While quadruple stars are not unheard of, what's truly astonishing is that there may be as many as ten stars in this system!

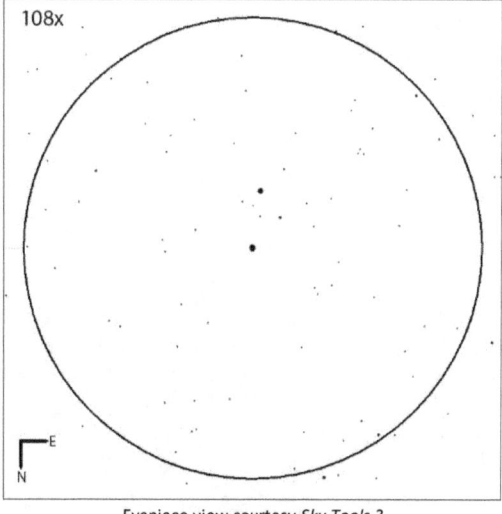

Eyepiece view courtesy *Sky Tools 3*

Sheliak

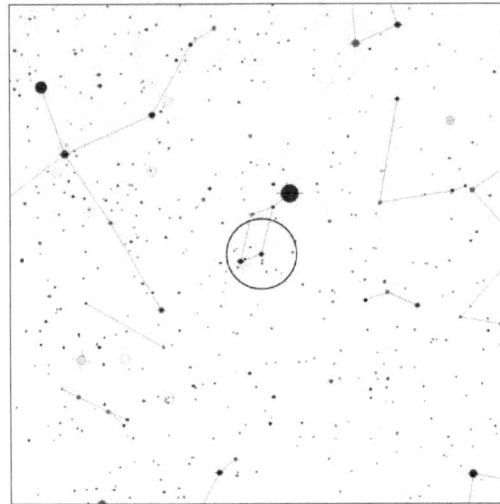

Map courtesy *Mobile Observatory*

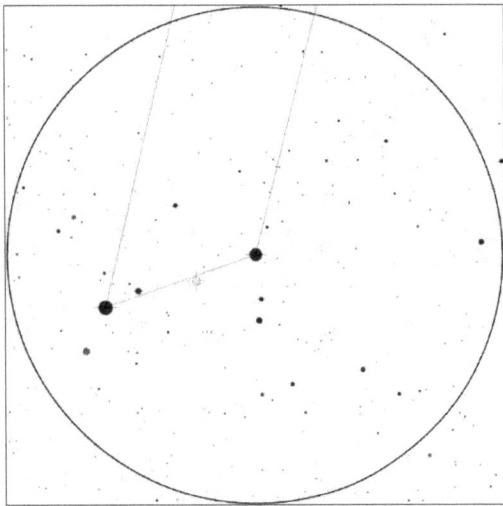

Finderscope view courtesy *Mobile Observatory*

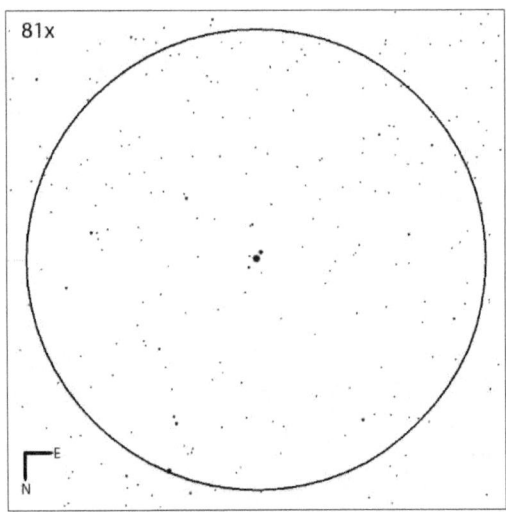

Eyepiece view courtesy *Sky Tools 3*

Designation(s):	Beta Lyrae
Constellation:	Lyra
R.A.:	18h 50m 05s
Declination:	+33° 21' 46"
Object Type:	Multiple Star
Location:	★ ★ ★
Rating:	★ ★
Best Seen:	Summer

Sheliak is a relatively wide double just beyond the range of regular binoculars. However, a small 'scope shouldn't have any problem resolving the pair at low magnification. The primary is pure white and about five or six times brighter than the blue companion.

Look carefully and you'll see the pair just off-center of a small triangle of faint stars. Increasing the magnification to around 100x will definitely make this easier to see.

Sheliak is one of those stars that makes you wish interstellar travel were possible. If you could visit this star, you'd see something quite amazing.

It lies about 950 light years away and its primary component is actually a very close double. The smaller of the two was once the giant star of the system but as it grew in size, its companion began to pull matter away from it.

In other words, the larger star was cannibalized by its smaller companion until the roles were reversed. The companion is now the dominant member of the pair and is surrounded by a disk of matter stripped away from the former giant. The two stars now orbit one another almost every thirteen days causing the star to dim slightly as seen from Earth. Try comparing its brightness to nearby stars and you may notice it fade and brighten over the intervening nights.

Designation(s):	Messier 57
Constellation:	Lyra
R.A.:	18h 53m 35s
Declination:	+33° 01' 45"
Object Type:	Planetary Nebula
Location:	★
Rating:	★ ★ ★
Best Seen:	Summer

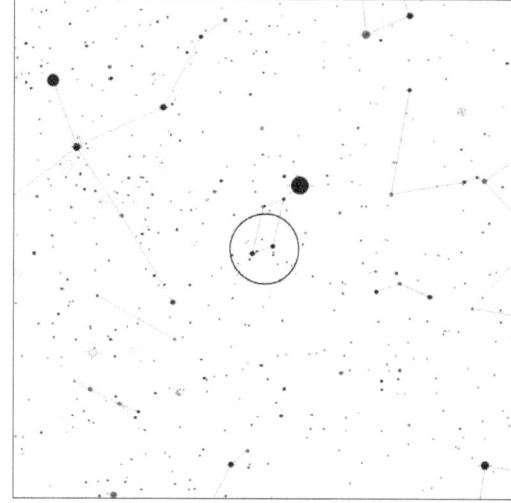

Map courtesy *Mobile Observatory*

A favorite of mine ever since I first saw it, it's very appropriately named because it truly looks like a smoke ring in space.

Surprisingly, I've been able to see it from both the city and also the suburbs. At 27x it appeared circular, dark grey, small and clearly not a star. However, you won't be able to see it as a ring until you increase the magnification.

At 70x I was barely able to see the central hole and the ring-shape, but I had to use averted vision to spot it. The shape becomes more apparent as you continue to increase the magnification but you'll need to always use averted vision to see the central hole.

At 161x I noted that it still appeared circular and cloud-like but that the nebula was well defined and the hole very apparent with averted vision. Focusing proved to be difficult while looking directly at it.

What you're seeing here is a dying star, some 2,000 light years away. The Ring Nebula is probably the best example of a planetary nebula – so-called because they look like planets when observed telescopically. It's actually a shell of gas thrown off from a giant star as it shrinks down to a white dwarf. You can actually see this star, in the center of the nebula, but you'll need a larger telescope to spot it.

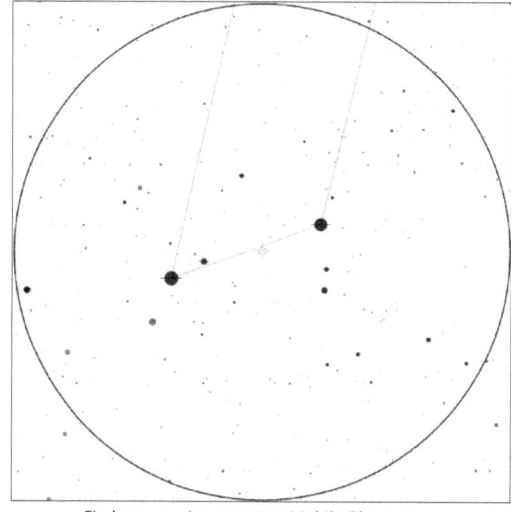

Finderscope view courtesy *Mobile Observatory*

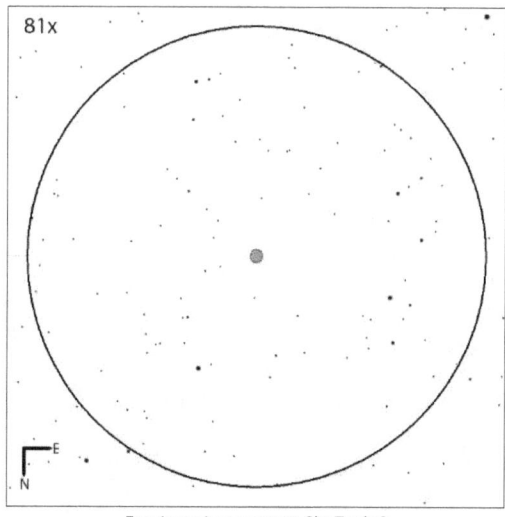

Eyepiece view courtesy *Sky Tools 3*

The Wild Duck Cluster

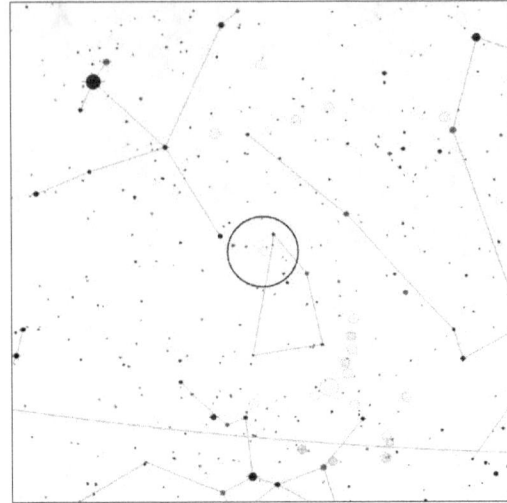

Map courtesy *Mobile Observatory*

Designation(s):	Messier 11
Constellation:	Scutum
R.A.:	18h 51m 06s
Declination:	-06° 16' 12"
Object Type:	Open Cluster
Location:	★ ★
Rating:	★ ★
Best Seen:	Summer

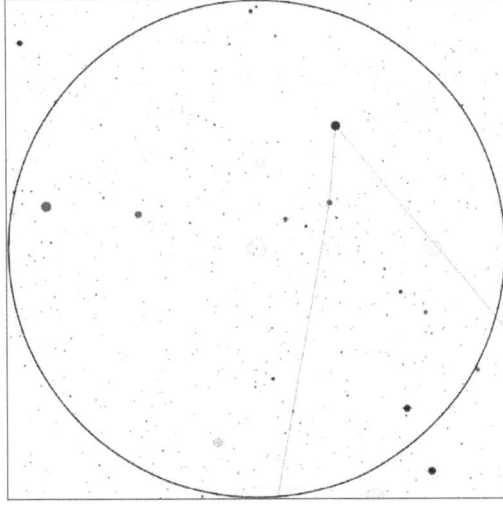

Finderscope view courtesy *Mobile Observatory*

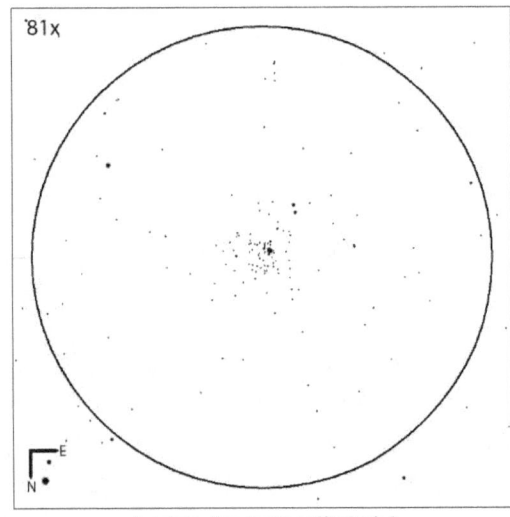

Eyepiece view courtesy *Sky Tools 3*

The Wild Duck Cluster is a small open star cluster found just to the south-west of Aquila, the Eagle. It can be easily seen with binoculars and can be found by following a slightly curved line of stars from Delta Aquilae, at the bottom of the constellation. (Alternatively, if you can easily locate Beta Scutum in your finderscope you should be able to spot the cluster within the same field of view.)

The cluster is disappointing from the city as it's small, faint and difficult to see at low power. You'll definitely need to use averted vision to really see anything of it.

If, however, you live away from the city lights, you're in for a treat. Through binoculars it appears small, compact, faint and globular – cloud-like, in fact – with a star-like point on the eastern edge.

Through a small 'scope at 35x you'll see a small, granular V shape with a single bright star at the tip. It's not hard to see how the cluster got its name as it looks like ducks flying south for the winter. You'll also see a few other bright stars appear within the same field of view, most notably a nearby double pair.

It's one of those clusters that definitely improves as you increase the magnification. Many more stars appear at just 54x with so many stars becoming apparent that it starts to lose its shape. The stars themselves are blue and are almost all of a uniform brightness

The Coathanger Cluster

Designation(s):	Collinder 399
Constellation:	Vulpecula
R.A.:	19h 26m 05s
Declination:	+20° 13' 14"
Object Type:	Asterism
Location:	★ ★
Rating:	★ ★
Best Seen:	Summer

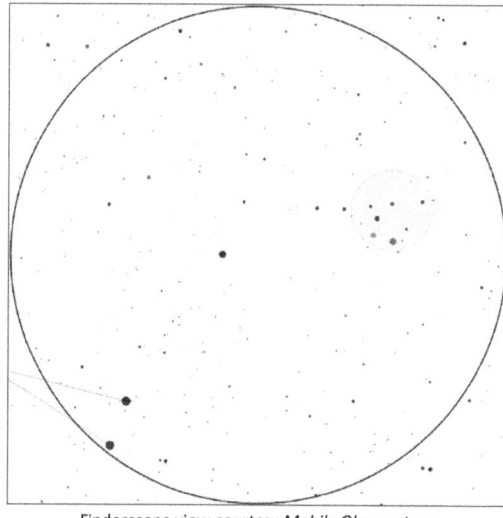

Map courtesy Mobile Observatory

The Coathanger Cluster is a large group of stars that's actually best observed with a very low power. In fact, you may prefer to observe it with either a pair of binoculars or even your telescope's finder.

To find it, you'll first need to locate the constellation of Sagitta, the Arrow, located a little under halfway between Albireo in Cygnus and Altair in Aquila. Look for the two stars that mark the feathers and place them at about the seven o'clock position in your field of view. The cluster should then be around the two o'clock mark.

The shape of the cluster should be quite apparent (albeit upside-down) through the finder or binoculars. When you turn your telescope toward it, make sure you're using your lowest power eyepiece first as you'll have difficulty fitting it into the field of view at even 35x.

The coathanger is formed by a chance alignment of eight or nine stars with the brightest forming the hook of the hanger.

The group has been known since antiquity, having first been recorded by the Persian astronomer al-Sufi around the year 964. The cluster also goes by the name Brocchi's Cluster, named for D. F. Brocchi, a stellar cartographer who mapped the cluster in the 1920's.

Finderscope view courtesy Mobile Observatory

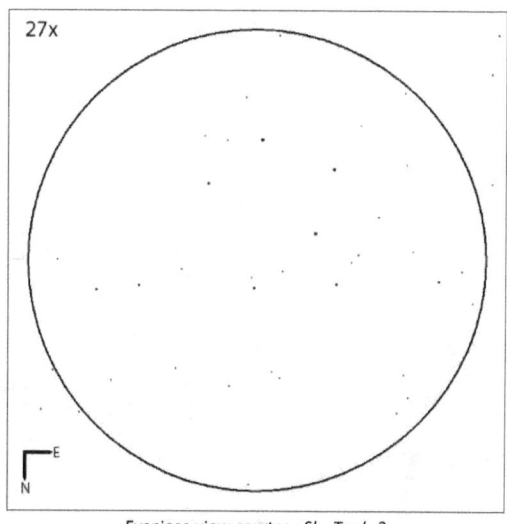

27x

Eyepiece view courtesy Sky Tools 3

The Dumbbell Nebula

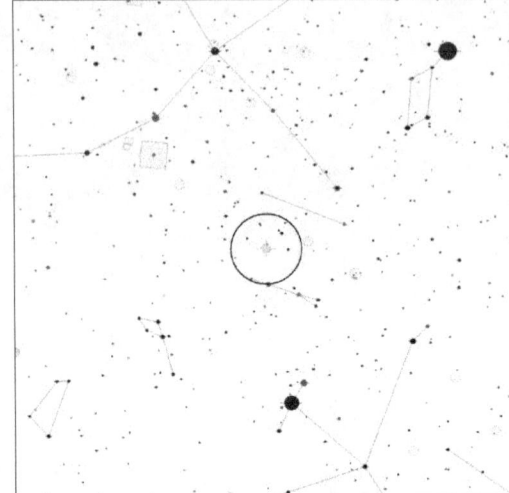

Map courtesy *Mobile Observatory*

Designation(s):	Messier 27
Constellation:	Vulpecula
R.A.:	19h 59m 36s
Declination:	+22° 43' 16"
Object Type:	Planetary Nebula
Location:	★
Rating:	★ ★
Best Seen:	Summer

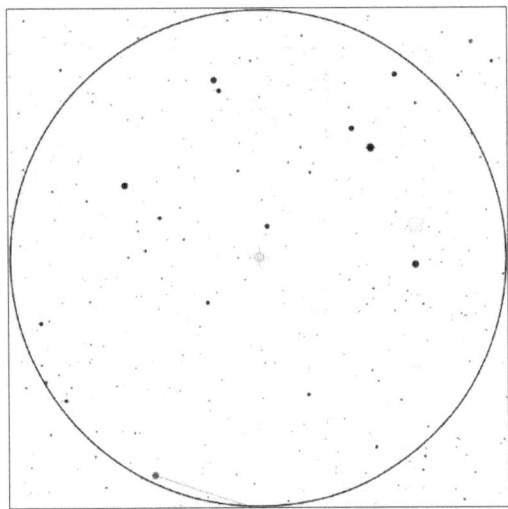

Finderscope view courtesy *Mobile Observatory*

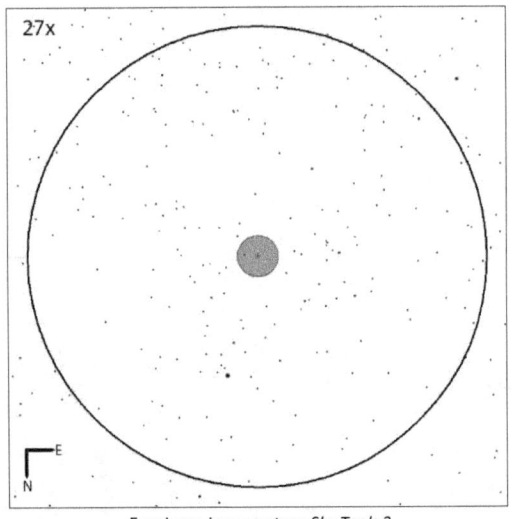

Eyepiece view courtesy *Sky Tools 3*

The Dumbbell Nebula, although faint, is actually relatively easy to find once you know where to look. You'll need to first find the constellation of Sagitta, the Arrow, about midway between Altair in Aquila and Albireo in Cygnus.

Look for Gamma Sagittae, one of the brighter stars that forms the body of the arrow itself. By placing it just on the edge or outside your finder's field of view (in the six o'clock position) you should be able to see a line of three stars near the center.

The Dumbbell Nebula is just to the south of the middle star, 14 Vulpeculae. I haven't seen it in a finder, but I've definitely snagged it with 8x30 binoculars under suburban skies. It was barely seen, circular, very misty and needed averted vision to be spotted.

If you're observing from the city, you'll probably need a telescope to see it at all. At 27x, it was seen but was extremely faint and needed averted vision. It appeared as a large, uniformly grey disk with no details seen. Increasing the magnification made observation easier.

It's much more apparent from suburban skies and should be easily seen at just 35x. It appeared rectangular at first but actually resembled a bat when observed with averted vision. Unlike the city, increasing the magnification will definitely improve the view and allow more details to be seen.

Designation(s):	Beta Cygni
Constellation:	Cygnus
R.A.:	19h 30m 43s
Declination:	+27° 57' 35"
Object Type:	Multiple Star
Location:	★ ★ ★
Rating:	★ ★ ★
Best Seen:	Summer

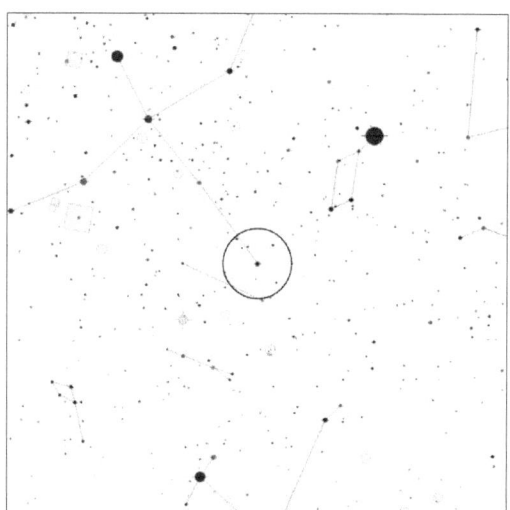

Map courtesy *Mobile Observatory*

Albireo is, by far, the most beautiful and popular double star in the entire night sky. This might sound like hyperbole, but it's really not. I challenge anyone to find a double that more easily found, easily split and with such a striking color combination.

It's not split with a regular pair of binoculars (say 10x50's) or a finderscope, but you won't need much power to reveal both components.

It's a very easy and fairly wide split at 27x with the primary being a pale gold and about 1½ times brighter than the pale blue secondary. I've noted that the strength of the colors will sometimes vary from night to night, so it's worth taking another look from time to time.

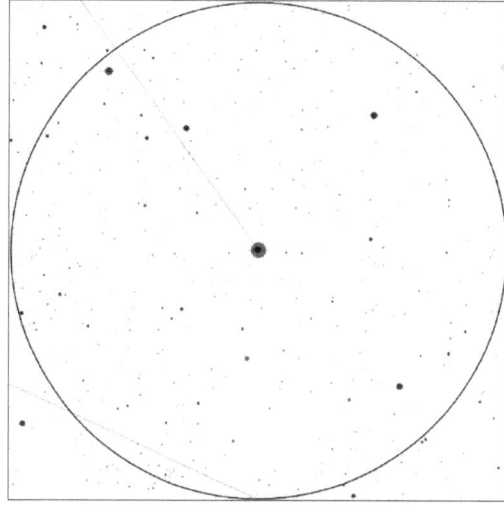

Finderscope view courtesy *Mobile Observatory*

You'll want to come back anyway – again and again – as it's a sight that's unparalleled throughout the rest of the sky. You'll show your family, your friends and your neighbors during summer barbecues and you'll anticipate their gasp – just as you did – when you first laid eyes on this beauty.

The pair lie about 430 light years away but it's not known for sure if the two components truly orbit one another. The primary, however, is itself a double but the stars are too close to be resolvable with amateur equipment.

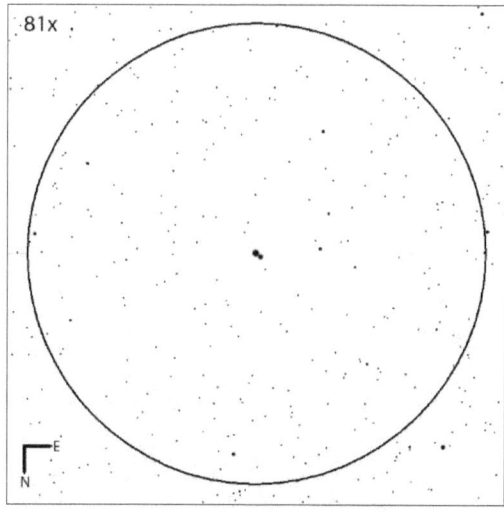

Eyepiece view courtesy *Sky Tools 3*

Messier 29

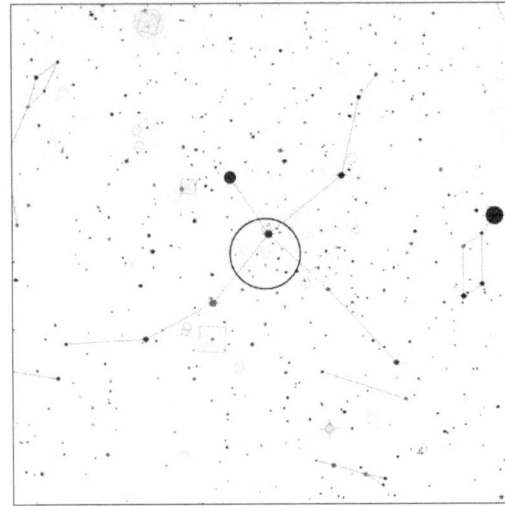

Map courtesy *Mobile Observatory*

Designation(s):	Messier 29
Constellation:	Cygnus
R.A.:	20h 23m 58s
Declination:	+38° 30' 28"
Object Type:	Open Cluster
Location:	★
Rating:	★ ★
Best Seen:	Summer

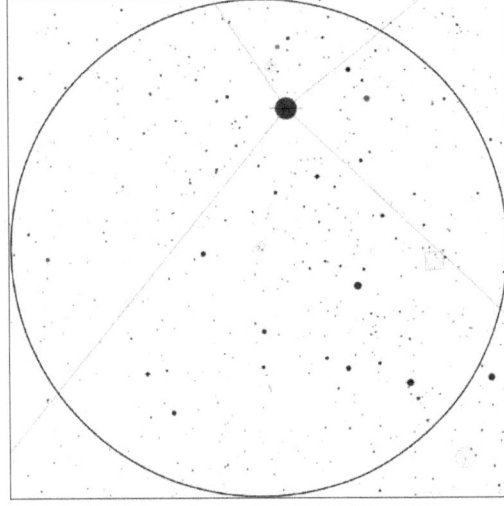

Finderscope view courtesy *Mobile Observatory*

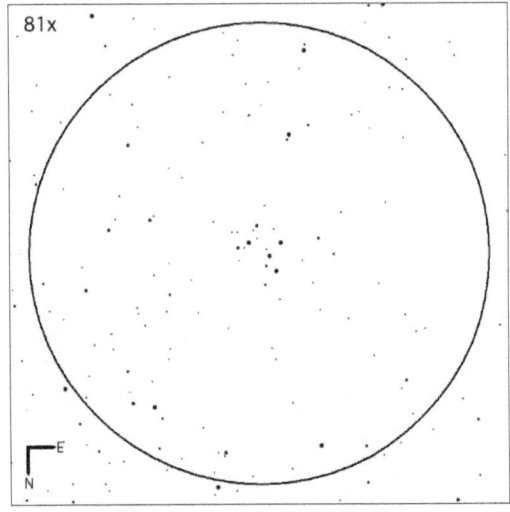

Eyepiece view courtesy *Sky Tools 3*

Everyone should have a favorite deep sky object. Whether it's a double star, an open or globular star cluster, a nebula or a galaxy, most observers have an object they feel a particular fondness for.

Mine is Messier 29. I can't explain why. It's not particularly large or spectacular and it doesn't contain hundreds of stars. It doesn't look like diamond dust on black satin. It doesn't have any golden orange stars nestled within it.

So why bother? Well, for one thing, it's easily found as Sadr, the central star in Cygnus the Swan, is a convenient location marker for it. If you place that star about halfway between the center and the edge of your finder's view (as depicted) you should have the cluster in the center.

You may not be able to see it in the finder and I've been unable to positively identify it with binoculars, but it should be readily apparent with a small telescope and a low magnification.

My first thought, upon "discovering" this cluster, was that it looked like a mini Pleiades, which came as a surprise. It's small, is surrounded by stars and has seven or eight bright stars, all white and about the same brightness.

Given its small size, it easily fits into the field of view, even at high magnification, but I've found it best at around 100x.

Designation(s):	Zeta Sagittae
Constellation:	Sagitta
R.A.:	19h 48m 59s
Declination:	+19° 08' 31"
Object Type:	Multiple Star
Location:	★ ★
Rating:	★ ★
Best Seen:	Summer

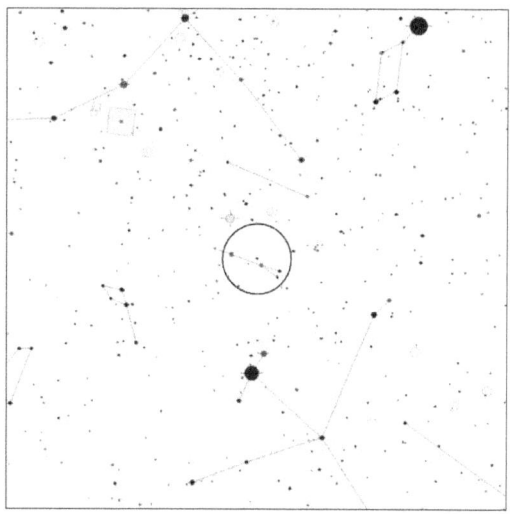

Map courtesy *Mobile Observatory*

Zeta Sagittae is a challenging double but provides a nice view for those who are able to split it. You'll need a higher power eyepiece and maybe even a barlow for this one.

Part of the challenge may also be in locating the star to begin with. It shouldn't be much of a problem if you live in the suburbs or under a dark sky as Sagitta isn't a difficult constellation to find. Just look midway between Albireo and Altair, the brightest star in Aquila the Eagle for the small, arrow shaped constellation.

If you live in a city, you'll need to use binoculars (or scan the area with your finderscope) because the chances are the stars will be too faint to see without optical aid.

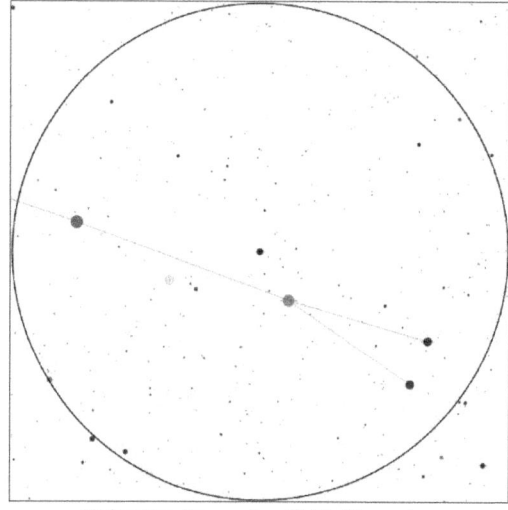

Finderscope view courtesy *Mobile Observatory*

Zeta can be seen just a little to the north of the main constellation itself and close to Delta Sagittae. You won't be able to split it with low power; 35x didn't do anything for me. At 54x, the secondary was just barely visible, but unless you've already spotted it at a higher magnification, it might elude you.

Try a magnification of about 100x and then work your way down. I was able to split the pair at 91x and saw a white primary and a tiny, faint blue secondary. Now drop the power down to a lower magnification and look for the secondary again. Keep doing this until the secondary disappears – how low can you go?

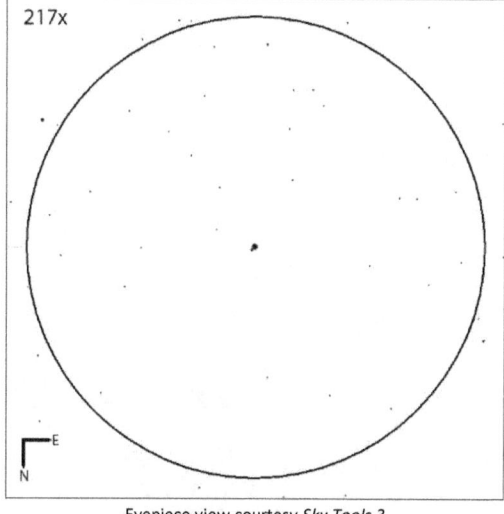

Eyepiece view courtesy *Sky Tools 3*

Messier 71

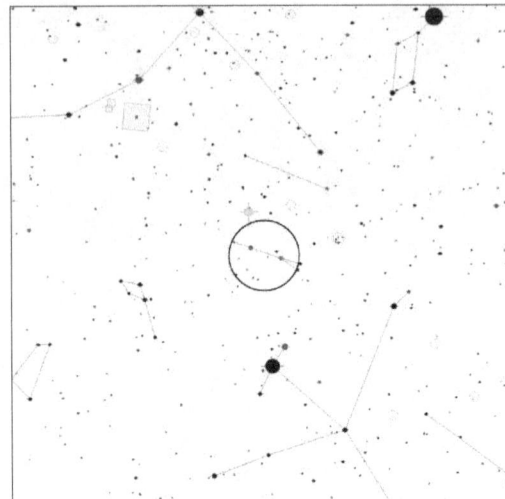

Map courtesy *Mobile Observatory*

Designation(s):	Messier 71
Constellation:	Sagitta
R.A.:	19h 53m 46s
Declination:	+18° 46' 45"
Object Type:	Globular Cluster
Location:	★
Rating:	★
Best Seen:	Summer

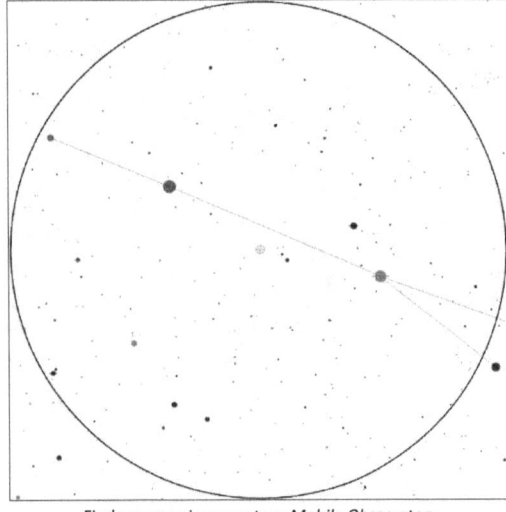

Finderscope view courtesy *Mobile Observatory*

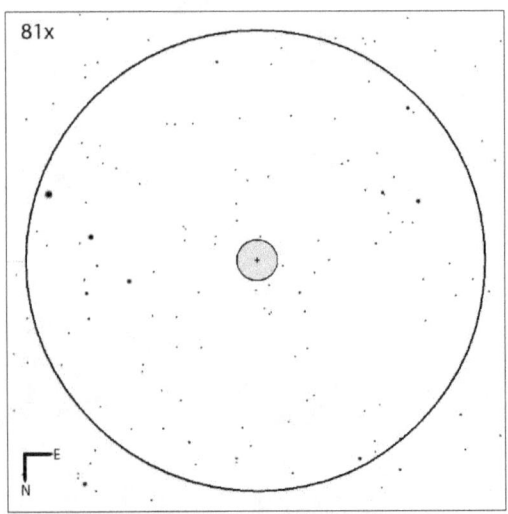

Eyepiece view courtesy *Sky Tools 3*

While you're in the vicinity of Sagitta, try your hand at Messier 71, another challenging sight for small 'scopes and located on just the other side of the constellation from Zeta.

It's too small and faint to be seen with in a finder but if you center the constellation in your field of view, the cluster should be visible through your telescope at low power.

This is a good example of an object you'd hunt down for the challenge, rather than for the eventual view. At 35x, it appears very faint, very small and very fuzzy. I've noted that it doesn't really look globular at all, but rather more like a nebula instead.

In fact, it doesn't even appear spherical. When I first observed it, I noted that it appeared rectangular and was uniformly dim with no bright core. Increasing the magnification to 70x improved the view a little but there was no resolution of the individual stars. (On the plus side, there's plenty of other stars in the field of view that will make it easy for you to focus.)

Messier 71 was discovered by Phillipe Loys de Chéseaux in 1746. The cluster lies about 27,000 light years away, is some 27 light years in diameter and was originally thought to be an open cluster.

Designation(s):	Alpha Capricorni
Constellation:	Capricornus
R.A.:	20h 18m 03s
Declination:	-12° 32' 41"
Object Type:	Multiple Star
Location:	★ ★ ★
Rating:	★ ★
Best Seen:	Summer

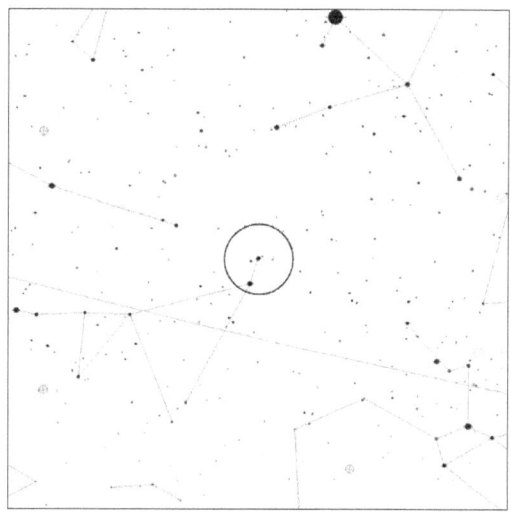

Map courtesy *Mobile Observatory*

Al Giedi is a wide, optical double, which means the two stars are not a true multiple star system.

It's a very easy object for binoculars or a finderscope with the western star appearing slightly fainter and with a hint of pale orange or cream color.

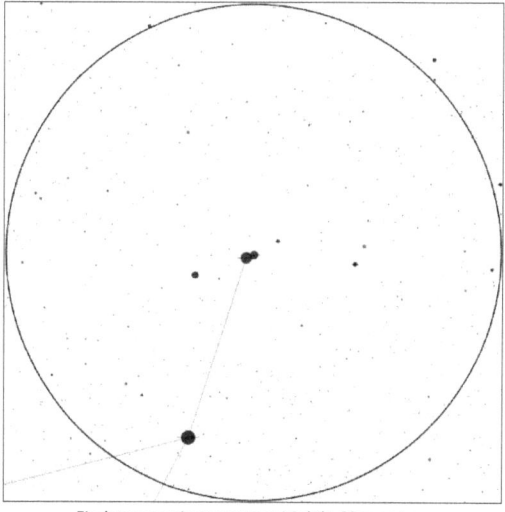

Finderscope view courtesy *Mobile Observatory*

Look through a telescope and the view improves somewhat. At 35x each star will reveal a faint companion. The western star, Alpha Capricorni[2], has a wide companion while the eastern star, Alpha Capricorni[1], also has a faint companion that appears much closer to the primary star.

(Only one of these companions may be visible from the city – you may need to increase the magnification to about 80x in order to see them both. Which do you see? The wide or the close companion?)

Al Giedi is the brightest star (or stars!) in the constellation of Capricornus, the Sea Goat. It's an ancient constellation, known to a number of cultures, with a rich mythology that goes back thousands of years.

Curiously, the constellation is almost always associated with water in some way. For example, in Greek mythology, it represents the god Pan. Part goat, the god panicked while escaping the monster Typhon and was only partly successful in turning himself into a fish before hitting the water!

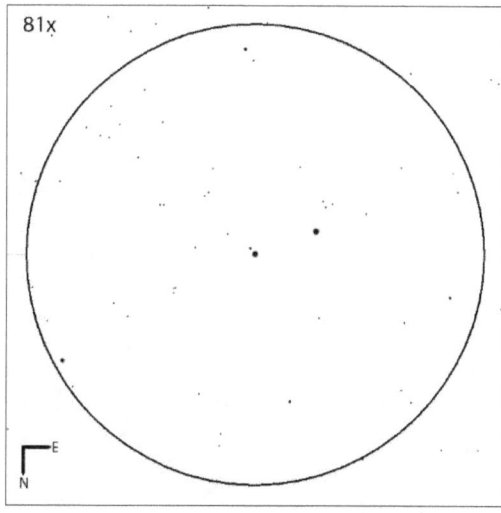

Eyepiece view courtesy *Sky Tools 3*

Gamma Delphini

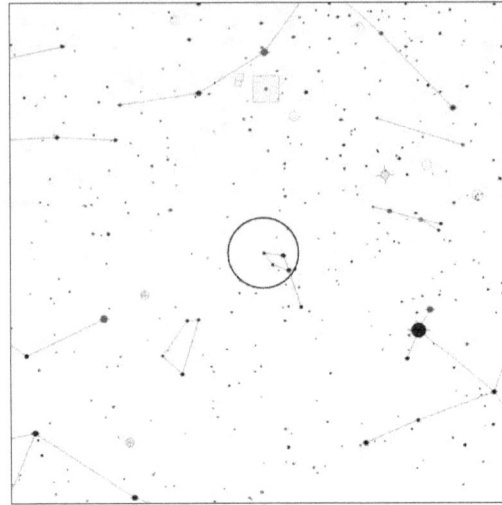

Map courtesy *Mobile Observatory*

Designation(s):	Gamma Delphini
Constellation:	Delphinus
R.A.:	20h 46m 39s
Declination:	+16° 07' 38"
Object Type:	Multiple Star
Location:	★ ★ ★
Rating:	★ ★
Best Seen:	Summer

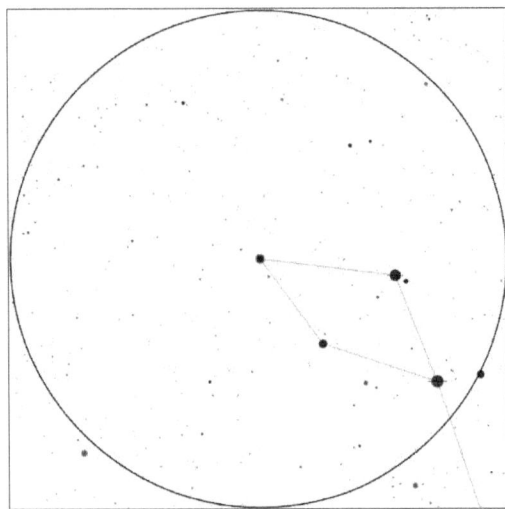

Finderscope view courtesy *Mobile Observatory*

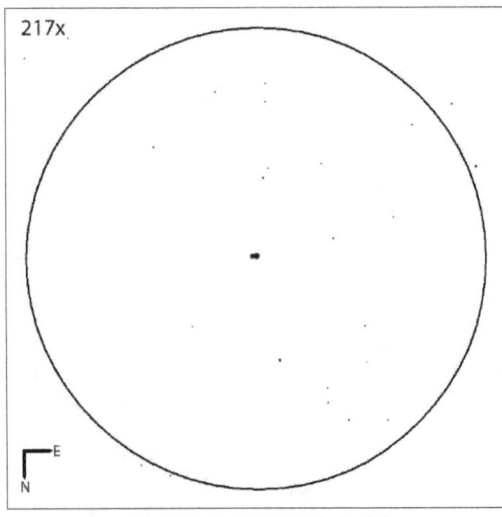

Eyepiece view courtesy *Sky Tools 3*

Gamma Delphini is a reasonably bright star that marks one corner of a star pattern (or *asterism*) known as Job's Coffin. It's also known to be a paler version of Albireo that's an easy split for small telescopes.

Binoculars (or a finderscope) won't reveal its companion but look out for Sualocin (Alpha Delphini) just to the west and within the same field of view. You'll see a wide pair of blue-white stars with the primary about twice as bright as the secondary. (Incidentally, try spelling the name of the star backwards. It's the name of an astronomer who named it after himself!)

Gamma is easily split at 35x and is a good example of your eyes showing you color that isn't necessarily there. As explained in the notes for Lambda Arietis, two stars may appear to be of slightly different colors as a result of the brightness contrast between them.

Take a look and note their colors and then increase the magnification. Do the colors change? Maybe they will and maybe they won't. Come back on the next clear night and try it again.

I usually see a very pale yellow and very pale blue pair of stars of almost equal brightness and increasing the magnification seems to make the colors stronger. But on other nights, one looks to be pale yellow while the other is clearly white. What do you see?

Designation(s):	Messier 15
Constellation:	Pegasus
R.A.:	21h 29m 58s
Declination:	+12° 10' 01"
Object Type:	Open Cluster
Location:	★
Rating:	★ ★
Best Seen:	Summer and Autumn

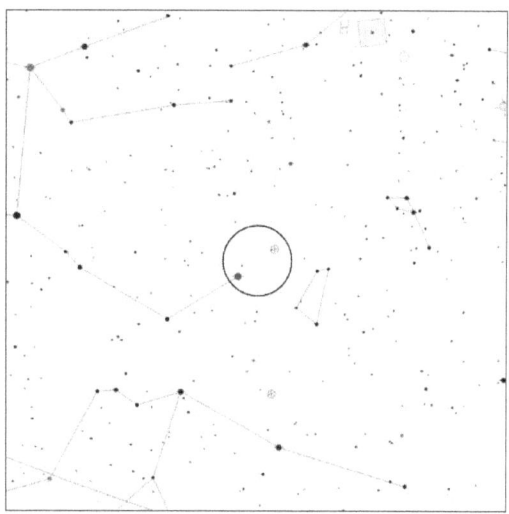

Map courtesy *Mobile Observatory*

Messier 15 is easily found, thanks to its close proximity to Enif, the brightest star in the constellation of Pegasus, the Flying Horse.

Visible in binoculars, draw a line through Baham (to the lower left of Enif) and Enif itself and look for a faint, fuzzy star next to a slightly brighter one. It's easy to mistake it for a star but careful observation will reveal its non-stellar nature.

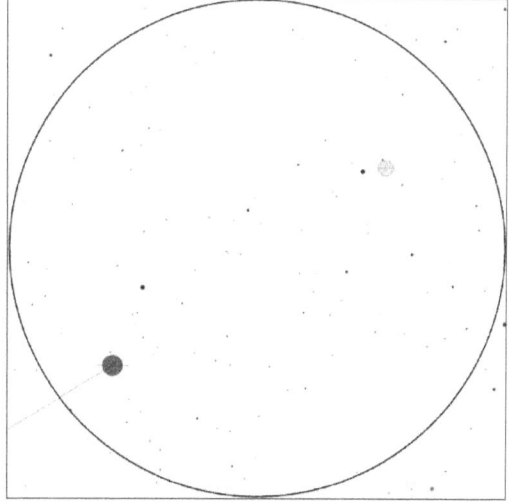

Finderscope view courtesy *Mobile Observatory*

In a finder, with Enif in about the eight o'clock position, the cluster may appear at around the two o'clock mark. Again, look out for the star that appears close to it.

Unsurprisingly, the cluster is better observed through a telescope. At low power (around 30x) you'll see a small, hazy "star" with four real stars within the same field of view. The stars formed a zig-zag pattern and the cluster appeared between two of the stars. Averted vision may help to make the cluster appear a little larger.

The view is improved at a higher magnification. Between 80x and 100x I noticed that the core appeared surprisingly bright and extended about two thirds of the way toward the edge of the cluster. At 107x I some resolution was hinted at and the edges seemed to extend a little further out.

At about 12 billion years old, this cluster is one of the most ancient known.

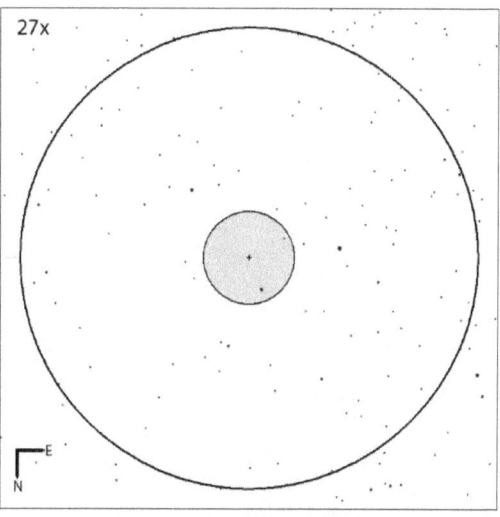

Eyepiece view courtesy *Sky Tools 3*

Observation Logs

Date	Time	Object

_____

_____

_____

_____

_____

Date	Time	Object

_____

_____

_____

_____

_____

Date	Time	Object

_____

_____

_____

_____

_____

Date	Time	Object

_____

_____

_____

_____

_____

Date	Time	Object

_____

_____

_____

_____

_____

Date	Time	Object

_____

_____

_____

_____

_____

Date	Time	Object

_____

_____

_____

_____

_____

Date	Time	Object

_____

_____

_____

_____

_____

Date	Time	Object

_____

_____

_____

_____

_____

Date	Time	Object

_____

_____

_____

_____

_____

Date	Time	Object

_____

_____

_____

_____

_____

Date	Time	Object

_____

_____

_____

_____

_____

Date	Time	Object

_____

_____

_____

_____

_____

Date	Time	Object

_____

_____

_____

_____

_____

Date	Time	Object

_____

_____

_____

_____

_____

Date	Time	Object

_____

_____

_____

_____

_____

Date	Time	Object

_____

_____

_____

_____

_____

Date	Time	Object

_____

_____

_____

_____

_____

Date	Time	Object

_____

_____

_____

_____

_____

Date	Time	Object

_____

_____

_____

_____

_____

Date	Time	Object

_____

_____

_____

_____

_____

Date	Time	Object

_____

_____

_____

_____

_____

Date	Time	Object

_____

_____

_____

_____

_____

Date	Time	Object

_____

_____

_____

_____

_____

Date	Time	Object

_____

_____

_____

_____

Date	Time	Object

_____

_____

_____

_____

Date	Time	Object

_____

_____

_____

_____

Date	Time	Object

_____

_____

_____

_____

Date	Time	Object

_____

_____

_____

_____

Date	Time	Object

_____

_____

_____

_____

Date	Time	Object

_____

_____

_____

_____

_____

Date	Time	Object

_____

_____

_____

_____

_____

Date	Time	Object

_____

_____

_____

_____

_____

Date	Time	Object

_____

_____

_____

_____

Date	Time	Object

_____

_____

_____

_____

Date	Time	Object

_____

_____

_____

_____

Appendix

The Easy Telescope Objects Listed by Constellation

Note: The co-ordinates have been slightly abbreviated and rounded up to the nearest minute for formatting reasons. Inputting these co-ordinates into a GoTo (with zero as the seconds) should still allow your telescope to find the object.

ANDROMEDA
Autumn and Winter

	Designation	R.A.	Dec.	Type	Location	Rating	Page
Pi Andromedae	Pi Andromedae	00h 37m	+33° 43'	Multiple Star	★★	★★	184
Andromeda Galaxy	Messier 31	00h 43m	+41° 16'	Spiral Galaxy	★★	★★	185
NGC 752	NGC 752	01h 58m	+37° 52'	Open Cluster	★★	★★	186
Almach	Gamma Andromedae	02h 04m	+42° 20'	Multiple Star	★★★	★★★	187

ARIES
Autumn and Winter

	Designation	R.A.	Dec.	Type	Location	Rating	Page
Mesarthim	Gamma Arietis	01h 53m	+19° 18'	Multiple Star	★★★	★★★	192
Lambda Arietis	Lambda Arietis	01h 58m	+23° 36'	Multiple Star	★★★	★★★	193

AURIGA
Winter

	Designation	R.A.	Dec.	Type	Location	Rating	Page
Messier 37	Messier 37	05h 52m	+32° 33'	Open Cluster	★	★★★	204

BOÖTES
Spring and Summer

	Designation	R.A.	Dec.	Type	Location	Rating	Page
Arcturus	Alpha Boötis	14h 16m	+19° 11'	Multiple Star	★★★	★	222
Delta Boötis	Delta Boötis	15h 16m	+33° 19'	Multiple Star	★★★	★★★	223

CANCER

	Designation	R.A.	Dec.	Type	Location	Rating	Page
Praesepe	Messier 44	08h 40m	+19° 40'	Open Cluster	★★	★★★	212
Iota Cancri	Iota Cancri	08h 47m	+28° 46'	Multiple Star	★★★	★★★	213
Messier 67	Messier 67	08h 51m	+11° 49'	Open Cluster	★	★★	214

CANIS MAJOR

	Designation	R.A.	Dec.	Type	Location	Rating	Page
Sirius	Alpha Canis Majoris	06h 45m	-16° 43'	Multiple Star	★★★	★	208
Messier 41	Messier 41	06h 46m	-20° 45'	Open Cluster	★★	★★★	209

CANIS MINOR

	Designation	R.A.	Dec.	Type	Location	Rating	Page
Procyon	Alpha Canis Minoris	07h 39m	+05° 13'	Multiple Star	★★★	★	210

CANES VENATICI

	Designation	R.A.	Dec.	Type	Location	Rating	Page
Cor Caroli	Alpha Canum Venaticorum	12h 56m	+38° 19'	Multiple Star	★★★	★★★	220

CAPRICORNUS

	Designation	R.A.	Dec.	Type	Location	Rating	Page
Al Giedi	Alpha Capricorni	20h 18m	-12° 33'	Multiple Star	★★★	★★	245

CASSIOPEIA

Autumn and Winter

	Designation	R.A.	Dec.	Type	Location	Rating	Page
Achird	Eta Cassiopeiae	00h 49m	+57° 49'	Multiple Star	★★★	★★★	188
Owl Cluster	NGC 457	01h 20m	+58° 17'	Open Cluster	★	★★★	189
Messier 103	Messier 103	01h 33m	+60° 39'	Open Cluster	★	★	190
NGC 663	NGC 663	01h 46m	+61° 13'	Open Cluster	★★	★★★	191

CORVUS

Spring

	Designation	R.A.	Dec.	Type	Location	Rating	Page
Algorab	Delta Corvi	12h 30m	-16° 31'	Multiple Star	★★★	★★	219

CYGNUS

Summer

	Designation	R.A.	Dec.	Type	Location	Rating	Page
Albireo	Beta Cygni	19h 31m	+27° 58'	Multiple Star	★★★	★★★	241
Messier 29	Messier 29	20h 24m	+38° 30'	Open Cluster	★	★★	242

DELPHINUS

Summer

	Designation	R.A.	Dec.	Type	Location	Rating	Page
Gamma Delphini	Gamma Delphini	20h 47m	+16° 08'	Multiple Star	★★★	★★	246

DRACO

Spring and Summer

	Designation	R.A.	Dec.	Type	Location	Rating	Page
Kuma	Nu Draconis	17h 32m	+55° 11'	Multiple Star	★★★	★★★	233

GEMINI

	Designation	R.A.	Dec.	Type	Location	Rating	Page
Messier 35	Messier 35	06h 09m	+24° 21'	Open Cluster	★★	★★★	205
Castor	Alpha Geminorum	07h 35m	+31° 53'	Multiple Star	★★★	★★★	206

HERCULES

	Designation	R.A.	Dec.	Type	Location	Rating	Page
Keystone Cluster	Messier 13	16h 42m	+36° 28'	Globular Cluster	★★	★★★	231
Rho Herculis	Rho Herculis	17h 24m	+37° 08'	Multiple Star	★★	★★	232

LEO

	Designation	R.A.	Dec.	Type	Location	Rating	Page
Regulus	Alpha Leonis	10h 08m	+11° 58'	Multiple Star	★★★	★	215
Adhafera	Zeta Leonis	10h 17m	+23° 25'	Multiple Star	★★★	★★	216
Algieba	Gamma Leonis	10h 20m	+19° 51'	Multiple Star	★★★	★★★	217
Denebola	Beta Leonis	11h 49m	+14° 34'	Multiple Star	★★★	★★	218

LEPUS

	Designation	R.A.	Dec.	Type	Location	Rating	Page
Gamma Leporis	Gamma Leporis	05h 44m	-22° 27'	Multiple Star	★★★	★★	203

LIBRA

	Designation	R.A.	Dec.	Type	Location	Rating	Page
Zuben Elgenubi	Alpha Librae	14h 51m	-16° 00'	Multiple Star	★★★	★★	224

LYRA

	Designation	R.A.	Dec.	Type	Location	Rating	Page
Double Double	Epsilon Lyrae	18h 44m	+39° 40'	Multiple Star	★★★	★★★	235
Sheliak	Beta Lyrae	18h 50m	+33° 22'	Multiple Star	★★★	★★	236
Ring Nebula	Messier 57	18h 54m	+33° 02'	Multiple Star	★	★★★	237

MONOCEROS

	Designation	R.A.	Dec.	Type	Location	Rating	Page
Beta Monocerotis	Beta Monocerotis	06h 29m	-07° 02'	Multiple Star	★★	★★	207

ORION

	Designation	R.A.	Dec.	Type	Location	Rating	Page
Mintaka	Delta Orionis	05h 32m	-00° 18'	Multiple Star	★★★	★★	199
Meissa	Lambda Orionis	05h 35m	+09° 56'	Multiple Star	★★★	★★★	200
Orion Nebula	Messier 42	05h 35m	-05° 23'	Nebula	★★★	★★★	201
Sigma Orionis	Sigma Orionis	05h 39m	-02° 36'	Multiple Star	★★★	★★★	202

PEGASUS

	Designation	R.A.	Dec.	Type	Location	Rating	Page
Messier 15	Messier 15	21h 30m	+12° 10'	Globular Cluster	★	★★	247

PERSEUS

	Designation	R.A.	Dec.	Type	Location	Rating	Page
Double Cluster	NGC 869 & 884	02h 22m	+57° 09'	Open Clusters	★★★	★★★	194
Messier 34	Messier 34	02h 42m	+42° 45'	Open Cluster	★★	★★	195

PUPPIS

	Designation	R.A.	Dec.	Type	Location	Rating	Page
Messier 93	Messier 93	07h 44m	-23° 51'	Open Cluster	★	★★	211

SAGITTA

	Designation	R.A.	Dec.	Type	Location	Rating	Page
Zeta Sagittae	Zeta Sagittae	19h 49m	+19° 09'	Multiple Star	★★	★★	243
Messier 71	Messier 71	19h 54m	+18° 47'	Globular Cluster	★	★	244

SAGITTARIUS

	Designation	R.A.	Dec.	Type	Location	Rating	Page
Messier 22	Messier 22	18h 36m	-23° 54'	Globular Cluster	★★	★★	234

SCORPIUS

	Designation	R.A.	Dec.	Type	Location	Rating	Page
Graffias	Beta Scorpii	16h 05m	-19° 48'	Multiple Star	★★★	★★★	225
Jabbah	Nu Scorpii	16h 12m	-19° 28'	Multiple Star	★★	★★★	226
Messier 80	Messier 80	16h 17m	-22° 59'	Globular Cluster	★	★	227
Messier 4	Messier 4	16h 24m	-26° 32'	Globular Cluster	★	★★	228
Butterfly Cluster	Messier 6	17h 40m	-32° 15'	Open Cluster	★★	★★★	229
Messier 7	Messier 7	17h 54m	-34° 48'	Open Cluster	★★	★★	230

SCUTUM

	Designation	R.A.	Dec.	Type	Location	Rating	Page
Wild Duck Cluster	Messier 11	18h 51m	-06° 16'	Open Cluster	★★	★★	238

TAURUS

	Designation	R.A.	Dec.	Type	Location	Rating	Page
Pleiades	Messier 45	03h 48m	+24° 10′	Open Cluster	★★★	★★★	197
Crab Nebula	Messier 1	05h 35m	+22° 01′	Supernova Remnant	★	★	198

URSA MAJOR

Spring and Summer

	Designation	R.A.	Dec.	Type	Location	Rating	Page
Mizar & Alcor	Zeta and 80 Ursae Majoris	13h 24m	+54° 56′	Multiple Star	★★★	★★★	221

URSA MINOR

All Year Round

	Designation	R.A.	Dec.	Type	Location	Rating	Page
Polaris	Alpha Ursae Minoris	02h 32m	+89° 16′	Multiple Star	★★★	★	196

VULPECULA

Summer

	Designation	R.A.	Dec.	Type	Location	Rating	Page
Coathanger Cluster	Collinder 399	19h 26m	+20° 13′	Open Cluster	★★	★★	239
Dumbbell Nebula	Messier 27	20h 00m	+22° 43′	Planetary Nebula	★	★★	240

Greek Alphabet

Alpha	α	Epsilon	ε	Iota	ι	Nu	ν	Rho	ρ	Phi	φ
Beta	β	Zeta	ζ	Kappa	κ	Xi	ξ	Sigma	σ	Chi	χ
Gamma	γ	Eta	η	Lambda	λ	Omicron	o	Tau	τ	Psi	ψ
Delta	δ	Theta	θ	Mu	μ	Pi	π	Upsilon	υ	Omega	ω

Elongations of Mercury and Venus, 2016-2025

	Mercury		Venus	
2016	February 6th	Morning Sky		
	April 17th	Evening Sky		
	June 4th	Morning Sky		
	August 16th	Evening Sky		
	September 28th	Morning Sky		
	December 10th	Evening Sky		
2017	January 18th	Morning Sky	January 11th	Evening Sky
	March 31st	Evening Sky	June 2nd	Morning Sky
	May 17th	Morning Sky		
	July 29th	Evening Sky		
	September 11th	Morning Sky		
	November 23rd	Evening Sky		
2018	January 1st	Morning Sky	August 16th	Evening Sky
	March 14th	Evening Sky		
	April 28th	Morning Sky		
	July 11th	Evening Sky		
	August 26th	Morning Sky		
	November 5th	Evening Sky		
	December 14th	Morning Sky		
2019	February 26th	Evening Sky	January 5th	Morning Sky
	April 11th	Morning Sky		
	June 23rd	Evening Sky		
	August 9th	Morning Sky		
	October 19th	Evening Sky		
	November 27th	Morning Sky		
2020	February 9th	Evening Sky	March 24th	Evening Sky
	March 23rd	Morning Sky	August 12th	Morning Sky
	June 3rd	Evening Sky		
	July 21st	Morning Sky		
	September 30th	Evening Sky		
	November 9th	Morning Sky		
2021	January 23rd	Evening Sky	October 29th	Evening Sky
	March 5th	Morning Sky		
	May 16th	Evening Sky		
	July 4th	Morning Sky		
	September 13th	Evening Sky		
	October 24th	Morning Sky		

	Mercury		Venus	
2022	January 6th	Evening Sky	March 19th	Morning Sky
	February 16th	Morning Sky		
	April 28th	Evening Sky		
	June 15th	Morning Sky		
	August 26th	Evening Sky		
	October 8th	Morning Sky		
	December 20th	Evening Sky		
2023	January 29th	Morning Sky	June 3rd	Evening Sky
	April 11th	Evening Sky	October 23rd	Morning Sky
	May 28th	Morning Sky		
	August 9th	Evening Sky		
	September 21st	Morning Sky		
	December 3rd	Evening Sky		
2024	January 11th	Morning Sky		
	March 24th	Evening Sky		
	May 9th	Morning Sky		
	July 21st	Evening Sky		
	September 4th	Morning Sky		
	November 15th	Evening Sky		
	December 24th	Morning Sky		
2025	March 7th	Evening Sky	January 9th	Evening Sky
	April 20th	Morning Sky	May 31st	Morning Sky
	July 3rd	Evening Sky		
	August 18th	Morning Sky		
	October 29th	Evening Sky		
	December 7th	Morning Sky		

Oppositions of Mars, Jupiter and Saturn, 2016-2025

	Mars		Jupiter		Saturn	
2016	May 21st	Scorpius	March 7th	Leo	June 2nd	Ophiuchus
2017			April 7th	Virgo	June 14th	Ophiuchus
2018	July 26th	Capricornus	May 8th	Libra	June 27th	Sagittarius
2019			June 9th	Ophiuchus	July 8th	Sagittarius
2020	October 13th	Pisces	July 14th	Sagittarius	July 20th	Sagittarius
2021			August 19th	Capricornus	August 1st	Capricornus
2022	December 7th	Taurus	September 26th	Pisces	August 13th	Capricornus
2023			November 2nd	Aries	August 27th	Aquarius
2024			December 7th	Taurus	September 7th	Aquarius
2025	January 15th	Gemini			September 20th	Pisces

Dates of Maximum Brightness for Selected Long Period Variables, 2017-2030

(Data calculated using *Sky Tools 3* by Skyhound, http://www.skyhound.com)

Mira in Cetus. Period 332 days, magnitude range approximately 3.0 to 10.0. See page 130.

March 26th, 2017	February 11th, 2018	January 9th, 2019	December 7th, 2019
November 3rd, 2020	October 1st, 2021	August 29th, 2022	July 27th, 2023
June 23rd, 2024	May 21st, 2025	April 18th, 2026	March 16th, 2027
February 11th, 2028	January 8th, 2029	December 6th, 2029	November 3rd, 2030

Chi Cygni in Cygnus. Period 408 days, magnitude range approximately 5.0 to 14.0. See page 138.

October 27th, 2017	December 9th, 2018	January 21st, 2020	March 4th, 2021
April 16th, 2022	May 29th, 2023	July 10th, 2024	August 22nd, 2025
October 4th, 2026	November 16th, 2027	December 28th, 2028	February 9th, 2030

Propus in Gemini. Period 233 days, magnitude range 3.1 to 3.9. See page 144.

August 21st, 2017	April 11th, 2018	November 30th, 2018	July 21st, 2019
March 10th, 2020	October 29th, 2020	June 19th, 2021	February 7th, 2022
September 27th, 2022	May 18th, 2023	January 6th, 2024	August 26th, 2024
April 16th, 2025	December 5th, 2025	July 26th, 2026	March 16th, 2027
November 4th, 2027	June 24th, 2028	February 11th, 2029	October 2nd, 2029
May 23rd, 2030			

R Leporis in Lepus. Period 427 days, magnitude range 5.5 to 11.7. See page 150.

May 7th, 2017	July 8th, 2018	September 8th, 2019	November 8th, 2020
January 9th, 2022	March 12th, 2023	May 12th, 2024	July 13th, 2025
September 13th, 2026	November 14th, 2027	January 14th, 2029	March 17th, 2030

Complete List of Constellations

Latin	Genitive	English Name	Abbreviation	Size
Andromeda	Andromedae	The Princess	And	19th
Antila	Antilae	The Air Pump	Ant	62nd
Apus	Apodis	The Bird of Paradise	Aps	67th
Aquarius	Aquarii	The Water Bearer	Aqr	10th
Aquila	Aquilae	The Eagle	Aql	22nd
Ara	Arae	The Altar	Ara	63rd
Aries	Arietis	The Ram	Ari	39th
Auriga	Aurigae	The Charioteer	Aur	21st
Boötes	Boötis	The Herdsman	Boo	13th
Caelum	Caeli	The Chisel	Cae	81st
Camelopardalis	Camelopardalis	The Giraffe	Cam	18th
Cancer	Cancri	The Crab	Cnc	31st
Canes Venatici	Canum Venaticorum	The Hunting Dogs	CVn	38th
Canis Major	Canis Majoris	The Large Dog	CMa	43rd
Canis Minor	Canis Minoris	The Little Dog	CMi	71st
Capricornus	Capricorni	The Sea Goat	Cap	40th
Carina	Carinae	The Keel	Car	34th
Cassiopeia	Cassiopeiae	The Queen	Cas	25th
Centaurus	Centauri	The Centaur	Cen	9th
Cepheus	Cephei	The King	Cep	27th
Cetus	Ceti	The Sea Monster	Cet	4th
Chamaeleon	Chamaeleontis	The Chameleon	Cha	79th
Circinus	Circini	The Compass	Cir	85th
Columba	Columbae	The Dove	Col	54th
Coma Berenices	Comae Berenices	Berenices' Hair	Com	42nd
Corona Australis	Coronae Australis	The Southern Crown	CrA	80th
Corona Borealis	Coronae Borealis	The Northern Crown	CrB	73rd
Corvus	Corvi	The Crow	Crv	70th
Crater	Crateris	The Cup	Crt	53rd
Crux	Crucis	The Southern Cross	Cru	88th
Cygnus	Cygni	The Swan	Cyg	16th
Delphinus	Delphini	The Dolphin	Del	69th
Dorado	Doradus	The Goldfish	Dor	72nd
Draco	Draconis	The Dragon	Dra	8th
Equuleus	Equulei	The Foal	Equ	87th
Eridanus	Eridani	The River	Eri	6th

Latin	Genitive	English Name	Abbreviation	Size
Fornax	Fornacis	The Furnace	For	41st
Gemini	Geminorum	The Twins	Gem	30th
Grus	Gruis	The Crane	Gru	45th
Hercules	Herculis	The Hero	Her	5th
Horologium	Horologii	The Clock	Hor	58th
Hydra	Hydrae	The Water Snake	Hya	1st
Hydrus	Hydri	The Lesser Water Snake	Hyi	61st
Indus	Indi	The Indian	Ind	49th
Lacerta	Lacertae	The Lizard	Lac	68th
Leo	Leonis	The Lion	Leo	12th
Leo Minor	Leonis Minoris	The Little Lion	LMi	64th
Lepus	Leporis	The Hare	Lep	51st
Libra	Librae	The Scales	Lib	29th
Lupus	Lupi	The Wolf	Lup	46th
Lynx	Lyncis	The Lynx	Lyn	28th
Lyra	Lyrae	The Lyre	Lyr	52nd
Mensa	Mensae	The Table Mountain	Men	75th
Microscopium	Microscopii	The Microscope	Mic	66th
Monoceros	Monocerotis	The Unicorn	Mon	35th
Musca	Muscae	The Fly	Mus	77th
Norma	Normae	The Carpenter's Level	Nor	74th
Octans	Octantis	The Octant	Oct	50th
Ophiuchus	Ophiuchi	The Serpent Bearer	Oph	11th
Orion	Orionis	The Hunter	Ori	26th
Pavo	Pavonis	The Peacock	Pav	44th
Pegasus	Pegasi	The Flying Horse	Peg	7th
Perseus	Persei	The Hero	Per	24th
Phoenix	Phoenicis	The Phoenix	Phe	37th
Pictor	Pictoris	The Painter's Easel	Pic	59th
Pisces	Piscium	The Fishes	Psc	14th
Piscis Austrinus	Piscis Austrini	The Southern Fish	PsA	60th
Puppis	Puppis	The Poop Deck	Pup	20th
Pyxis	Pyxidis	The Compass	Pyx	65th
Reticulum	Reticuli	The Net	Ret	82nd

Complete List of Constellations (cont.)

Latin	Genitive	English Name	Abbreviation	Size
Sagitta	Sagittae	The Arrow	Sge	86th
Sagittarius	Sagittarii	The Archer	Sgr	15th
Scorpius	Scorpii	The Scorpion	Sco	33rd
Sculptor	Sculptoris	The Sculptor	Scl	36th
Scutum	Scuti	The Shield	Sct	84th
Serpens	Serpentis	The Serpent	Ser	23rd
Sextans	Sextantis	The Sextant	Sex	47th
Taurus	Tauri	The Bull	Tau	17th
Telescopium	Telescopii	The Telescope	Tel	57th
Triangulum	Trianguli	The Triangle	Tri	78th
Triangulum Australe	Trianguli Australis	The Southern Triangle	TrA	83rd
Tucana	Tucanae	The Toucan	Tuc	48th
Ursa Major	Ursae Majoris	The Great Bear	UMa	3rd
Ursa Minor	Ursae Minoris	The Little Bear	UMi	56th
Vela	Velorum	The Sails	Vel	32nd
Virgo	Virginis	The Virgin	Vir	2nd
Volans	Volantis	The Flying Fish	Vol	76th
Vulpecula	Vulpeculae	The Fox	Vul	55th

Glossary

Aperture

The aperture of a telescope is the width of the lens on a refractor telescope or the width of the open end of a reflector telescope tube. In both cases, this is the end that points up toward the sky.

See also *Reflector Telescope* and *Refractor Telescope*

Astronomical Unit

One Astronomical Unit (AU) is the distance from the Earth to the Sun. This is taken as 149,597,870 kilometers or 92,955,806 miles. To make life (and polite conversation as your astronomical society) easier, this is taken as 150 million kilometers or 93 million miles.

See also *Light Year*

Averted Vision

Averted vision is the "trick" of looking at an object out of the corner of your eye. As your eyes are more sensitive to light (rather than color) with averted vision, you're more likely to see a particularly faint object. Averted vision can be particularly useful in observing the details in a nebula.

Barlow Lens

A barlow lens is an accessory that allows you to effectively double (or, with some, triple) the magnification of any eyepiece used with it. A barlow is considered by many to be an essential tool for the amateur astronomer.

Declination

Just as geographical co-ordinates use longitude and latitude, stellar cartographers use right ascension and declination, respectively. Declination, like latitude, is measured in degrees and indicates an object's position in the sky relative to the northern and southern hemisphere. For example, Polaris has a declination of almost +90° and is therefore almost overhead from the North Pole. An object with a declination of 0° will pass overhead when observed from the equator. There are no bright stars with a declination of -90° and, therefore, no star marking the celestial South Pole.

See also *R.A. (Right Ascension)*

Focal Length

Focal length is the distance that light travels from the telescope's entry point (i.e., the lens on a refractor telescope or the aperture on a reflector) to the eyepiece. It is measured in millimeters. The longer the focal length, the higher the possible magnification of the telescope.

See also *Aperture, Reflector Telescope* and *Refractor Telescope.*

Globular Clusters

There are two types of star cluster: open and globular. Globular star clusters are like huge balls of stars, with each one containing thousands of stars. These stars are packed into a relatively small space, often with only a few light years between each one. The clusters form a halo around the galactic center and mostly contain older stars, often billions of years old.

See also *Open Clusters*

Greatest Eastern/Western Elongation

As both Mercury and Venus orbit the Sun at a closer distance than the Earth, they never stray very far from the Sun in the sky. However, there comes a time when each planet has moved the furthest it can before it starts to move back towards the Sun again. This point is called its greatest elongation.

Greatest eastern elongation means the planet is at its furthest point to the east of the Sun, thereby making it visible in the evening sky (confusingly, in the *west* after sunset.) When the planet is at its greatest western elongation, it's visible in the morning sky in the *east* before sunrise.

See also *Opposition*

Light Year

A light year is the distance that light travels in a year. Light travels at an approximate speed of 186,000 mph per *second*. In other words, at a distance of roughly 92 million miles, light from the Sun takes about eight minutes to reach us.

Given that the distance from the Earth to the Sun is one *Astronomical Unit*, it can be said that light travels 7 ½ AUs every hour, 180 AUs every day, 1,260 AUs every week, about 5,400 every month and 65,520 every year. The closest star is Proxima Centauri, which, at a distance of 4.24 light years, is some 277,804 AUs away. (And this is why we don't measure every distance in miles and kilometers.)

See also *Astronomical Unit*

Nebula (pl. nebulae)

A nebula is a huge cloud of gas and dust in space. These clouds are light years in diameter and are the birthplaces of the stars themselves. As the gas and dust particles move through space, they collide with other particles and gradually begin to grow in mass. Eventually, the clump of gas becomes so massive that a gravitational collapse begins, causing nuclear fusion and the birth of a star. Most of the nebulae are located away from bright stars (and may require a little practice to locate) but the Orion Nebula is easily seen with unaided eyes on a winter's night.

At the other end of a star's life are planetary nebulae. A planetary nebula is the last dying gasp of a star. It's the shell of gas and dust that the star ejects as it collapses and dies, causing the shell to expand out into space. They're named *planetary* nebulae because many look like the tiny disc of a planet when observed through a telescope.

The most famous planetary nebula is the Ring Nebula, which looks just like a smoke ring in space. This can easily be seen with a small telescope but you might need a slightly larger 'scope (around 100mm or more) to see the central hole itself. You can see the Ring Nebula during the summer months.

If the star goes supernova, you may well be able to see the remnants in the night sky. Many of these are faint (such as the Veil Nebula, which actually looks like the remains of an exploded star) but one, the Crab Nebula, should also be visible on a winter's night.

Open Clusters

Open clusters are a type of star cluster. Whereas globular clusters appear small, spherical and often faint, most open clusters are quite the opposite. Many contain tens or hundreds of stars of varying brightness, scattered across an area of sky sometimes larger than the full Moon. There are a number that are easily observed in a small telescope with the Pleiades, Praesepe and Butterfly clusters being fine examples.

See also *Globular Clusters*

Opposition

A planet (or other solar system body, such as an asteroid) is said to be at opposition when it's directly opposite the Sun in the sky. When this occurs, it rises at sunset and sets at sunrise and is therefore visible throughout the entire night. This is when the planet is at its best for observation, as it will be at its closest to the Earth and will appear largest and brightest in the night sky.

Opposition can only occur if the body orbits the Sun at a greater distance than the Earth. For example, the planets Mars, Jupiter and Saturn all periodically reach opposition but the planets Mercury and Venus will always appear close to the Sun in the sky.

See also *Greatest Eastern/Western Elongation*

Optical Binary

An optical binary is a chance alignment of two stars in the sky, thereby giving the illusion that the pair are a true multiple star system. In reality, the stars may be hundreds of light years apart and moving in different directions through space.

See also *Primary Component* and *Secondary Component*

Planetary Nebulae

See *Nebula*

Primary Component

The primary component of a multiple star system is the brightest with the next brightest being the secondary. For example, the famous double star Albireo has a golden primary with a sapphire blue secondary.

See also *Optical Binary* and *Secondary component*

R.A. (Right Ascension)

On star maps and charts, R.A. is the celestial equivalent of longitude and enables astronomers to easily locate an object in the night sky. Unlike longitude, it's measured in hours, minutes and seconds with 00h 00m 00s holding a special significance. When the Sun reaches that point in its movement across the sky, the March (vernal) equinox occurs. Spring then begins in the northern hemisphere and while autumn begins in the south.

See also *Declination*

Reflector Telescope

A reflector telescope is one that works using mirrors. The light enters through the open end (the *aperture*) and then hits the primary mirror at the bottom of the telescope tube. The light is then reflected back up towards the aperture, where it hits a smaller secondary mirror and is directed into the eyepiece. This has the benefit of having the eyepiece conveniently located near the top of the tube and also allows for a longer *focal length* without increasing the length of the tube itself.

See also *Aperture, Focal Length* and *Refractor Telescope.*

Refractor Telescope

A refractor telescope uses lenses to magnify the image for the observer. It comprises of a long tube with a primary lens at the aperture, where the light enters the telescope. The light then travels all the way down the tube and then exits through the eyepiece at the bottom.

See also *Aperture, Focal Length* and *Reflector Telescope.*

Secondary Component

The secondary component of a multiple star system is typically the second brightest star, with the brightest being the primary.

See also *Optical Binary* and *Primary Component*

Supernova Remnant

See *Nebula.*

Recommended Resources

Facebook Groups
- *Astronomy for Beginners (A4B)*
- *Astronomy for Fun*
- *Astronomy Workfile*
- *Online Astronomy Society*
- *Space Science and Astronomy for Home Educators*
- *Telescope Addicts – Astronomy & Astrophotography*
- *UK Astronomy*

Books
- *365 Starry Nights* – Chet Raymo
- *Astronomy Hacks* – Robert Bruce Thompson & Barbara Fritchman Thompson
- *Binocular Highlights* – Gary Seronik
- *Celestial Harvest* – James Mullaney
- *Celestial Sampler* – Sue French
- *Cosmic Challenge* – Philip S. Harrington
- *Deep Sky Observing with Small Telescopes* – David J. Eicher
- *Deep-Sky Wonders* – Sue French
- *Deep Sky Wonders* – Walter Scott Houston
- *Double Stars for Small Telescopes* – Sissy Haas
- *Illustrated Guide to Astronomical Wonders* – Robert Bruce Thompson & Barbara Fritchman Thompson

- *Observer's Sky Atlas, The* – E. Karkoschka
- *Observing Handbook and Catalogue of Deep-Sky Objects* – Christian B. Luginbuhl and Brian A. Skiff
- *Planet Observer's Handbook, The* – Fred W. Price
- *Pocket Sky Atlas* – Roger W. Sinnott
- *Star-Hopping for Backyard Astronomers* – Alan M. MacRobert
- *Star Names – Their Lore and Meaning* – Richard Hinckley Allen
- *Star Watch* - Philip S. Harrington
- *Touring the Universe Through Binoculars* – Phillip S. Harrington
- *Turn Left at Orion* – Guy Consolmagno and Dan M. Davis

Software

- *Mobile Observatory* by Wolfgang Zima (http://zima.co.)
- *Sky Tools* by Greg Crinklaw (http://www.skyhound.com)
- *Stellarium* (Open Source software – http://www.stellarium.org)